Air Transport Policy and National Security

AIR TRANSPORT POLICY AND NATIONAL SECURITY

A Political, Economic, and Military Analysis

By FREDERICK C. THAYER, JR.

The University of North Carolina Press
Chapel Hill

Manufactured in the United States of America
Library of Congress Catalog Card Number 65-25600
Printed by Rose Printing Company, Inc.

Preface

AN IMPORTANT SERIES OF DEVELOPMENTS in the late 1950's gradually brought to varying degrees of public attention the general subject of this book: the interrelationships of the political, military, and economic aspects of both military airlift policy and commercial air transport policy. Several ranking Army generals called attention to the subject by retiring during the second term of the Eisenhower administration and then writing books sharply critical of the president's defense policy emphasis upon the Air Force's Strategic Air Command.[1] To the Army generals, this emphasis had caused the Air Force to dangerously neglect its other assigned responsibilities, among them the provision of airlift for the Army. Prominent scholars in the defense policy field echoed the protests which then were included in the political attack on Eisenhower's defense policies during the 1960 presidential campaign.

A second but related controversy emerged at about the same time but from another quarter. Several spokesmen for the U.S. commercial airlines repeated charges made by the 1955 Hoover Commission that the Military Air Transport Service (MATS, operated by the Air Force but with some Navy participation) was much too large. The commercial operators argued that the government was, in effect, operating its own airline in unfair competition with them. While the companies wanted to get for themselves a share of the Cold War military traffic carried by MATS, the question they raised was part of an emerging congressional debate about the role of commercial airlines in war or other emergencies. This aspect of the over-all controversy, however, received less attention than the questions raised by the Army generals and played no part in the 1960 campaign.

1. Matthew B. Ridgway and H. H. Martin, *Soldier: The Memoirs of Matthew B. Ridgway* (New York: Harper & Brothers, 1956); James M. Gavin, *War and Peace in the Space Age* (New York: Harper & Brothers, 1958); Maxwell D. Taylor, *The Uncertain Trumpet* (New York: Harper & Brothers, 1959, 1960).

If one focuses on the phenomenon of airlift itself, it becomes quickly apparent that these controversies were much more intimately related than usually indicated in studies directed either to purely military questions or to problems of government regulation of the airline industry. The efforts made since World War II to institutionalize government and airline industry planning for emergency use of commercial airlift capabilities were tied closely not only to the old question of "government competition with private enterprise" but also to questions of defense mobilization, military strategy, and the position occupied in world markets by U.S. international carriers. When the special problem of MATS entered the congressional arena in the 1950's, congressmen having strong ties with the Army were largely responsible for expanding the discussion of the issues. Two former Army generals elected in 1959 to participate directly in the controversy over commercial versus military airlift services. They suggested that the Army's airlift needs could be met as easily by the airlines as by the Air Force.

In mapping a research strategy, there seemed little choice but to attempt an interdisciplinary approach that would include all of the important considerations. Political scientists interested in defense policy have analyzed some of the military aspects of these issues but, in their preoccupation with broad strategic questions, have overlooked the economic problems and even some of the precise military questions. Especially there is no shortage of economic studies of airline regulation and the market analysis tools used by the Civil Aeronautics Board (CAB) to make individual route decisions, and subsequent suggestions about how the CAB might more effectively promote competition. Most such studies, however, use the CAB's birth date (1938) as the point of departure, do not question the premises underlying the CAB's creation, and bypass almost completely any questions of foreign policy or military strategy. It thus seemed appropriate to attempt an integrated approach (subject of course to the author's ability to work within several disciplines) that would treat the whole subject in all of its complexity.

The decision to be interdisciplinary has had several consequences, some of which add to the problems facing the reader. It has been necessary, for one thing, to retrace some four decades of the political and economic history of U.S. airlines, principally because most of what has happened in more recent years is traceable to attitudes shaped in 1934 and these, in turn, are understandable only when earlier events are analyzed. Second, the book includes material that some may think too technical for a political study, because the military side of the controversy concerned detailed questions of strategy, tactics, and equipment that cannot be comprehended solely by using the word "airlift." A major difficulty of the 1950's was that congressmen, administrators, and generals did not make clear what kind of airlift they had in mind and this confusion, it should be noted, spilled over into the scholarly studies of defense policy. Congress, moreover, ultimately developed its own airlift program in 1960, by making technical choices between the types of military transports being proposed. I have found it necessary, finally, to include specific suggestions for refinements of economic theory, revisions to the airline regulatory process, and comments on specific military strategies, even if these may make the book appear unduly prescriptive. If the suggestions and the evidence supporting them seem complex, it bears repeating that airlift problems entered the congressional arena in this same complex, interdisciplinary manner.

If a single thread runs through this book, it is that the U.S. has too quickly rejected reasonable and pragmatic solutions in favor of less satisfactory policies. Several major examples are analyzed in this book, as well as some minor ones. Among the former are the Roosevelt administration's cancellation of all air mail contracts in 1934, the concessions granted foreign airlines in the 1950's by the Eisenhower administration, and some of the Kennedy administration's innovations in military airlift organization. Current military airlift and commercial airline policies are the product of these decisions and others. The situation is not a desirable one, but it cannot be blamed upon any particular individual, interest group, or political party.

I have used a historical framework, even if it would have been equally valid, especially for political analysis, to begin with the most recent controversies and then separately trace each of them back to the beginning. It seemed more useful, however, to focus on the interrelationships from the outset than to subject the controversies to a more definite separation. A major premise of the book is that airlift policy already has been compartmentalized for much too long a time. Because of the ebb and flow of particular issues, not to mention discontinuities between war and peace, there is some overlapping in the book and at least the appearance of repetition. In one section or another it has been necessary to go back and pick up threads which were not important enough to warrant extended analysis when mentioned earlier. Conversely, the analysis of some questions is carried beyond the boundary of a particular section.

These statements apply especially to Parts Two and Three of the book. The former (Chapters 3 through 6) deals with military airlift and the airlines during World War II, the significant policy changes following the war, and the subsequent interest group activity that triggered congressional interest. The analysis in these chapters includes most of the Eisenhower years so as to adequately portray the interest groups themselves. Part Three (Chapters 7 through 9) focuses on congressional policy-making in the second Eisenhower administration and on the roles played by the interest groups in this specific arena. In a sense, Part Two describes the actors while Part Three recounts the drama itself.

Two other observations may serve to put this book in proper perspective. One is that strategic military airlift and international air transportation on a truly massive scale are relatively new phenomena. In the military field the U.S. did not even attempt until 1960 to use airlift instead of sealift for the transoceanic deployment of large troop concentrations. Transoceanic air travel has been really significant only since the mid-1950's. These forms of aviation are still in their infancy at a time when other forms of manned flight, e.g., strategic bombardment, may be declining in significance. From this point of view the airlift controversy was the outcome of attempts by many individuals and groups to cope with a new phenomenon difficult to accommodate within traditional models of economic competition and military organization. The most relevant of the political factors, finally, has been hidden behind necessary veils of government secrecy, as when the administration used the commercial airlines to counter Axis penetration of Latin America in the 1930's and when the government maintains liaison with the airlines as they fly the Berlin Corridor of the 1960's. This secrecy, coupled with the fledgling nature of international operations, has led many to assume too quickly that international airlines do not represent a special problem.

THE NATURE OF LONG-HAUL AIRLIFT SYSTEMS

The paragraphs that follow are a brief introduction to the basic factors associated with the operation of a fleet of transport aircraft. The explanation is principally for those who have had no previous association with the concepts of commercial or military airlift operation. Some of the factors are important also to an understanding of the principal types of military airlift.

Transport aircraft are designed to carry passengers, cargo, or some combination of the two. The efficiency of a single aircraft is generally proportional to the total amount of goods carried within any given time period. In a simplified example, assume that the task at hand is to airlift cargo between New York and Chicago. Other things being equal, an airplane making three such trips per week is more efficient than one that can make only two. In commercial operation, more total revenue will be accrued from the three trips, and the marginal cost of the third trip will be less than the cost of each of the first two trips.

If this relationship is carried to its logical conclusion, the most efficient airplane stays on the ground only long enough to discharge its cargo, take on a new load, and refuel. The rest of its time is spent in the air. While any airplane is on the ground, however, it is likely to need repairs if safe operation is to be maintained. The salaries paid the mechanics are, of course, a part of the operating cost. If an airplane is on the ground for say, two hours, four mechanics may be able to accomplish the average amount of repairs required. If the attempt is made to compress to one hour the time on the ground, more mechanics will be needed, perhaps as many as eight or ten. The point can be reached where the marginal cost of further reducing the ground time exceeds the additional revenue that can be earned.

In a commercial operation, it is important that airplanes operate on schedule; this promotes public confidence and encourages others to use the service. The same consideration affects scheduled military airlift; there must be assurance that defense goods will arrive according to plan. As operators gain experience with particular types of airplanes, they find cause-effect relationships between particular mechanical problems and the amount of time spent on the ground. One type of airplane, for example, may frequently develop small oil leaks within its engines. The working space around the leaks may be sufficient for only one mechanic and, whatever his level of skill, an average leak may require two hours to repair. If the airplane is scheduled for only one hour on the ground, and if this type of leak occurs on half the flights, then this half will not be able to adhere to the schedule. The other half may be on schedule, but the total result will be a loss of public confidence or late delivery of half the defense goods being shipped.

A premium is thus placed upon achieving the best equilibrium between time in the air, time on the ground, maintenance costs, and the percentage of flights that operate on schedule. There are, of course, other factors in the equation. Perhaps automated cargo terminals can load and unload aircraft more quickly and at less cost. An airplane carrying passengers may be scheduled to arrive at its destination at two in the morning. If it is then scheduled to leave on a new flight only one hour later, passengers will ignore it in favor of one leaving at a more comfortable time. The airplane arriving at two in the morning will thus be held for departure at a later hour and this, in turn, will cause some loss in over-all efficiency.

All of these factors, and others, make up a basic management tool for airlift systems, commonly known as aircraft utilization, defined as the average number of hours spent in the air each day by a single airplane.[2] In the example at hand, it may take four hours to fly from New York to Chicago, and four hours to return. If one airplane makes a single round trip each day, the airplane has a utilization rate of eight hours. The definition can be broadened to include the total number of aircraft operated by a single company, a group of companies, or a military organization. If the time frame covers a month or a year, it will include time spent in major overhaul. Taken together, all these considerations produce an average that is an index of efficiency and productivity.

Commercial airlines generally attain daily utilization rates of eight to twelve hours on their fleets. Strategic military transports operated by MATS fly at five-hour utilization in the expectation that the figure can be increased to eight or ten during emergencies. This introduces an additional consideration; if the utilization rate is doubled, the net effect is that of having doubled the airlift fleet without purchasing additional aircraft. More mechanics and more spare parts may accompany the increase, but the marginal cost is less than the cost of doubling the number of planes.

The utilization concept is closely related to one other that might be called the "airlift conveyor belt." Transports are most productive when they operate in an unbroken sequence. At a given time, some items on the conveyor belt are at one end, some at the other,

2. For a more detailed analysis of utilization and its mathematical relationship to total productivity, see R. Dixon Speas, *Technical Aspects of Air Transport Management* (New York, Toronto, London: McGraw-Hill Book Company, Inc., 1955), pp. 79–82.

and the remainder are spaced along the belt. If a fleet of transports is to be used at its maximum efficiency within the assumed environment, some airplanes will be in Chicago, some in New York, and some in between.

Suppose further that the Chicago airport is so small that there is room on the ground for only one airplane, and that it takes two hours to offload cargo and refuel a single plane. If twelve aircraft leave New York at the same time, there will be room at Chicago for only one, and the other eleven will have to await their turns. If the twelve airplanes leave New York at two-hour intervals, on the other hand, each will arrive at the appropriate time for its landing. In addition, the first one will be able to return to New York and begin another trip to Chicago, all within the same day. The conveyor belt thus makes it possible to use fewer airplanes to accomplish the same amount of work.

This maximum productivity, it will be noted, is quite different from most forms of combat aviation. The total effectiveness of fighters and bombers generally depends upon the number in the air at the same time. They take off at about the same time and return to base at the same time. The concentration of effort results in much less total flying for all of the aircraft than in a conveyor belt situation. The utilization concept is thus not very useful for the measurement of combat efficiency. Strategic Air Command bombers, for example, fly at the much lower utilization rate of one to three hours per day.

Three other characteristics of transport aviation deserve special mention. The conveyor belt, first of all, does not operate in midair without underpinning. Spaced along its length are various mechanisms to assist in keeping the belt in operation, and the transports themselves generally must land at several airfields before reaching their final destination. At each of the bases, there must be people to refuel the plane, mechanics to fix it, cargo handlers to offload it if need be, plus various other specialists. Taken together, all of the bases and supporting personnel comprise a strategic airlift system or an international airline complex. In this sense, the system is an entity with an extremely broad geographic base.

A second characteristic concerns the number of flight crews needed to operate the system. Standards have been developed over the years for the number of hours that flight crews can fly within any given time period. One such rule might be that pilots should fly no more than sixty hours each month. If the figure is exceeded, pilot proficiency deteriorates and the subsequent fatigue can lead to accidents. If a fleet flies at a two-hour utilization rate, each aircraft flies sixty hours per month and the number of crews is the same as the number of planes. If the rate is four hours, two crews are needed for each plane. The "crew ratio" in transport aviation rises in proportion to the utilization, with commercial airlines employing as many as four to six crews for each aircraft.

The third characteristic involves the manner in which these crews help achieve the higher utilization of transport fleets and the overall effect of the conveyor belt. Because there is more than one crew for each plane, the additional crews can be located at bases along the route of flight. When an airplane arrives at one of these bases, its crew can be replaced by another placed there in advance, and the airplane can continue its flight while the first crew gets some rest. The crew subsequently takes a later aircraft on to the next base along the way, and so on through the system until the destination is reached. While crews alternate periods of flying with rest periods, the airplane and its cargo continue on their way. The speed of cargo movement is thus geared to the airplane but not to the crew. If there were only one crew for each aircraft, it would be necessary to delay the cargo movement to allow the crew to rest.

An airlift system then, is an organizational complex of some magnitude. It is a necessity for effective operation but, once established, airplanes cannot operate outside its limits except at a loss to efficiency; the airplane moves at its maximum rate only when it stays on the conveyor belt. If the plane lands at a base with no mechanics to make minor repairs and no crew to continue the flight to the next base, the system breaks down. In this sense, an airlift system is relatively inflexible. Many people not familiar with these factors look upon an airlift system as something involving only airplanes, crews, and something to carry. It appears on the surface that airplanes can head in any direction with equal efficiency.

Notice will be taken later in this book of the interplay of some of these factors with particular problems of international airlift, whether commercial or military. Some important military tasks, for example, cause deviations from a conveyor belt operation, a reduction in the utilization rate, and a loss in total efficiency. Tactical military airlift within a combat theater is subject to constraints which prevent efficiency. The battlefield situation can be quite fluid, airfields can be primitive, spare parts can be destroyed by enemy action, and so forth. Yet another major difference will get much attention in subsequent pages; when tactical military airlift involves massive airborne operations and thousands of paratroopers, the resultant system departs completely from the conveyor belt concept. In the operational sense, paratroop operation is quite similar to traditional bomber operations; all planes take off simultaneously, fly close formation to the target, and return to base together. These differences are crucial to an understanding of the military aspects of this book.

Similarly, the length of international commercial routes often is accompanied by a relatively sparse travel market that does not permit operation of very many flights along the route. A certain number of crews must be stationed along the way, and all the associated facilities must be maintained; the result is a relatively high overhead cost.

Acknowledgments

THIS BOOK STARTED OUT AS A DISSERTATION for the University of Denver's Graduate School of International Studies, one of those advanced research centers where an interdisciplinary approach flourishes under the sensible guidance of Josef Korbel. I owe a very special debt to Vincent Davis as a trenchant critic, a rigorous adviser, a supplier of valuable reference materials, and a wholehearted supporter during the necessary revision for publication. Joe R. Wilkinson helped me develop the relationships between international politics and economics, particularly with regard to the special position of the Export-Import Bank. Glenn Snyder sharpened my awareness of decision-making mechanisms for national security policy. My long-time employer, the United States Air Force, and especially Colonel Wesley Posvar of the Air Force Academy, made it possible for me to leave military duties temporarily in favor of advanced study.

Samuel Huntington, John Galbraith, and Thomas Schelling were kind enough to comment on the manuscript. I especially thank Gardiner Means who discussed at length with me the applicability of his own studies to the confused field of domestic and international airline regulation. If there is any value to the economic approach in Chapter 11, it is traceable to his influence. One intellectual debt is of long standing and is acknowledged here: Harvey Mansfield introduced me to the complexities of the administrative process. Earlier research done under his supervision enabled me to gain some understanding of congressional investigations, an important aspect of this book.

I owe much to people in government, some of whom (for what I hope are obvious reasons) must remain unnamed in this book. But I must thank Mrs. Dorothy Cross and Mrs. Bessie Moore (both of Headquarters United States Air Force) for making available to me voluminous research and clipping files maintained for Washington policy-makers. J. B. Townsend supplied some similar material. On the other hand, I never can repay numerous Army, Navy, and Air Force associates for what I learned from them during eleven years of operational and staff experience with military airlift. I include also airline operating officials providing contract augmentation to the Military Air Transport Service.

The usual statement applies: only the author can be held responsible for what appears in this book.

Washington, D.C.
September 19, 1965

Contents

Tables

List of Abbreviations

AN ATTEMPT HAS BEEN MADE in this study to minimize the use of any type of jargon. It has been found necessary, however, to use a few abbreviations to avoid endless repetition of long official titles. The list below contains the abbreviations used in the text. The full name is used the first time each agency appears in the text.

ACC	Air Coordinating Committee
ALPA	Air Line Pilots Association
AMEX	American Export Air Lines
ATA	Air Transport Association
ATC	Air Transport Command
BOAC	British Overseas Airways Corporation
CAA	Civil Aeronautics Administration
CAB	Civil Aeronautics Board
CASF	Composite Aerial Strike Force
CAT	Civil Air Transport
CRAF	Civil Reserve Air Fleet
DATA	Defense Air Transportation Administration
DSA	Defense Supply Agency
DTMS	Defense Transportation Management Service
FAA	Federal Aviation Agency
IATA	International Air Transport Association
ICC	Interstate Commerce Commission
JCS	Joint Chiefs of Staff
JMTC	Joint Military Transportation Committee
MATS	Military Air Transport Service
MSTS	Military Sea Transport Service
PAA	Pan American World Airways
RAF	Royal Air Force
SAC	Strategic Air Command
STRAC	Strategic Army Corps
STRICOM	Strike Command
TAC	Tactical Air Command
TWA	Trans-World Airlines
VTOL	Vertical Take-Off and Landing

PART ONE

National and International Airline Policy before World War II

[1]

Airlines As An Issue of Party Politics

THE FIRST FORMAL INVOLVEMENT of the United States government in air transportation occurred in 1916 when the Post Office Department signed an air mail contract with private operators.[1] During the next two years U.S. intervention in World War I gave a great boost to all phases of aviation development in the country and encouraged some people to see relationships between military aviation and government support for commercial airlift. The National Advisory Committee on Aeronautics, an Army general among its members, proposed that the air mail system be organized as a training ground for Army pilots.[2] Air mail flights using this arrangement began in May, 1918.

The relationship between U.S. military aviation and non-military air transport may thus be dated from the World War I period. At that time and during the subsequent decade, the relationship was rudimentary at best. Aside from a notion entertained briefly in World War I that U.S. troops might be dropped by parachute behind enemy lines, little thought was given to any sort of military airlift until World War II. Technological advance made it possible for that war to provide the same impetus to transoceanic civil and military airlift that World War I had given earlier forms of commercial and military aviation. Even so, international commercial aviation and strategic military airlift have attained a truly significant status only in recent years. Commercial airlift has been the primary

1. Henry Ladd Smith, *Airways: The History of Commercial Aviation in the United States* (New York: Alfred A. Knopf, 1942), p. 54. The terms "airlift" and "air transportation" are used interchangeably in this book. In that sense, both terms encompass both civil and military operations, even if the word "airlift" usually has only a military connotation.

2. John H. Frederick, *Commercial Air Transportation* (Homewood, Illinois: Richard D. Irwin, 1955), p. 78.

means of transoceanic travel for only a few years, and it was not until almost the end of the Eisenhower administration that anyone gave serious attention to using airlift for the strategic deployment of troops.

When air transport policy is analyzed in a way that emphasizes the interaction between world politics, foreign economic policy, and military strategy, one of the first things that comes to mind is that these areas are the direct political responsibilities of the government. Outside groups can be asked for assistance but they cannot bear explicit responsibility for the policy-making outcomes. The political dimension of airlift policy thus takes on a significance well beyond that accorded "pure" economic or political analysis, and one is disposed to search for the political trends and attitudes that have shaped the government's response to its political, economic, and military responsibilities.

The striking fact is that no administration since Hoover's has explicitly accepted responsibility for a national airlift policy. Whatever one's evaluation of how that administration performed, any study of the subject that excludes what happened before 1933 overlooks the only experiment of its type. The Hoover administration took on the job only after more than a decade of experimentation had failed to provide a solution. The Roosevelt administration took office committed to a general condemnation of everything done by its predecessor, especially in the economic arena. Using evidence provided by an aggressive investigating committee, the new administration canceled all air mail contracts on grounds that its predecessor and the commercial operators had conspired to stifle competition and to extract millions of dollars in unjustified or illegal subsidy and mail payments. This administrative action and the subsequent public opprobrium visited upon the industry led to permanent and mutual hostility between its managers and liberal Democrats.[3]

Several major policy decisions made since 1934 have reflected the strained relationships between government and industry. The

3. The term "liberal Democrats" is not used either pejoratively or flatteringly, but only to designate in shorthand, noncommittal fashion an important segment of the Democratic party traditionally thought unfriendly to big business. In a general way, this segment is about the same as the "presidential Democrats" described by James M. Burns, but this does not imply agreement with his central argument. James MacGregor Burns, *The Deadlock of Democracy: Four-Party Politics in America* (Englewood Cliffs, N.J.: Prentice-Hall, Inc., 1963), pp. 324–29.

basic regulatory legislation enacted in 1938, for example, was written so as to give what might now be called undue emphasis to the principle of free competition. All recent administrations have found it all but impossible to form a meaningful partnership with the international airlines, even while using them to achieve specific policy objectives, as in the pre-World War II period. After the war, the Truman administration decided that U.S. airlines must compete against each other on many international routes. Congressional Democrats, many of them persistently opposed to the subsidies paid the larger airlines, became the principal supporters of the nonscheduled airlines. These small companies, organized after the war thanks to government sale of surplus aircraft, spent the subsequent years searching for a market that could be created only by taking it from the major companies.

Finding no market, the "nonskeds" became in the 1950's the loudest supporters of turning over to the commercial industry the traffic carried on military aircraft. Several Democrats joined forces with these small companies but, as a necessary compromise, agreed to terms proposed by the major companies to which both the nonskeds and the Democrats had for so long been opposed. The result, between 1956 and 1960, was a most curious political alliance between liberal Democrats and various types of business lobbies, in general opposition to a Republican administration generally sympathetic to the business world. When Army spokemen joined the same alliance, Democrats found themselves advocating both an increase and decrease in military airlift capacity—a highly confusing arrangement which this book undertakes to explain.

The preceding discussion should not be taken as an assertion that the history of airlift policy, commercial or military, can be understood simply by putting in a bad light a major segment of a political party. Current airlift problems are not the fault of any single individual, interest group, or political party. The Eisenhower administration, for example, appears to have been much too generous in granting concessions to foreign airlines. Democrats, however, have controlled the executive branch for most of the period since 1933, and most of their decisions reflect the atmosphere that surrounded the Hoover-Roosevelt transition.

It seems relevant to this analysis, therefore, to examine the politico-economic considerations that went into the Democratic decision of 1934 and the compromise regulatory process adopted in 1938. The remainder of this chapter and all of the next will explore the interwar decades, focusing first on the domestic and then on the international aspects of airlift policy during that period. The long-term problems depend so much upon the 1934 policy reversal that one must ask why the Hoover administration took on the responsibility for an industry that had been struggling along since the World War I beginnings mentioned earlier.

THE STRUGGLE IN THE 1920'S TO AVOID THE SUBSIDY ISSUE

World War I experience encouraged the Post Office to press for an expanded air mail system free of military connections. By 1918 the Department had won permission to organize its own flying service solely for mail carrying purposes. By 1920 the routes extended all the way west to San Francisco, and the Post Office was in the airlift business in a fairly major way; it remained until 1934 the one government agency significantly involved in domestic and international operation.[4]

The Harding administration surprisingly agreed to continued expansion of air mail service as an exception to a general policy of reducing government expenditures.[5] Both the administration and Congress let it be known, however, that the mail service would be surrendered to private enterprise as soon as practicable. The railroads encouraged the decision by arguing that the service was unfair competition with other forms of transportation.[6] The administration and Congress gradually came to realize also that passenger travel, already known to be the ultimate objective, could not by itself finance airline growth until air travel gained public acceptance. The earliest attempts to establish passenger airlines floundered because the job was beyond the capability of private enterprise. Other groups, particularly within the military services, advocated some form of subsidy on grounds that the Europeans were forging ahead. Without government support, they argued, the U.S. might not realize the economic and military potential of aviation.[7]

4. E. P. Warner, an early member of the Civil Aeronautics Board, ascribes the beginning of commercial air transportation to the "spontaneous attraction of the airplane and the Post Office for one another." Quoted in Frederick, *op. cit.*, p. 77.

5. The president reportedly was impressed by the overnight flights undertaken to deliver mail from San Francisco to New York in 33 hours. Smith, *op. cit.*, p. 71.

6. *Ibid.*, p. 84.

7. *Ibid.*, p. 88.

It was quite convenient for the U.S. government to use air mail for the promotion of an airline industry. Within such a large country, a significant increase in speed was immediately significant to those using the mails. Had the U.S. been confined within the smaller areas known to Europeans, for example, the subsidy issue might not have been obscured. Air mail made it possible for the U.S. to give the surface appearance of not using subsidy, despite the virtual impossibility of properly allocating costs and expenditures between passenger and mail operation within a single transport fleet. Something of a tradition emerged; government assistance always should be hidden behind the fiction that it is not needed.

Congress drafted the first air mail statute, the Kelly Act of 1925, so as to avoid the subsidy issue.[8] The contracts awarded that year were only for small connecting routes, the Post Office continuing to operate its own transcontinental service until 1927. Nevertheless, the modern airline system can be dated from the Kelly Act, because the carriers awarded those initial four-year contracts eventually became parts of United, American, and Trans World, the largest of the permanent domestic companies.[9]

Attempts to avoid the appearance of subsidy while retaining its substance made difficult the drafting of suitable amendments to the Kelly Act. A 1926 amendment contained a complex rate structure which paid carriers the same amount for trips of three hundred or one thousand miles. The companies understandably sought to develop short route patterns giving them the greatest return for the shortest flight.[10] A 1928 amendment lowered postage rates to encourage wider use of air mail and made it possible for the first time for mail payments to exceed revenues. The cleverest operators began to air mail all sorts of miscellaneous items back and forth across their assigned routes. The difference between the postage rates and mail payments then became clear profit. Some companies even worked out rebate schemes with manufacturers who agreed to ship heavy equipment by air.[11] These experiences demon-

8. Samuel B. Richmond, *Regulation and Competition in Air Transportation* (New York and London: Columbia University Press, 1961), p. 4. Commercial operators were allowed a maximum of four-fifths of the revenues collected for air mail; the rate was quite high (four cents per ounce) and the postmaster general could not exceed the revenue limitations. Francis A. Spencer, *Air Mail Payment and the Government* (Washington: The Brookings Institution, 1941), pp. 30–31.
9. Smith, *op. cit.*, p. 104.
10. Spencer, *op. cit.*, p. 35.
11. *Ibid.*, pp. 38–39. Smith, *op. cit.*, pp. 125–26.

strated that it might not be possible to provide for sensible development of passenger transportation without creating self-defeating loopholes.

The Republican administrations of the 1920's gradually moved toward clearer definitions of the responsibilities of the government and the infant industry. The Air Commerce Bill of 1926 set up a Bureau of Aeronautics within the Department of Commerce, thus establishing government supervision of safety, licensing of crews and aircraft, and the building of national airways. The legislation had two important effects, both often overlooked. Federal licensing and inspections made it inevitable that the basic operational elements of all airlines would be virtually identical, thus giving passengers few reasons for choosing one airline over another. The national airways were, secondly, aerial highways serving much the same function as had land grants to railroads. Radio stations, light beacons to guide pilots at night and over mountains, and weather forecasting facilities all emphasized the relatively fixed nature of airlift systems.[12]

The postmasters general of the 1920's consistently used all of their authority to promote passenger travel, despite statutory difficulties and active opposition. Because legislation mentioned only air mail delivery, it was difficult to award contracts to companies having airplanes suitable for passenger travel but more expensive to operate than austere transports designed only for mail carriage.[13] One postmaster general decreed that contracts be awarded only to responsible bidders having capital, equipment, pilots and experience, even though his action eliminated all but a few companies.[14] When he used this rule to award a contract to a company which had not submitted the lowest bid, the comptroller general objected strenuously before finally agreeing to the criterion of "responsibility." Nevertheless, the industry's growth pattern demonstrated the need for more direct measures to rationalize the airline map.

The Post Office adopted the apparently sensible practice of

12. See the Preface to this book.

13. Spencer, *op. cit.*, p. 31. Richmond, *op. cit.*, p. 4.

14. Smith, *op. cit.*, p. 104. This was a subject of much dispute. William Boeing, for example, built single-engine mail planes that were cheap to operate but did little for passenger operation. One competitor was William Stout, inventor of what became known as the "Ford Tri-motor," a three-engine, all-metal transport that significantly advanced passenger travel, even if it was more expensive to operate. *Ibid.*, pp. 105–6, 119–20.

gradually withdrawing from the air mail service in favor of private industry. Initially, for example, the New York-San Francisco route was divided in two at Chicago. The piecemeal withdrawal of the Post Office and the necessity to follow the pattern of bids submitted by existing companies were not enough to provide for logical development of a transcontinental route system. To those operating the companies it was clear that coast-to-coast flights would become a reality and, inevitably, each company flying a subdivided route sought to expand its route system and to improve its competitive position. Thus, the first signs of attempted mergers appeared by the end of 1928. An internecine corporate struggle followed, complicated by the connection of each of the largest operators (through holding companies) to an aircraft or engine manufacturer.[15] The press of events virtually forced the Hoover administration to decide whether existing companies should be left to their own devices to shape the national airline map or whether the government should assert a public interest in the outcome.

The range of alternatives available in 1929 has remained about the same throughout the intervening years, although international and military considerations have become more significant. The choices can be summarized in a few rhetorical questions. Can direct subsidy be forever avoided? Is indirect subsidy, e.g., the fostering of improvements in passenger travel through the device of air mail payments, too complex an undertaking to fit within the governmental process? Should an established enterprise be favored over the small operator trying to get started, especially if passengers put great store in past records of safe performance? Is competitive bidding useful in situations where the lowest price may not best serve the ultimate objective? The Hoover administration made the only attempt to answer such questions within a framework of accepted political responsibility for the outcomes.

THE HOOVER ADMINISTRATION'S DECISION AND THE NEW DEAL REACTION

The administration's bold step toward government regulation seems to have had two immediate causes. The first was a breakdown in rate negotiations between the Post Office and the carriers, resulting from some of the difficulties mentioned earlier. The precise issue was the extent to which losses on passenger operation

15. *Ibid.*, pp. 130–55.

should be absorbed by government air mail payments. A second cause was the powerful Republican newly appointed as Postmaster General, Walter Folger Brown. He had no aviation experience, but he proved himself an energetic administrator eager to learn about his new tasks. He immersed himself in the study of air mail and airlift matters until, in the words of one airline official, "he knew more about it than anyone in the industry."[16] His studies convinced him that anarchic expansion of the air routes and the airlines had to be abandoned.[17]

The 1930 McNary-Watres Act gave Brown enough authority to act as a "one-man public utility commission" and to "unify" the industry, and he did not hesitate to use it.[18] He transformed the mail payment system into one that was close to outright subsidy, but not so close as to provoke congressional opposition.[19] The Post Office began to pay by the "space-mile" rather than the "pound-mile," i.e., carriers were paid for the space made available regardless of the Post Office's ability to fill the airplane with mail. Simply stated, this was a charter system; the government paid for the entire airplane, so to speak, not for just the mail being airlifted. This provision was, of course, designed to encourage the purchase of large passenger airplanes. Brown also got control over bidding conditions, service standards, systems of accounts, and crew personnel, e.g., he could force airlines to use cabin attendants to care for passengers. He could extend the routes of existing companies if he thought it in the public interest to do so and, as before, he could issue long-term certificates. He even managed a statutory provision eliminating carriers with less than six months' experience over 250-mile routes.[20] He met with one significant defeat; Congress refused to explicitly authorize him to abandon competitive bidding.

16. A president of Eastern Air Lines, quoted in *ibid.*, p. 185.

17. In attempting to avoid the problem of passenger travel while concentrating on air mail profits, the industry created an odd situation. It was calculated, for example, that a 150-pound passenger would pay $1,003.12 for a coast-to-coast trip at air mail rates. *Ibid.*, p. 157.

18. Spencer, *op. cit.*, p. 43. Frederick, *op. cit.*, p. 81.

19. U.S. Congress, House, Committee on the Judiciary, *Report of the Antitrust Subcommittee (Subcommittee No. 5) Pursuant to H. Res. 107 on Airlines.* 85th Congress, 1st Session, Committee Print (Washington: Government Printing Office, 1957), p. 10 (hereinafter *Celler Report*).

20. The statute was credited to Brown and three representatives of industry; one had been an Assistant Secretary of Commerce and another had been chief of the Post Office's air mail service. Smith, *op. cit.*, p. 158.

Brown moved immediately to eliminate what he considered chaotic conditions, principally by forcing existing companies to merge. He called the major airlines together and asked them to produce some new route systems, specifying that each of the three transcontinental routes he wanted must be operated by a company having no connection with the other two. In effect, Brown handed the companies an airline map, then told them to reorganize so as to fit the map.

When meetings began in the spring of 1930 the industry deferred to Brown, because each participant felt unable to surrender his company's interests to the larger objective of a new system. Brown accepted the challenge, decided who should operate each of his route systems, announced the decision to the companies, and then withheld mail contracts until the mergers were accomplished. The meetings dragged on through the summer as Brown successfully fended off the comptroller general once again; that official argued that Brown's extension of existing route systems had exceeded the statutory authority.[21] Using all his authority and perhaps more, Brown prevailed over the comptroller general, brought to heel the major companies, and then acted in concert with them to produce the new airline map. The very nature of the process made Brown's decisions distasteful to the small operators who fell victim to his grand design.

Brown's treatment of one such company is significant because it led directly to the Roosevelt administration's rejection of the entire arrangement. This small airline had set up a low-fare schedule between New York, Philadelphia, and Washington; it had cut costs to the minimum by selling tickets and picking up passengers by taxicab at railroad stations. Its bid for an air mail contract, however, was overruled in favor of Eastern Air Transport (later Eastern Air Lines), because Brown's master plan included yet a fourth major trunk line between New York and Florida. The plan left no room for the small company that could operate only between the largest cities and whose revenues would not permit it to buy new aircraft.[22]

Brown insisted that companies forced out of business be paid a fair price for their assets, but his abrupt insistence on conformity encouraged the losers to seek political redress. Chief among them was Tom Braniff, founder of the airline of that name, who had

21. *Ibid.*, pp. 156–96; 239–48.
22. *Ibid.*, pp. 218–23.

been operating along a part of the routes serviced by United and American. Denied a mail contract, Braniff took his case to people who stood to be influential if the Hoover administration left the scene.[23]

The dispute came to Congress through the friendship of Fulton Lewis, Jr. (then a reporter for the Hearst newspaper chain) with the operator of the small New York-Washington airline. Hearst refused to print the story of how a low bidder had lost out, but Lewis took his evidence to the staff investigator of a special Senate Committee headed by Hugo Black; the committee initially had been formed to investigate the Hoover administration's handling of merchant marine affairs.[24] With Lewis acting as committee adviser, the focus of the investigation shifted to Brown's airline machinations.

Black persuaded President Roosevelt early in 1934 that existing contracts should be canceled on grounds of illegality.[25] Postmaster General James A. Farley suddenly issued an order in February, to be effective within only ten days, canceling all domestic contracts. In nine lengthy specifications, Farley charged that his predecessor had abetted wild stock promotions, illegally extended both contracts and routes, destroyed incriminating evidence, and acted in collusion with a few favored companies. Farley added that Brown had wasted millions by not using competitive bidding to achieve the lowest price to the government.[26] The cancellations left the U.S. without an air mail system, and the administration's attempt to provide a replacement led to some unfortunate consequences.

The Army took on the task of flying the mail and wrote a tragic chapter in American aviation. Unprepared for the job and some extremely severe weather, hampered by inexperienced

23. *Ibid.*, pp. 259–61.

24. Air mail was added to the committee's agenda after one member became concerned over Brown's denial of some bids from United Air Lines. *Ibid.*, pp. 214–17; 226. For the general position of investigations during this era, see M. Nelson McGeary, *The Development of Congressional Investigating Power* (New York: Columbia University Press, 1940), p. 13.

25. Arthur M. Schlesinger, Jr., *The Coming of the New Deal* (Boston: Houghton Mifflin Co., 1958), p. 451.

26. Smith, *op. cit.*, p. 212. It seems likely that the initiative for the cancellations came from outside the Post Office Department. Smith was puzzled by Farley's appearance before the Black Committee only twelve days before the cancellation was announced; Farley said nothing of illegality at the time. Farley, moreover, participated in conferences with the airlines in June, 1933—on the subject of rearranging the routes. Presumably, Farley had not thought these conferences were illegal. *Ibid.*, pp. 243, 264.

pilots and airplanes with inadequate instruments, the Army suffered a series of accidents.[27] Arthur Schlesinger, Jr. cites this event as the first major setback for the Roosevelt administration; it appeared responsible for an unnecessary loss of lives. The administration swiftly reversed itself and ended the Army's mission. The Post Office, determined to avoid using the Brown-inspired McNary-Watres Act, advertised for new bids under an old wagon-road law not originally designed for competitive bidding.[28]

The government awarded new temporary contracts in April, 1934, only two months after the cancellations, and inadvertently renewed the airlift system first created in 1925. Despite the administration's attempt to encourage low bidding by small operators, the major companies out-guessed their competitors. Feeling they could risk absurdly low bids just to get re-established, the big companies regained most of their routes despite Farley's edict barring the airline officials involved in Brown's 1930 conferences. In circumventing the edict, the airlines took on the names they have retained to this time. The north-south route went to Eastern Air Lines, with United, American, and Transcontinental (later TWA) taking on the coast-to-coast systems.[29]

This important series of events in 1934, in which all airline operators were branded as co-conspirators with a discredited administration, created the permanent barrier between those operators and liberal Democrats. The administration's rejection of the Hoover approach converted the airline system itself into a political issue. Brown's meetings became known as "secret spoils conferences," and Black himself never wavered from the position that his investigation had been part of a "righteous crusade" against Hoover administration injustices.[30] The House Anti-trust Subcommittee, chaired by Democrat Emmanuel Celler, investigated the airlines in 1957, accepted the 1934 charges at face value, reproduced them with approval in its report, and implied that Farley had been lax in permitting the major airlines to recapture the mail contracts.[31] Schlesinger dismisses Brown as an individual who mis-

27. *Ibid.*, pp. 249–58. Frederick, *op. cit.*, pp. 82–83. *Celler Report*, pp. 12–13. Spencer, *op. cit.*, pp. 59–60.

28. *Ibid.*, p. 61.

29. Smith, *op. cit.*, p. 279. Frederick, *op. cit.*, p. 83 n. This stands as the only instance in which purely competitive bidding was used. Lucile Sheppard Keyes, *Federal Control of Entry into Air Transportation* (Cambridge: Harvard University Press, 1951), p. 68.

30. Quoted from an interview in Smith, *op. cit.*, pp. 224–25.

31. *Celler Report*, p. 13.

takenly equated the sheer size of a company with its efficiency.[32] Thus, from a relatively partisan point of view, the Roosevelt administration's drastic actions were fully justified. From other vantage points, however, it is not quite so easy to adequately evaluate Brown's handiwork and his successor's rejection of it.

The abrupt dismissal of Brown's regulatory scheme may have been caused by the interaction of an emotionally charged political climate, the zeal of an aggressive investigating committee, general unfamiliarity with the nature of the new industry, and misunderstandings generated by the disguising of outright subsidy. The Roosevelt administration took office under circumstances which made its predecessor's activities seem little less than criminal. The congressional investigating committees of the time performed as public tribunals determined to inflict social punishment upon transgressors. Among the most unpopular groups of the era were the businessmen widely alleged to be responsible for the depression. Investigators found it easy to detect conspiracies, much as later investigators discovered Communist conspiracies within a later Democratic administration. The methods used by the committees of the 1930's were like those used by practically all of the more spectacular committees dedicated to the exposure of alleged wrongdoing.[33] Whatever their philosophical orientation, such committees usually depend for their sustenance upon a "devil theory" of government within which the political opposition seems part of an evil conspiracy. Perhaps the greatest distortion in this case was the continued charge that Brown's conferences were "secret," made long after careful research had shown otherwise.[34] The political environment, then, all but prevented the administration from accepting even the soundest of its predecessor's policies and has clouded all subsequent discussions of airlift policy.

32. Schlesinger, *op. cit.*, p. 450.

33. Walter Lippmann, "The Senate Inquisition," *Forum*, 84 (1930), 129–32. McGeary, *op. cit.*, pp. 23–50.

34. The Post Office issued press releases for all of the meetings. Unfortunately, not all newspapers thought the subject important enough in 1930 to warrant complete coverage. Trade journals, however, covered all the meetings and Brown's intention to alter the route structure. The fact was later noted in the decision of the Court of Claims concerning airline attempts to recover losses. Smith, *op. cit.*, p. 243. Paul W. Godehn and Frank E. Quindry, "Air Mail Contract Cancellations and Resulting Litigation," *Journal of Air Law and Commerce*, 21 (Summer, 1954), 273. The authors of the *Celler Report* used Smith as an authority for some statements made in the report, but they chose to overlook his discussion of the so-called "spoils conferences." *Celler Report*, p. 10.

Nobody in the Roosevelt administration (at least in 1934) seemed to understand the complexity of the new industry, especially its relationship to the question of subsidy. Schlesinger notes, for example, that Black favored both competitive bidding and subsidy, a position with which Farley presumably concurred.[35] The two notions are, of course, somewhat contradictory; subsidies are authorized only when continued operation of a given enterprise is deemed to be in the public interest, in which case any losses incurred from competitive bidding ultimately will be absorbed within the subsidies. It must be admitted, however, that one not associated with the Post Office throughout the decade of experimentation before 1930 might neither have comprehended nor accepted that department's responsibility to build a passenger airlift system.

The ultimate legal and political disposition of Farley's charges did not support him. Henry Ladd Smith notes how Senator Warren Austin, a minority member of the Black Committee, demonstrated that Farley was unfamiliar with the subject and that the Post Office's own solicitor did not think Brown had acted illegally.[36] The only litigation ever undertaken was on the initiative of the companies whose contracts were canceled, and the government compensated some of them.[37] Despite the official banishment of the airline officials, moreover, one of them became an Air Force general during World War II and later was appointed to head the Defense Air Transportation Administration within the Commerce Department.[38]

Dispassionate analysts of interwar airlift problems conclude that the Roosevelt administration reversed what had been, on the whole, a satisfactory representation of the public interest. Henry Ladd Smith thinks Brown was a man of vision, especially in view of the mechanics of the problems he faced. Brown's extensive testimony before the Black Committee revealed an intelligent administrator, intimately familiar with the details of his job.[39] Francis Spencer concludes that Brown was successful in designing a more rational

35. Schlesinger, *op. cit.*, p. 450.

36. Smith, *op. cit.*, pp. 241–47.

37. An examiner of the Court of Claims exonerated Brown, but the Court itself took a middle position. It held that Brown had acted in concert with the larger companies to keep out the small companies; the Court, nonetheless, awarded United Airlines more than $350,000 in compensation. *Ibid.*, pp. 276–78. Godehn and Quindry, *op. cit.*, pp. 274–76. By 1936 the Attorney General had awarded $600,000 in settlements. McGeary, *op. cit.*, p. 90.

38. Ray W. Ireland, an official of United Airlines.

39. Smith, *op. cit.*, pp. 271–76.

airlift system and in giving the public both an attractive passenger service and improvements in safety. Unfortunately, continues Spencer: "In an organizational set-up giving to a political appointee quasi-legislative and quasi-judicial powers usually reserved for independent commissions, it is to be expected that when decisions are made contrary to those sought by special interest groups, such groups will accuse the administrative officer of bias or downright fraud, regardless of how good his intentions or how correct his decisions may be."[40]

One might speculate that without Brown's intervention and his ability to comprehend the need for large airlines, the industry might not have advanced beyond the limits set by the railroad map; there never has been a transcontinental railroad. In any event the cancellations created not only a lasting hostility between government and industry, but a four-year policy vacuum as well.

THE POLICY VACUUM: 1934–38

Cancellation of air mail contracts left the U.S. with no permanent apparatus for regulating either air mail or passenger airlift. Farley could not very well return to statutes he had discredited, nor could the administration argue immediately that it needed a broad grant of regulatory authority. Inasmuch as Brown had been attacked for exercising too much centralized governmental power, the administration seemed obliged to ask for less power and to deny or ignore government responsibility for the airline system. While this was an unusual position for an administration with a broad philosophy of the government's responsibility toward the economy, its policy statements left it little other choice. Airlift policy floundered in a vacuum until 1938; the four-year period combined patchwork air mail regulation, uncontrolled airline operation, mutual hostility, and lack of public confidence in an industry condemned by a generally popular administration.

Of forty air mail bills introduced after the cancellations and, while the temporary contracts were in force, Congress enacted a composite bill proposed by Senator Black.[41] The statute required competitive bidding but included loopholes permitting extension of the temporary contracts. This authority, plus the fact that

40. Spencer, *op. cit.*, pp. 57–58.

41. This marked the beginning of complex and detailed statutory provisions for airline regulation. Pre-1938 legislation is reproduced in *Ibid.*, pp. 319–76.

Brown's route system remained the foundation of the air mail system, made it certain that things would be left as they were.

Three agencies shared the responsibility; the Post Office designated schedules and routes but only in terms of air mail needs, the Interstate Commerce Commission (ICC) fixed rates, and the Commerce Department licensed planes and crews. Initially at least, there were several restrictions; operators and manufacturers had to cut any corporate ties, limits were put on the number of contracts to be given one company and the length of route extensions, and the ICC could reduce rates but not raise them.[42] All but the first of these restrictions were gradually relaxed, but the ICC and the Post Office could not agree on the ground rules for rate determinations. The former wanted to include both mail and passenger considerations, but the Post Office insisted that it would bear only the costs associated with the mail.

The policy vacuum simply did not allow for a logical development of the airline system. Anyone could start a small company if he could license his aircraft and crews. If he took business away from the air mail companies, their losses were covered by whatever change in rates the Post Office accepted. Yet the government had to preserve the air mail system, by now an integral part of the economy, and the emergence of new aircraft promised significant breakthroughs in passenger travel.

The controlled mail system and the uncontrolled passenger system had to come together within some sort of policy framework. The Federal Aviation Commission, a temporary study group formed under the provisions of the Black bill, said as much and suggested a new regulatory agency to control competition through economic regulation. Airlines, said the commission, should operate only when issued certificates of public convenience and necessity.[43]

The year 1938 was the deadline for new legislation, for the 1934 statute provided that payments could not exceed postal revenues after that time.[44] The airlines could not have afforded the newer and larger airplanes within such limitations. Senator Pat McCarran of Nevada was the leader of congressional attempts to produce a

42. *Ibid.*, pp. 79–88. Smith, *op. cit.*, pp. 286–88.

43. Summarized in U.S. Congress, Committee on the Judiciary, *Monopoly Problems in Regulated Industries*, Hearings before Subcommittee, 84th Congress, 2nd Session, on Airlines (Washington: Government Printing Office, 1956), II, 927 (hereinafter *Celler Committee Hearings*).

44. Spencer, *op. cit.*, pp. 92–93.

new law, but outside groups were quite active. The older airlines formed the Air Transport Association (ATA) in 1936, a typical trade association. One of its founders was Braniff Airways, now the partner of companies it had accused of conspiring with Brown. ATA worked with the ICC during 1937 and 1938 in writing proposed bills. The Post Office and Commerce departments initially opposed the attempts, but an interdepartmental committee overruled them and put the administration's weight behind the project.[45]

The four-year policy vacuum thus ended in something of a compromise. The administration admitted the need to return, at least in part, to Brown's system, and it tacitly admitted that something of an airline oligopoly might have to be tolerated. The industry, on the other hand, would have opposed an attempt to return it to regulation by executive agency.

THE NEW NATIONAL POLICY: REGULATION BY INDEPENDENT AGENCY

The basic legislation was passed in 1938 and re-enacted in 1938 as the Federal Aviation Act.[46] This is the policy declaration of the 1958 statute, changed only in minor details from the 1938 original:

> In the exercise and performance of its powers and duties under this ACT, the Board shall consider the following, among other things, as being in the public interest, and in accordance with the public convenience and necessity:
>
> (a) The encouragement and development of an air transportation system properly adapted to the present and future needs of the foreign and domestic commerce of the United States, of the Postal Service, and of the National Defense;
>
> (b) The regulation of air transportation in such manner as to recognize and preserve the inherent advantages of, assure the highest degree of safety in, and foster sound economic conditions, such transportation, and to improve the relations between, and coordinate transportation by, air carriers;
>
> (c) The promotion of adequate, economical, and efficient service by air carriers at reasonable charges, without unjust discriminations, undue preference or advantages, or unfair or destructive competitive practices;
>
> (d) Competition to the extent necessary to assure the sound

45. Smith, *op. cit.*, pp. 300–5. *Celler Committee Hearings* II, 932–33. *Celler Report*, pp. 137, 280–83.

46. Section 102, Public Law 725, 85th Congress.

growth of an air-transportation system properly adapted to the needs of the foreign and domestic commerce of the United States, of the Postal Service, and of the National Defense;

(e) The promotion of safety in air commerce; and

(f) The promotion, encouragement, and development of civil aeronautics.[47]

This new policy recognized subsidy as legitimate support for an industry important to the achievement of broad policy objectives, an unusual policy declaration.[48] Nevertheless, the new policy contained some apparently unreconcilable contradictions. References to a high degree of safety, sound economic conditions, and destructive bidding practices seemed to favor companies already in business. The advocacy of competition, on the other hand, encouraged formation of new companies or more direct competition between existing ones. In setting up objectives that could not be simultaneously maximized, the statute gave the new regulatory agency considerable freedom of action in choosing between objectives.[49] Any critic of the agency could find within the general policy statement proof that his favorite objective had been slighted.

The airlines were turned over to detailed regulation by the classic device of the independent agency having quasi-executive, quasi-judicial, and quasi-legislative powers. The law set up the five-man Civil Aeronautics Board (CAB), each member appointed by the president to a six-year term and removable only for inefficiency or malfeasance. At first, the CAB headed a larger agency containing both the regulatory apparatus and the licensing and safety functions formerly within the Commerce Department. Reorganizations later separated the regulatory and licensing mechanisms; the latter remained within the Commerce Department between 1940 and 1958 and then went to an independent agency, the Federal Aviation Agency (FAA).[50] The CAB's regulatory powers have been relatively static since 1940, and quite extensive:

The regulation of fares and rates for carriage of persons and property;

47. *Ibid.*, Sec. 101.

48. Keyes, *op. cit.*, p. 60.

49. Richard E. Caves, *Air Transportation and its Regulators,* (Cambridge: Harvard University Press, 1962), pp. 126–27.

50. The reorganization is traced in *Celler Report, op. cit.*, p. 45. The 1958 re-enactment of the legislation created the Federal Aviation Agency as a new organization.

The fixing of subsidy and service mail rates;

The guarantee of loans to certain classes of carriers for purchases of new equipment;

The enforcement of the economic provisions of the Federal Aviation Act;

The approval or disapproval of mergers and control and interlocking relationships, and of intercarrier agreements affecting transportation;

The regulation of air carrier accounting practices and the development of air carrier reporting systems;

The maintenance of public records of tariffs, schedules, and other material required to be filed by air carriers;

The licensing of domestic air routes, and with approval of the President, of international air routes operated by U.S. and foreign air carriers;

Participation in the negotiation of air agreements between the United States and other governments covering the exchange of air rights;

Authorization of the navigation of foreign civil aircraft in the United States;

Assuring protection of the public by (1) requiring the performance of safe and adequate air carrier service, and (2) eliminating rate discrimination and unfair competition or unfair and deceptive practices in air transportation;

Investigation and determination of probable causes of aircraft accidents; and

Adjudication of appeals from safety enforcement decisions of the Administrator of the Federal Aviation Agency, and participation in safety rulemaking proceedings of the Administrator as appropriate.[51]

Despite this broad grant of authority, the legislation contained significant limitations on the CAB's powers. Existing air mail companies were authorized to stay in business (the origin of the term "grandfather carriers"). Some companies were exempted from regulation, principally those in short-haul charter, or air-taxi operations; the nonscheduled airlines later claimed this provision as their own. The needs of international relations forced a significant compromise with the principle of complete independence for the regulatory agency; the president retained authority over international route decisions.

51. U.S. Civil Aeronautics Board, *Annual Report, 1959* (Washington: Government Printing Office, 1960), p. 1 (hereinafter *CAB Annual Report*). Duties are in Richmond, *op. cit.*, p. 16.

The provision for "grandfather carriers" and the political necessity to keep subsidy appropriations as low as possible made it inevitable that the CAB would promote the economic viability of existing companies. Aircraft accidents just prior to 1938, moreover, indicated a connection between experience, economic health, and safety in operation. The general notion was that only a reasonable profit level insured that carriers performed all necessary mechanical repairs and did not otherwise cut corners.[52] Between 1938 and 1951, then, the CAB managed the industry with what might be called the "remainder" method. Having licensed a carrier to fly certain routes, the CAB computed the operating costs, estimated the passenger revenues, then included the difference and the necessary profit in the mail payments which were, in turn, charged to the Post Office budget. Since 1951 the subsidy and mail payments have been administratively separated.[53]

It is worth noting that the industry turned over to the CAB was only slightly different from the one Brown had organized in 1930. The "big four" (United, American, Trans-World, Eastern) dominated the list but there also was a group of smaller carriers, and it is this smaller group that has varied in size during the subsequent years.[54]

AN EVALUATION

The first twenty years of the airline industry included a progression from direct governmental promotion and operation, to haphazard expansion and uncontrolled corporate warfare, to abrupt disavowal of executive branch responsibility and, finally, to regulation by ostensibly independent agency. It was mentioned at the start of this chapter that analysis of the pre-1938 period is essential if one is to include the only instance in which the government explicitly accepted responsibility for a national airline system. To

52. Keyes, *op. cit.*, p. 90.

53. The relationship between subsidy and mail payments is complex and need not be detailed in this book. Suffice it to say there is no wholly satisfactory way to determine the exact amount of subsidy when an airline carries both government mail and private passengers. While subsidies and mail payments have been listed separately since 1951, it can be argued that subsidy still is concealed within the mail payments. Frederick, *op. cit.*, p. 243. Caves, *op. cit.*, pp. 254–59, 411–18.

54. Initially, the smaller group numbered fifteen, but this has been reduced to these seven: Braniff, Continental, Delta, National, Northeast, Northwest, Western.

say only that one administration expressed the public's interest while another did not, however, is surely not enough. The purpose of retracing old ground must be to ask new questions and to provide new insights.

There is considerable political irony in the chronology of events that led to the birth of the CAB. By 1938 political scientists, economists, and administrative lawyers were in the process of discovering that independent regulatory commissions needed systematic scholarly attention. The increased interest dates from the 1937 report of the President's Committee on Administrative Management (Brownlow Committee), which coined the phrase "headless fourth branch of government" to describe the collectivity of such agencies. The committee's recommendation, quite close to the Roosevelt administration's general philosophy, was that both the agencies and the industries they regulated should be made more responsive to Presidential policy direction and the national interest. Had this point of view been prominent in 1934, of course, the administration might not have rejected the Hoover approach, for it certainly incorporated policy control by the executive branch.

Professor Robert E. Cushman's staff experience with the Brownlow Committee inspired him to produce his pioneering study of the subject.[55] Since that time, scholars and government study groups have tended, for the most part, to discuss these agencies as an entity and to concentrate on their similarities instead of upon the differences between the industries they regulate. As the 1949 Hoover Commission put it, "The most basic fact about the commissions is the similarity of the situations with which they . . . deal."[56] This collective approach is something less than satisfactory, at least with respect to airline regulation.

The approach has led political scientists to argue for policy control by the executive, administrative lawyers to ask for more judicialization of the process so as to discourage contact between government and industry, and economists to demand that all such industries conform to the same competitive pattern. These are by no means rigid categories, but they are definable enough and significant enough to warrant comment.

Quite a few political scientists argue that the independent com-

55. Robert E. Cushman, *The Independent Regulatory Commissions,* (New York, London, and Toronto: Oxford University Press, 1941), pp. vii–viii.

56. Quoted in Leon I. Salomon (ed.), *The Independent Federal Regulatory Agencies* (New York: The H. W. Wilson Company, 1959), p. 30.

missions, designed originally to put important industries beyond the range of "politics," have in more recent years become the captives of the industries they are supposed to regulate.[57] One solution suggested has been to turn away from the principle of regulatory independence and return the commissions to executive policy control.[58] The collective approach used by administrative lawyers and economists has had a more demonstrable effect upon the CAB.

The essence of the regulatory process is the quasi-judicial approach, with the regulators sitting as judges who "decide controversies between government and citizen or among citizens."[59] The procedures include pre-hearing examinations, pre-hearings, analyses, conferences, oral arguments, cross-examination; as *Time* puts it, every citizen has the right to be heard, and lawyers have made an art of dragging out cases while being paid substantial fees.[60] To use the words of James M. Landis, the procedures are both "cumbersome" and "wasteful."[61] Perhaps the most important aspect of this judicialism is the degree to which the procedure, hence the activity of any commission, depends upon argument between two litigants. In its early days the CAB even used the device of

57. The argument generally is that commissions are created in an attempt to control the evils associated with monopolistic practices. The desire for reform gradually is corrupted as the commissioners come in closer and closer contact with the members of the industry. Tempted by extracurricular favors and the prospect of lucrative jobs after they end their government service, the regulators take an ever narrower view of their responsibilities and equate the public interest with the industry's interest. For a compilation of views on the subject, see Samuel Krislov and Lloyd D. Musolf *The Politics of Regulation* (Boston: Houghton Mifflin Company, 1964), p. 261, and Salomon (ed.), *op. cit.* Probably the most critical of recent books is Bernard Schwartz, *The Professor and the Commissions* (New York: Alfred A. Knopf, 1959).

58. Political scientists, to be sure, hedge their recommendations for increased policy control. Professor Emmette S. Redford, for example, says that policy guidance should be limited to general statements in areas that the agencies cannot resolve. The agencies then would use the guidance to make day-to-day decisions. He adds, however, that all delegations of power to the president to approve specific decisions should be removed. Taken as a whole, Professor Redford's ideas would both increase and decrease executive control because, among other things, the president would lose authority over international airline cases. Emmette S. Redford, "The President and the Regulatory Commissions," Report Prepared for the President's Advisory Committee on Government Reorganization (Mimeographed: Washington, 1961) pp. 5, 25–28.

59. Henry J. Friendly, *The Federal Administrative Agencies* (Cambridge: Harvard University Press, 1962), p. 156.

60. *Time*, March 17, 1958, quoted in Salomon (ed.), *op. cit.*, p. 96.

61. U.S. Congress, Senate, Committee on the Judiciary, *Report on Regulatory Agencies to the President-Elect*, 86th Congress, 2nd Session Committee Print (Washington: Government Printing Office, 1960), p. 41.

the public counsel, a staff attorney who ostensibly represented the consumer and thus opposed the airlines.[62] The long-term results of this approach have been to encourage more direct competition between airlines, so that any route question always will involve two companies, and to promote the implicit assumption that any request from an airline is a setback for the public interest. Scholars have argued that this process can be depoliticized by separating the administrative and judicial functions of the agencies.[63]

The intellectual pressures mentioned thus far have combined with a third, contributed this time by economists. The cost of airplanes and their necessary appurtenances was low enough in 1938 to make it appear that airlines would forever be quite easy to organize. It followed that the public interest would be served best by having as many suppliers of the service as possible, that competitive price bidding should be used for all government contracts, and that the ultimate objective should be free and unregulated competition between carriers. These arguments turned on the old economic question of whether the airline industry should be considered a "natural monopoly."[64] By treating all regulatory agencies on the same basis, then, scholars have asked for more executive policy guidance, the freest possible competition among the greatest number of companies, and an ever increasing judicialization of the entire process.

Those who have studied the specific problems of the airlines, without reference to regulatory agencies which supervise other industries, have come up with different answers. It has been mentioned that those who thoroughly analyzed Brown's activities tended to praise him. The Federal Aviation Commission, whose recommendations started the ball rolling toward the CAB, suggested the use of indirect competition as the best solution for the national airline system.[65] Other students of aviation have agreed

62. The CAB encouraged the counsel to oppose the arguments advanced by industry. Spencer, *op. cit.*, pp. 247–48. Henry Ladd Smith, *Airways Abroad* (Madison: University of Wisconsin Press, 1950), pp. 45, 275.

63. See summary in Krislov and Musolf (eds.), *op. cit.*, pp. 149–83.

64. A "natural monopoly" exists when the most efficient size of the firm is so large relative to the market that normal competition cannot be relied upon to protect the public from unfair pricing and discrimination. Keyes, *op. cit.*, pp. 74–75, 99.

65. "If an airline running from coast to coast acquires faster and more comfortable airplanes, it takes but a little time for the patrons of a line running up and down the Mississippi Valley to complain if it fails to make the same advances." Quoted in Keyes, *op. cit.*, p. 101.

that airlines were more similar (even in 1938) to public utilities than to anything else, that competitive bidding has proved unsatisfactory in an industry needing subsidy, and that freedom from regulation could not be relied upon to provide transportation in travel markets that were not inherently profitable.[66]

Most important to this book is the degree to which the treatment of regulatory agencies as virtually identical to each other has obscured from the start the crucial question of international aviation. This was understandable during the interwar period, for the Atlantic and Pacific oceans still were formidable barriers to an international airline industry. The statutory provisions for international regulation seemed almost an afterthought when the U.S. was not overly conscious of its world position. While the president had to approve international decisions, such cases followed the same procedural route as domestic cases. This had the unfortunate effect of making it appear as though a single system of regulation was in use for both types of cases and, under the circumstances, the CAB and the executive agencies concluded that international cases also should involve two contending parties. It was quickly forgotten that international aviation could not be left to the managerial control of an agency without political responsibility.

66. Frederick, *op. cit.*, p. 85. Richmond, *op. cit.*, pp. 7–9. Smith, *Airways*, pp. 302–3, Spencer, *op. cit.*, p. 227.

[2]

The Distinctive Nature of International Operations

THE U.S. AIRLINE INDUSTRY grew to prominence within a political, economic, military, and geographic environment that seemed to make it unnecessary to search for an important relationship between the international airlines and world politics. To the extent that international operations were considered at all, they were viewed as ordinary extensions of the domestic system. Scholarly studies of airline regulation exhibited the same indifference and viewed the CAB as a regulatory agency no less qualified to supervise international operations than to arrange for the airlift of domestic air mail. This chapter is designed to show that even before World War II, however, there was evidence to demonstrate the important differences between domestic and international operations.

The first such evidence was a realization by European countries that their geographic environments made any sort of airline industry an immediate item of foreign policy. For U.S. policymakers the New York-Washington route was a purely domestic affair but the London-Paris route, while almost identical in length, was a matter for international negotiation. This forced Europeans to make all their decisions in the light of the political, economic, and military aspects of the airline industry. Some U.S. military officers argued in the 1920's that the U.S. should adopt the system already prevalent in Europe, but the trend toward nationalization was not entirely relevant to the American context. Government and industry thus had to evolve their own distinctive relationship, and in a way that ultimately produced results far more satisfactory than those realized in Europe. Even while achieving success, however, the U.S. solution succumbed to the notion that the international airlines must conform to the economic and procedural model of CAB regulation.

The "chosen instrument" concept is the key to an understanding of the European and American experiences. The concept has traditionally meant that a common carrier is granted an exclusive franchise over certain routes and is guaranteed a "fair rate of exchange" for servicing those routes.[1] The carrier, in return, makes its planes and pilots available for military use whenever necessary. The Europeans generally have used the "chosen instrument" principle. The American tendency, largely an outgrowth of the 1934 air mail cancellations, has been to deny that the international carriers have any place in foreign policy, despite persistent evidence to the contrary.

Whatever the environmental differences, the U.S. and European systems evolved within a shared agreement on certain points of international law: (1) each state has sovereignty over its own airspace; (2) each state can admit or not admit aircraft into its airspace; and (3) airspace over the high seas, or any other location not subject to a state or other jurisdiction, is free to all.[2]

THE EUROPEAN TREND TOWARD NATIONALIZATION

The most significant aspect of the European experience was the degree to which those countries were conscious from the beginning of the importance of the international airlines to national prestige, influence, and imperial unity.[3] Most European countries had overseas territories or colonies reachable by air, even with the short-range airplanes of that time. This direct relationship caused all of the European countries to nationalize their airlines in one way or another, and to accept the necessity for direct subsidy.[4]

The Germans concentrated on commercial aviation because they were initially denied a military role after World War I. Even in the 1920's they developed a geopolitical theory of airlift which envisaged a Pan-European airline expanding to South America and the Far East. This was to be an instrument of the German search

1. Oliver James Lissitzyn, *International Air Transport and National Policy* (New York: Council on Foreign Relations, 1942), p. 209.

2. *Ibid.*, p. 365. Robin Higham, *Britain's Imperial Air Routes, 1918 to 1939* (London: G. T. Foulis and Co., Ltd., 1960), p. 19.

3. Lissitzyn, *op. cit.*, pp. 56–58, 64–67.

4. It is worth noting that the nature of colonial routes also made it unlikely that heavy traffic would be available. See *ibid.*, pp. 115–21.

for lebensraum and an economic pincers to overcome British and U.S. competition.[5] The vision did not materialize, but there were several instances of close co-operation between European airlines; the French and Belgians formed a single company for African operations, and the French and Germans (and later the Italians) joined forces in South America.[6] The Germans provided technical assistance to several countries in South America and had a direct interest in airlines formed in Chile and Colombia.[7]

The British unified military and civil aviation within a single governmental ministry, with the latter being left to "fly by itself," in Winston Churchill's phrase.[8] The attitude persisted only until 1921, when the first British cross-channel airline was forced out of business by subsidized continental competitors.[9] The British then concluded that it made little sense to subsidize companies in direct competition against each other. After a two-year experiment in which four airlines flew separate and non-competitive routes, the British decided in 1923 to form a single chosen instrument.[10] By this time the Air Ministry had complete control of all aviation, and the pilots of the chosen instrument were in the military reserve. Civil aviation needs had lower priority than did a strategic bombardment force and squadrons for the Royal Navy, and this was disastrous in a period of peace.[11]

The British decided in the 1930's that the existing chosen instrument was too conservative and unimaginative, given the absence of any competition. Conservative members of Parliament sought a new solution that would provide for indirect competition between companies having different routes. In the confusion attending the attempts at reorganization, the British adopted an even more rigid solution than before. Just prior to World War II, the new company emerged with the title of British Overseas Airways Corporation

5. *Ibid.*, pp. 93, 336–37; William A. M. Burden, *The Struggle for Airways in Latin America* (New York: Council on Foreign Relations, 1943), pp. 38–39.
6. *Ibid.*, pp. 93, 336–37; Lissitzyn, *op. cit.*, pp. 334–35, 396.
7. Burden, *op. cit.*, pp. 38–46. It might be noted that much of South America is as close to Europe as to the U.S.
8. Churchill was then Secretary of State for War and Air. Higham, *op. cit.*, p. 23.
9. *Ibid.*, pp. 24–25; Lissitzyn, *op. cit.*, pp. 174–78.
10. The decision was made hurriedly by a Conservative government just about to leave office. The Labor Party had demanded complete nationalization. Higham, *op. cit.*, pp. 49, 67–69.
11. Higham describes British commercial transport as having been a "stepchild" during this period. *Ibid.*, p. 40.

(BOAC).[12] This was a public corporation and, by any reasonable standard, a completely nationalized enterprise.

Students of interwar international airline activity conclude that the Europeans probably went too far toward complete unification and nationalization. Had each government organized one other company along largely noncompetitive routes, it could have compared one company against the other. This would have been especially appropriate in the British case, given the geographical extent of the empire.[13] Moreover, says Oliver Lissitzyn, the European system had distinct foreign policy disadvantages. For one thing, private operation seemed more in the tradition of the Western nations. It would have been advantageous, moreover, to have overseas companies at least partially independent of the government, and not totally identified as instruments of foreign policy.[14]

Despite all the attempts to make airlift subordinate to government policy control, then, the Europeans were much less successful than the U.S., with its vague and pragmatic approach toward unofficial partnership between government and airlines. In one respect, however, the U.S. approach was anything but vague; the government had to face quite early the issue of nationalization.

THE AMERICAN REJECTION OF NATIONALIZATION AND MILITARY CONTROL

The decision to completely separate commercial and military aviation was the most significant U.S. decision in the first decade of the airline industry. This proved a highly satisfactory solution and gave the U.S. a position of world leadership in commercial aviation. To achieve this outcome it was first necessary to deny the bid of military airmen for immediate adoption of the British model. This aspect of the relationship seems to have been overlooked in some analyses of U.S. military history.[15]

The central figure in U.S. military aviation in the 1920's was General William "Billy" Mitchell, ultimately punished for carry-

12. The British intended to create three companies, one of which would be owned jointly by the other two. The companies were joined into one at the insistence of a former head of the British Broadcasting Corporation who was called in to manage one of the companies. *Ibid.*, p. 264.

13. *Ibid.*, p. 312; Lissitzyn, *op. cit.*, pp. 282–83.

14. *Ibid.*, pp. 132–33.

15. There is a notable exception. This discussion amplifies but does not alter the findings of one pioneer study. Lawrence J. Legere, "Unification of the Armed Forces" (unpublished doctoral dissertation, Harvard University, 1950), pp. 93–127.

ing to the point of insubordination his protests against what he believed the neglect of military aviation. His approach often is described as having involved only the narrow military issue of whether aviation (especially bombardment) should be removed from the Army and Navy and organized as a separate military service. One of his biographers, for example, describes the struggle as one between: " . . . farseeing political leaders, progressive industrialists, prophetic airmen, independent generals, and rugged sea dogs of vision, and a press . . . which envisioned aviation in its true scope . . . and high office-holders with narrow views, the entrenched bureaucrats in the war and navy departments who could not endure the thought of a rival service, conservatives and skeptics who were always in opposition to anything new."[16] Bernard Brodie describes Mitchell's contribution as largely in his "quality and conduct as a man and officer rather than in his stature as a military philosopher."[17] Yet Mitchell's theories ranged far and wide, well beyond narrowly military considerations. A typical example was his view of how the U.S. airline industry should be organized.

He was quite specific in recommending the unification of all forms of aviation, civil and military. On top of his suggested organizational pyramid would have been a department of defense headed by a cabinet officer and containing three subordinate departments of Air, Army, and Navy. In line with the British model, the aviation department would have been responsible for national progress in all forms of aviation, and it would have set up and operated experimental airlines. Flying airplanes developed internally, government corporations would produce financial and operational data concerning the efficiency of specific airplanes. The most efficient would then be made available to those private operators permitted to organize their own airlines. The process would have been continuous, with the government always testing new aircraft and operating experimental aircraft along the same routes as private operators having earlier models.[18]

It is difficult to see how this could have led to anything but a

16. Isaac Don Levine, *Mitchell, Pioneer of Air Power* (Cleveland and New York: World Publishing Company, 1943), p. 173.

17. Bernard Brodie, *Strategy in the Missile Age* (Princeton: Princeton University Press, 1959), p. 77.

18. William Mitchell, *Winged Defense* (New York and London: G. P. Putnam's Sons, 1925), pp. xviii–xix, 89, 114–15.

totally nationalized industry. Mitchell's airline proposals remain largely ignored, a factor which has led to misinterpretations of of the recommendations made by important government study groups of the 1920's, as they attempted to produce a national aviation policy.

One such group was the Crowell Commission, a civilian advisory body which Secretary of War Newton D. Baker appointed to study European experience after World War I. One paragraph of the Commission's 1920 report often is interpreted as an unequivocal endorsement of Mitchell's proposals: " . . . the concentration of the air activities of the United States, military, naval, and civilian, within the direction of a single Government agency, created for the purpose, co-equal in importance with the Department of War, Navy, and of Commerce, to be called in this report . . . the National Air Service."[19]

Later in the same report, however, the Crowell Commission advocated the use of air mail to create a healthy industry organized and operated as a private enterprise.[20] Taken as a whole, then, the Commission's report favored separation of military and commercial aviation; the reference to civil aviation within a National Air Service seems only to have meant miscellaneous functions of government aviation, e.g., aircraft used by the agency charged with licensing pilots and aircraft.

The Lampert Committee, a select committee formed in 1925 by the House of Representatives, concentrated on military aviation. Air Force histories generally credit the committee with having been sympathetic to Mitchell and his colleagues, and with having advocated both a unified air force and a single department of defense.[21] The committee favored a single defense department, but it recommended only that the Army and Navy provide more adequate representation of aviation interests at the higher staff levels

19. Wesley Frank Craven and James Lea Cate (eds.), *Plans and Early Operations, January 1939 to August 1942* (*The Army Air Forces in World War II* [Chicago: The University of Chicago Press, 1948]), I, 24. See also Mitchell, *op. cit.*, p. 117; Levine, *op. cit.*, p. 178.

20. Reproduced in U.S. Congress, House, Select Committee of Inquiry into Operations of the United States Air Services, *Testimony of Major Edward V. Rickenbacker*, Hearings, 68th Congress, 2nd Session (Washington: Government Printing Office, 1925), II 1229–30.

21. Alfred Goldberg (ed.), *A History of the United States Air Force, 1907–1957* (Princeton: D. Van Nostrand Company, Inc., 1957), p. 36; Craven and Cate (eds.), *op. cit.*, I, 29; Robert E. Cushman, *The Independent Regulatory Commissions* (New York: Oxford University Press, 1941), p. 390.

—there was no mention of a separate air force: " . . . the Air Services of both the Army and Navy should . . . be adequately represented on the General Staff of the Army and the General Board of the Navy by members who will firmly support the full and complete use of Army and Navy aviation."[22]

The Morrow Board, appointed in 1925, was perhaps the most important group to study aviation policy during the interwar period. Unlike the Lampert Committee and the Crowell Commission, the Morrow Board was generally thought hostile to Mitchell.[23] Its report was the first to emphasize export sales of aircraft as a means of maintaining a production mobilization base without large government expenditures.[24] This particular aspect of government activity has been one of the most significant, if least understood, components of national aviation policy, and it will reappear in later chapters of this book. The Morrow Board was initially significant, however, because it bluntly stated that military and civil aviation should be kept completely separated.[25] Henry Ladd Smith concludes that the board set the definitive national policy on the subject.[26]

One might conclude that the area of political agreement on aviation policy was much broader than is indicated in accounts based chiefly upon organizational struggles within the military. Had military airmen had their way, civil aviation would have been subordinated to its military counterpart. As in Great Britain the result might have been disastrous in the interwar period. In retrospect Mitchell and his colleagues had much less support than they thought, perhaps because some of their proposals ran counter to the traditional American notion that the government, especially

22. U.S. Congress, House, Select Committee of Inquiry into Operations of the United States Air Services, *Inquiry into Operations of the United States Air Services*, 68th Congress, 2nd Session H. Report 1653 (Washington: Government Printing Office, 1925), p. 9.

23. Mitchell considered the report one that "retarded our progress in aviation for ten years," as among the "blackest pages ever written into our Government's records," and as a surrender to the "air trust," accused of building unsafe airplanes for military use in World War I. Craven and Cate (eds.), *op. cit.*, I, 28; H. H. Arnold, *Global Mission* (New York: Harper & Brothers, 1949), p. 120; Emile Gauvreau and Lester Cohen, *Billy Mitchell* (New York: E. P. Dutton, 1942), pp. 128–31, 163, 169.

24. *Ibid.*, p. 170.

25. U.S. *Report of the President's Aircraft Board* (Washington: Government Printing Office, 1925), p. 6.

26. Henry Ladd Smith, *Airways* (New York: Alfred A. Knopf, 1942), pp. 98–99.

the military, should neither organize nor dominate any important segment of the national economy.

THE EMERGENCE OF AN AMERICAN INTERNATIONAL INDUSTRY: PARTNERSHIP AND ITS REJECTION

Until the eve of World War II the U.S. had an international airlift policy, not spelled out in detail, which included a useful partnership between government and industry. All of this gradually changed, however, when the CAB entered the picture. The Roosevelt administration's attempts to deny any partnership and to increase direct competition in the international field led to permanent hostility between the government and Pan American (PAA), the only U.S. international carrier prior to World War II. While the partnership was active, however, the U.S. achieved quite a few specific foreign policy objectives.

The military provided the initial impetus for international operations. Major Henry H. Arnold, later to become chief of the Army Air Forces in World War II, noticed in the mid-1920's the expansion toward the Panama Canal of a German-controlled Colombian airline. Stationed in Washington at the time, he consulted with the postmaster general who advised that unless someone organized an American company to carry air mail to Latin America, he might have to award the Colombian airline an air mail contract. Arnold and Major Carl Spaatz drew up plans for an American company and found an individual willing to try to raise the capital. On the basis of only a prospectus for future operation of a company to be named "Pan American," the postmaster general denied the Colombian request for a contract.[27] This action set off the rush to organize American companies for operation in Latin America.

The most successful of the prospective entrepreneurs was Juan Trippe, organizer of the company that ultimately became the present-day Pan American. When Postmaster General Brown reorganized the airline map he chose the company as the single overseas mail carrier, largely because Trippe had obtained exclusive landing concessions from the Cuban government. Through negotiations with other Latin American governments, sometimes

27. Arnold, *op. cit.*, pp. 115–16.

helped along by the State Department, Trippe gradually expanded his route system within the hemisphere. By first securing landing rights from the governments involved, Trippe made certain that none of his potential competitors could provide air mail service. He could then submit air mail bids at the highest allowable rate, knowing he had no real competition.[28]

Most of Pan American's initial route expansion occurred during the Hoover administration; once Brown settled on PAA as the overseas carrier, its expansion conformed to his notions of how the airlines should expand. He thus warned PAA and domestic companies to stay out of each other's markets.[29] The fact that Latin America was the natural geographic area in which the U.S. first operated an international airline led directly to the informal partnership between government and industry.

Whatever its feelings about airlines in general, the Roosevelt administration had little choice but to co-operate with the international industry in an arrangement that proved highly successful. In a departure from its condemnation of the industry, the administration abstained from canceling PAA's contracts; the action would have hindered the "Good Neighbor" policy, given PAA's extensive contacts in that area. When PAA turned to the Pacific in the 1930's, it found an ally in the Navy Department, equally interested in building up the Pacific islands (Wake, Midway, Guam) as outposts of U.S. military and economic strength. The Navy gave whatever assistance it could, within the limits set by sparse military appropriations. Farley himself demonstrated the national interest in the program by presiding at the inauguration in 1935 of PAA's famed "Clipper Service" across the Pacific.[30]

When the administration became concerned over Axis penetration into Latin America, PAA rather quickly became a secret instrument of foreign policy. In 1939 the government supplied the company with funds to develop new and expanded services in Latin America as an offset to Italian and German operations. By 1941 the Reconstruction Finance Corporation had set up a subsidiary to disburse funds for the same purpose. The economic pressure ultimately forced the German-controlled Colombian airline to sell out to PAA, and the company then replaced the Ger-

28. Henry Ladd Smith, *Airways Abroad* (Madison: University of Wisconsin Press, 1950), p. 59.
29. *Ibid.*, p. 32.
30. *Ibid.*, pp. 34–35.

mans still working with the company.[31] PAA also became the contractor for the secret Airport Development Program, building or improving some fifty airfields in the Latin American area.[32] These airfields, in turn, made up the route system along which the U.S. delivered supplies and aircraft to Great Britain before American entry into World War II. It is difficult to describe any of these activities as anything but generally useful and necessary to the attainment of important policy objectives.

Both the company and the government, however, contributed to a continuously strained relationship. PAA took a long time to remove German influence from the Colombian airline, even though a complex problem of severance pay prolonged an undesirable situation. The government suspected the company of having gained too much economic advantage from building Latin American airfields. There was evidence also that PAA, relatively secure in its position in Latin America, suppressed technological advancement in that area in the 1930's, e.g., the company was slow to begin flying at night.[33] This apparent failure to exploit the Latin American market occurred while PAA was concentrating its attention on the Pacific, and before the administration became interested in Axis activity. The CAB's public counsel and the Post Office jointly attacked PAA's subsidy payments, pointing out that the company had been paid as much as all the domestic companies combined.[34]

It is difficult to evaluate each of these assertions, but the combined attack upon PAA was questionable on several grounds. PAA's subsidy payments, for example, could not logically be compared with those paid domestic companies. Government air-

31. *Ibid.*, pp. 66–69; Lissitzyn, *op. cit.*, pp. 331–32; Burden, *op. cit.*, pp. 63–79.

32. U.S. Congress, House, Committee on the Judiciary, *Report of the Antitrust Subcommittee (Subcommittee No. 5) Pursuant to H. Res. 107 on Airlines*, 85th Congress, 1st Session Committee Print (Washington: Government Printing Office, 1957), pp. 261–62 (hereinafter *Celler Report*).

33. This was the evaluation of a technical adviser to the American Embassy in Brazil, made in early 1942. He had been an airline executive and became an Air Force general during World War II. U.S. Congress, House, Committee on the Judiciary, *Monopoly Problems in Regulated Industries*, Hearings before Subcommittee, 84th Congress, 2nd Session, on Airlines (Washington: Government Printing Office, 1956), IV, 2940–41 (hereinafter *Celler Committee Hearings*).

34. Burden, *op. cit.*, p. 117. Between 1929 and 1939, PAA received some $47 million in payments, all other airlines receiving about $50 million. Smith, *Airways Abroad*, p. 44.

fields and airways were not available in many overseas areas; the company provided them for itself, and this made its operations much more expensive than the domestic norm. Even while attacking PAA, moreover, the government was providing the company with funds for counter-Axis activity.[35] In the midst of this combination of public attack and secret support, the administration apparently decided to establish a new and more explicit national policy.

The opportunity arrived when an American steamship company decided to organize its own international airline. The Navy lent some of its pilots to American Export Airlines (AMEX) so the company could survey the Atlantic route which promised to be the most lucrative of all, if the most difficult to traverse because of weather conditions. By the time the AMEX petition for an Atlantic route reached the Civil Aeronautics Board, World War II had sealed off most of Europe; the decision turned only on the question of direct competition on the single route to Lisbon. Government agencies supported AMEX and testified to the emergency needs of the government in a world in crisis. The Post Office admitted it would have to increase mail payments to offset the cost of supporting two companies, but it argued that the emergency made the increase acceptable.[36] Other agencies apparently assumed that two companies on the same route automatically would provide increased airlift capacity.[37] The CAB, in its first major international decision, relied upon the testimony from the government agencies and approved the AMEX application. When the president approved the decision, a new partisan contest began.

AMEX could not begin operation until assured that mail payments would be forthcoming, and Congress refused to appropriate the funds. PAA had acquired considerable support within Congress from its long record of pioneering in aviation, and the company called upon all its support.[38] Even though the AMEX certifi-

35. It might be argued that the Roosevelt administration had to conceal its association with PAA, especially where the activity involved neutral countries. Even if this argument be conceded, however, the attacks on the company seem to have been unduly harsh.

36. *Ibid.*, pp. 46–61; *Celler Report*, pp. 240–41, 252; *Celler Committee Hearings*, IV, 2901–2; Lissitzyn, *op. cit.*, pp. 243–51.

37. On any given route, the total number of aircraft operating is a function of the traffic available. Yet, many government agencies always have assumed that more airlines mean more airplanes.

38. Smith, *Airways Abroad*, p. 59.

cate had been issued on emergency grounds, Congress chose to deal with the other complicating issues involved in the case.[39] The increase in mail appropriations was distasteful, of course, and the case also raised the question of whether one form of transportation (steamships) should operate in a competing field. Most importantly, PAA had proved itself the political master of the administration, an unhappy situation for both parties.

In choosing to ignore or overlook the accomplishments of the government-industry partnership, the administration perpetuated the deficiencies of the policy statements of 1934. The partnership had lacked an explicit delineation of what it entailed, but it had worked; it had given the U.S. the world lead in international air transport. The AMEX petition, on the other hand, emerged from the framework of economic and administrative theories that prescribed the need for two contending parties to appear before a court of equity. Had a more detailed analysis been undertaken, a less drastic solution might have suggested itself.

PAA's expansion had occurred in three distinct stages, each involving a separate geographic region. Once the company established itself in one region, the government might have insisted that a new company be formed for each of the other regions. The administration, of course, would have had to acknowledge its responsibility for the desired outcomes. Unfortunately, PAA's expansion beyond Latin America occurred during the four-year policy vacuum that followed the 1934 cancellations.

The hostility resulting from the AMEX case was, if anything, more intense than had followed the 1934 action. The administration suffered a direct political defeat, and this could only have increased its determination to ultimately reverse the congressional decision. It will be seen in the next two chapters that the administration went so far as to form a new alliance with the domestic airlines to break PAA's seeming monopoly in the international field, even if the action was delayed by the war. Perhaps the most significant result of the AMEX case, however, was the almost totally confused pattern of international regulation that emerged.

Since this first explicit international decision, the question of

39. The AMEX case also demonstrated the lack of agreement as to Congressional intentions when the Civil Aeronautics Act was passed. Senators Harry Truman and Pat McCarran had been equally active in the passage of the statute; Truman voted for the AMEX payments and McCarran voted against them. *Congressional Record*, May 7, 1941.

responsibility for the status of the international airlines has been unclear. In making its case for direct competition the administration relied upon foreign policy arguments and thus admitted that the industry was something other than a purely commercial enterprise. The CAB, in turn, relied upon the administration's arguments, admitting that only the president could evaluate the needs of foreign policy. The president then approved the CAB recommendation, thus making it appear he was concurring in a nonpolitical decision reached by an independent agency. This established the permanent pattern of international regulation, with the CAB and the executive agencies holding each other responsible for the decisions finally made in each instance; neither side has been willing to accept final responsibility. Yet any realistic analysis indicates that the CAB is anything but an independent agency when it acts in international cases.

The CAB functions only as a normal agency of the executive branch, and the president is no more bound to follow its advice than he is the advice of any other executive agency. The U.S. has been left with a wholly unreal arrangement in which an ostensibly independent agency functions as a foreign policy agency of the executive branch, the members of the agency are not subject to removal at the pleasure of the president, and the entire process takes place within a quasi-judicial framework. The procedural atmosphere effectively eliminates the possibility of meaningful partnership between government and industry, simply because judges must maintain a respectful distance from litigants.

AN EVALUATION

This chapter attempts to demonstrate that the decision to manage both domestic and international airline systems within a single regulatory framework was of dubious validity. Acceptance or rejection of the argument, however, depends upon one's intellectual approach, or assumptions. It should be possible to evaluate the U.S. and European experiences by applying a brief test to several such approaches.

It might be argued, for example, that a U.S. international carrier is only a normal commercial enterprise and that it should have to take its chances within a truly competitive environment. But if one uses such an argument he should be prepared to accept either of two major consequences. One is that the airline may not succeed in overcoming its foreign competitors, it may then have to go out of business (as the British did in 1921), and the U.S. may be left with no international carrier of its own. A second consequence is that the government cannot expect the purely commercial enterprise to function as an instrument of national policy except in time of war and, moreover, the government must be prepared to tolerate whatever activities are undertaken by the airlines of other countries. Given the necessities of international politics, these consequences appear serious enough to warrant immediate rejection of this approach.

A second approach is exemplified by the Roosevelt administration's decision to promote additional competition on the Atlantic routes. It might be argued that while an international airline is, in some respects, an instrument of national policy, each airline will perform more effectively if threatened by competition from another U.S. airline as well as from airlines of other countries. There seem to be three replies to the argument, one of which is that multiple competition among U.S. companies requires that each maintain a costly set of facilities at each of the airfields used along the way. A second rejoinder follows from the first; it is likely that the cost of subsidy will be higher than with a single company. Yet a third rejoinder involves a new question: what values and interests are at stake in international competition?

If the U.S. promotes two international companies in competition against each other, both companies are (presumably) sometime instruments of national policy. Each company's competitive objective, however, must be elimination of the other. If each must consider all its competitors as threats, then its fellow countrymen appear little different than the airlines of friendly or unfriendly governments. The competive objective thus puts each company at odds with the national interest, because it was national policy which set up multiple competition in the first place. Neither of two U.S. companies can afford to be self-effacing in its view of the national interest for fear its U.S. opponent will gain an insurmountable competitive advantage. There is an apt analogy here to theories of international alliance behavior. Each U.S. company is tempted to ally itself with a foreign competitor if this will eliminate a third competitor who also is American. In sum, direct competition between airlines of the same country tends to substitute narrowly competitive values and interests for the national interest. It can be argued, then, that if one admits that international airlines are or can become direct instruments of foreign policy, direct competition between airlines of the same country is, at best, self-defeating.

Still another approach might suggest that the government needs fairly direct control if it is to be certain that an international airline will respond to the needs of foreign policy. It seems reasonably clear that if the government wants absolute assurance of immediate and satisfactory response, the only way to get it is through something that approaches nationalization. When the Roosevelt administration introduced direct competition into the international industry and further separated the industry from direct contact with the government itself, it virtually prevented the industry from being more responsive to executive policy direction. If nationalization is out of the question, and if some partnership is the only answer, then a certain amount of tension and dispute seems acceptable in the interest of maintaining the strength of the partnership.

The vague interwar partnership between government and industry seems to have been the only reasonable solution open to the U.S., and it turned out to be a better one than produced elsewhere. No matter its success, however, the administration chose to reject it. This is not to argue that the U.S. should have permitted a single company to operate all the international routes. Taken together, European and American experience suggests that the U.S. might have tried the regional system mentioned earlier. The U.S. would have had to accept indirect competition, some withdrawal of the international airlines from the CAB regulatory process, and explicit recognition of the connection between the airlines and foreign policy. It is worth repeating that, given the relationship between the CAB and the president in international cases, all of this was within the power of the executive branch.

PART TWO

World War II and its Aftermath

[3]

World War II: The Traditions and Dimensions of Military Airlift

U.S. ENTRY INTO WORLD WAR II halted the political and economic conflicts discussed in previous chapters, everyone now looking for ways in which air transport might contribute to the national effort. The war marked the beginning of airlift as a significant aspect of military operations.

Strategic airlift (between the U.S. and overseas areas) was entirely new and, until the war, virtually unknown. It developed from a general theory of mobilization, in which all commercial transportation systems were to work for the government during emergencies. The airlines themselves built and operated the wartime strategic airlift organizations (the Army's Air Transport Command and the Naval Air Transport Service); these units, while in the military services, were yet another form of government-industry partnership.[1] In a departure from military traditions, the Army and Navy retained in Washington the control and management of these airlift systems, not delegating those functions to

1. Airlift usually is defined as being strategic or tactical, but the distinctions are not clear. The terms "intertheater" and "intratheater," respectively, also are used upon occasion, the former connoting operations between the U.S. and overseas theaters, or between theaters. The "Hump" operation between India and China in World War II, in which Chinese and American forces were supplied by air, is called "the birthplace of strategic airlift," and the forerunner of the Berlin Airlift, even though it occurred wholly within an overseas theater. Wesley Frank Craven and James Lea Cate (eds.), *Services Around the World* (*The Army Air Forces in World War II* [Chicago: The University of Chicago Press, 1958]), VII, 151. The Hump and the Berlin Airlift are known as strategic operations apparently because of their massive, sustained, and scheduled nature, and perhaps because of their political importance. The terms "strategic" and "tactical" are used here in their most common form which, in retrospect, seem based upon the degree of direct association with the Army.

theater commanders.[2] In a sense, World War II strategic airlift was on the fringes of the military services and their more traditional associations with the purely combat functions of military aviation.

Wartime tactical airlift was, on the other hand, more in the nature of a purely military function, operated and organized in much the same manner as any other form of combat aviation. One principal task was distribution of goods and people within a theater of operations. A second task was the dropping of paratroopers employed in airborne assault operations. These tasks conflicted with each other, each requiring its own type of airlift organization. This conflict ultimately had much to do with the airlift controversy of the 1950's, because the leading Army spokesmen of that later period were generals who gained their airlift experience in the airborne operations of World War II and Korea. Whatever the tasks of tactical airlift operations, however, they were controlled by theater commanders, e.g., General Eisenhower in Europe.

The next four chapters are concerned with the addition to airlift policy issues of these two military dimensions, and the postwar policy changes that led so directly to the controversies of the 1950's. This chapter deals with wartime experiences in both strategic and tactical airlift, the government-industry partnership that characterized the former, and the disputes between high and low levels of command that marked the latter. Tactical airlift experience raised an important question that never has been faced nor answered: Exactly how effective are massive airborne assaults?

Chapter 4 details the changes in commercial airlift structure that were direct outcomes of World War II. While the U.S. made certain that its domestic airlines would extend their routes around the world and become direct competitors against each other and especially against Pan American, it had to accept an international agreement that made such competition more difficult. U.S. international carriers ultimately came to feel the government had sacrificed their interests to those of their foreign competitors, and the

2. General Eisenhower controlled Allied bombers during preparations for the World War II invasion of Europe and until September, 1944. After that time, the Combined Chiefs of Staff (Britain and U.S.) controlled them, acting through an agency located in Britain, but Eisenhower retained some effective authority. Dwight D. Eisenhower, *Crusade in Europe* (Garden City: Permabooks, 1952), pp. 256–57, 346. The Air Force in Washington controlled the B-29's which bombed Japan late in the war.

companies were quite willing to join any movement that seemed to promise them relief from economic difficulties.

Chapter 5 analyzes the postwar nonscheduled airlines, the result of the government's decision to sell off surplus aircraft. These airlines never captured a significant share of the market, but they became politically identified as enterprising small businesses important to Cold War strategy, especially the Berlin Airlift and the Korean War.

Chapter 6 traces the postwar evolution of military airlift, to demonstrate how both strategic and tactical operations contributed to disputes within the military services and between them and the industry. The Army, arguing that the Air Force had neglected tactical airlift, submitted numerous requests for assistance. Strategic airlift, on the other hand, became the target of external groups, e.g., the Hoover Commission, which accused it of unfair competition with private enterprise.

TRADITIONS OF MILITARY AVIATION: PARATROOPERS, TWO AIR FORCES, AND AN AIRLINE RESERVE

While Mitchell's theories contemplated total unification of all aviation, the Air Force's combat traditions envisaged operational separation of civil and military aviation. To the military airmen, the airplane was a combat weapon; military responsibilities were limited to development of combat potential. Using this logic military airmen generally ignored noncombat functions of aviation. Their first thought of air transport was as a combat weapon within a broader framework of what can be described here as two military air forces.

Mitchell established one tradition when he proposed during World War I that airplanes drop Army troops behind German lines. General Pershing approved the plan and it would have been the first paratrooper operation in history, but a change in the battlefield situation led him to cancel the project.[3] The proposal was significant in that an operation of that sort would have been like all other forms of combat operation. Like the fighter planes of World War I and the later bombers, transports carrying troops would have left an airfield near the front lines, flown to the battle

3. H. H. Arnold, *Global Mission* (New York: Harper & Brothers, 1949), p. 86.

area, dropped the troops, and returned to the same airfield. All the airplanes would have been in the air at the same time and would have returned together. Airlift systems organized for supply purposes, of course, could not have operated in the same manner.[4] Mitchell apparently felt that paratrooper transports would be a part of one of the air forces he proposed at about the same time.

Mitchell's World War I suggestion for a two-way division of air power established the dominant Air Force tradition. He proposed to Pershing that one air force include "large aeronautical groups for strategical operations against enemy aircraft and enemy materiel"; this was to be a "truly independent mission," not controlled by local ground commanders.[5] A second air force was to be attached to ground armies, corps, and divisions, and controlled by the commanders of those units. Besides the airplanes used to drop paratroopers, this second air force presumably was to include observation planes and other miscellaneous types. The strategic air force would be directly controlled by Pershing, a concept clearly the forerunner of strategic bombardment theory.

Military airmen did not achieve an organizational solution they wanted until at least 1935 but, in the meantime, they clung to the notion of two air forces. General Mason Patrick, chief of the Air Force during the 1920's, suggested labeling the two organizations the "air force" and the "air service." The former, an offensive striking force, would have comprised 80 per cent of total air strength and would have been under the control of General Headquarters (GHQ). The other 20 per cent would have been the "air service," associated with ground forces. The War Department approved the recommendation in 1923, but it was vetoed by Navy members of the Joint Army-Navy Board.[6]

The unfortunate air mail crashes of 1934 stimulated a 1935 reorganization that finally gave the bombardment theorists much

4. Any airlift system designed for supply purposes would operate in the "conveyor belt" arrangement described in the book's introduction.

5. Alfred Goldberg (ed.), *A History of the United States Air Force 1907–1957* (Princeton: D. Van Nostrand Company, Inc., 1957), p. 21.

6. Wesley Frank Craven and James Lea Cate (eds.), *Plans and Early Operations, January 1939 to August 1942* (*The Army Air Forces in World War II* [Chicago: The University of Chicago Press, 1948]), I, 28. For analyses of the evolution of military aviation doctrine between wars, see Craven and Cate (eds.), *op. cit.*, I, 33–54; also Gordon Turner, "Air and Sea Power in Relation to National Power" in Turner and Richard D. Challener (eds.), *National Security in the Nuclear Age* (New York: Frederick A. Praeger, Inc., 1960, pp. 235–36.

of what they wanted. A new GHQ Air Force, described by General Arnold as a "highly mobile" organization, contained only offensive air power concerned principally with bombardment. Aside from attack and fighter units needed to protect the bombers, the remainder of the Air Force was organized into "observation" squadrons attached to Army Corps Areas.[7] The GHQ Air Force foreshadowed the Strategic Air Command (SAC), the dominant element of the U.S. Air Force created in 1947.

The 1935 reorganization resulted in such primacy for strategic bombardment that the next four years were, in the words of an Air Force historian, a period of "overemphasis on the long-range bomber." The Air Force did not overcome until 1944 a development lag in fighter and interceptor aviation.[8] Significantly, strategic bombardment theory required centralized control of the bombers, but Mitchell and his successors seemed content to leave permanently with ground commanders the other functions of military aviation. At the very least, this dual tradition left tactical airlift with a low priority within the Air Force.

The vague notion that the airlines were something of a national reserve was in line with the World War I connection between military and civil aviation, and with the general tradition that transportation systems have some contribution to make to military preparedness. A conspicuous example of the latter was the railroads' contribution to the evolution of strategy in the 19th century.[9] Edward V. Rickenbacker, World War I aviator and later an airline executive, considered airline pilots as reservists, and advo-

7. Major General H. H. Arnold and Lieutenant Colonel Ira C. Eaker, *This Flying Game* (New York and London: Funk and Wagnalls Company, 1936), pp. 151–53.

8. Goldberg (ed.), *op. cit.*, pp. 42–46.

9. Railroads greatly influenced the strategy of the U.S. Civil War and the mobilization theories of the Prussians and Germans. It seemed originally that each new improvement in mobility made it possible to use fewer troops and to substitute speed of movement for mass. The results were just the opposite; by making it possible to move great numbers of troops, railroads helped expand wars from engagements involving small troop concentrations to massive conflicts involving entire populations. Edward Meade Earle, "Adam Smith, Alexander Hamilton, Friedrich List: The Economic Foundations of Military Power" in Earle (ed.), *Makers of Modern Strategy* (Princeton: Princeton University Press, 1943), pp. 148–49; Hajo Holborn, "Moltke and Schlieffen; The Prussian-German School" in *ibid.*, pp. 177, 181, 199–201; Walter Millis, *Arms and Men* (New York: Mentor Books, 1958), pp. 102–3, 111–12; B. H. Liddell Hart, *Strategy* (rev. ed.; New York: Frederick A. Praeger, 1960), pp. 143–45.

cated government operation of flying schools.[10] In the 1930's, the ICC and the Federal Aviation Commission noted the relationship between airlines and the mobilization base, fast wartime communications, and emergency needs for qualified technicians.[11] President Roosevelt described the industry as a "reservoir" of men and machines always available for the defense effort.[12]

The Air Transport Association suggested the wartime arrangement with the airlines that ultimately prevailed; its president established liaison with military agencies, looking ahead to a contractual arrangement during emergencies. Even if diverted completely to military traffic, industry hoped to maintain each company's integrity within a framework of negotiated prices.[13] This was rudimentary planning at best, since few could have foreseen a need for wartime transoceanic military airlift operations.[14]

These early traditions of military aviation thus postulated three separate layers of air power. One was the strategic bombardment force, identified as an organization that should be separated both from other military services and from all other forms of military aviation. The second layer included the remaining forms of military air power, less important than the first but still within the military sphere. Tactical airlift, to the extent anyone thought about it beyond Mitchell's original paratrooper proposal, was in this second layer. A third layer, finally, was outside the Air Force in a military reserve not in uniform. Even before World War II began, this third layer became a necessity, and showed that it might become an unusual form of military aviation.

STRATEGIC AIRLIFT AS A DEPARTURE FROM TRADITIONAL ORGANIZATION

The first requirement for something approaching a strategic airlift system grew out of the U.S. decision to help the British in the period after 1939. It became apparent that the best way to

10. Henry Ladd Smith, *Airways* (New York: Alfred A. Knopf, 1942), pp. 89–90.

11. Oliver James Lissitzyn, *International Air Transport and National Policy* (New York: Council on Foreign Relations, 1942), pp. 88–91, 416–19.

12. *Ibid.*, pp. 73, 92; Lucile Sheppard Keyes, *Federal Control of Entry into Air Transportation* (Cambridge: Harvard University Press, 1951), p. 84.

13. Craven and Cate (eds.), I, 351; Goldberg (ed.), *op. cit.*, pp. 50–51; Henry Ladd Smith, *Airways Abroad* (Madison: University of Wisconsin Press, 1950), p. 74.

14. To the military, the airlines' sole objective was to provide comfortable transportation for the public, not to be confused with the military task of producing "vehicles of destruction." Arnold and Eaker, *op. cit.*, p. 179.

deliver airplanes to the British was to fly them, because of German submarine activity. The limited range of the airplanes of that time made necessary the use of an extensive network of refueling bases. A joint civil-military organization was set up to handle the operation.

When President Roosevelt asked Air Force chief General Arnold to begin delivering aircraft, Arnold's first thought was to turn over the job to an airline. Pan American, Arnold, and the British arranged a contract between the latter and a PAA subsidiary, with the company charged with ferrying aircraft through Brazil to Africa and the Middle East. PAA constructed the bases under the provisions of the Airport Development Program mentioned in the previous chapter. The company controlled the bases, flew the aircraft and, by the end of 1942, had delivered 464 aircraft across this southern route system.[15] The ferry mission gradually became an enterprise involving American civilian and military pilots, as well as the British themselves. Americans delivered aircraft to what were called points of "ultimate takeoff" from the Western Hemisphere, e.g., Montreal, the British then taking over. The ostensibly civil nature of the enterprise made it possible to assist the British within a framework of official neutrality; thus began the strategic airlift system.[16]

A transport network evolved out of the necessity to shuttle crews back and forth, to establish a flow of spare parts for the aircraft already delivered, and to provide transportation for diplomatic representatives. The ferry system itself became a much larger enterprise than originally envisaged; 30,000 planes were delivered in 1942, 72,000 in 1943, and 108,000 in 1944.[17] In the first year or so, increasing involvement with the British and Russians made it necessary to airlift government emissaries back and forth. With no long-haul transports available (airlines had been using flying boats), bombers were modified to fit into the first scheduled military transport service.[18] Air transportation and aircraft delivery ultimately merged into a single global system.

15. Arnold, *op. cit.*, pp. 204, 210; Craven and Cate (eds.), I, 322–24; U.S. Congress, House, *Report of the Antitrust Subcommittee (Subcommittee No. 5) of the Committee on the Judiciary, Pursuant to H. Res. 107 on Airlines*, 85th Congress, 1st Session, Committee Print (Washington: Government Printing Office, 1957), pp. 261–62 (hereinafter *Celler Report*).

16. Craven and Cate (eds.), I, 310–12; Oliver LaFarge, *The Eagle in the Egg* (Boston: Houghton Mifflin Company, 1949), pp. 10–11, 17.

17. Craven and Cate (eds.), VII, 19.

18. See note 16 above.

The explicit government-industry partnership followed official U.S. entry into the war. The president authorized the Secretary of War to commandeer aircraft he thought necessary, and the Army took over more than half of the existing airline fleet.[19] At a series of Washington meetings, the government agreed to lease the aircraft back to their operators.[20] This left the companies intact so as to use their experience with maximum efficiency. The flying equipment available for long-haul operation was meager; eleven converted bombers, nine Boeing clippers, two Martin flying boats, and five Boeing stratoliners began an operation that ultimately expanded to more than one thousand four-engine transports.[21] Two Army commands contracted with the airlines; the Ferrying Command handled overseas arrangements, and the Air Service Command was responsible for domestic contracts.[22] The Navy, on the other hand, organized its own system, largely by awarding contracts to PAA, its partner of the 1930's.[23]

To the wartime chairman of the CAB, L. Welch Pogue, this seemed an unduly complicated arrangement, because the three operations tended to overlap. He asked for a single military airlift system, but Navy opposition could not be overcome. The Air Force agreed with Pogue, however, and established in 1942 the single organization known as Air Transport Command (ATC). Key airline executives occupied important posts in ATC, the most notable assignment being that of C. R. Smith, President of American Airlines, as Deputy Commander (Major General).[24] The result was a highly successful enterprise in which the idea of civil-military partnership took on new dimensions.

Not only did leading airline officials join ATC (some PAA executives joined the Navy organization), but even the ATA took on a quasi-official status. Given an office in the Commerce Building, it acted as a command post for the industry. It issued directives to the airlines for the War Department, managed the domestic

19. Reginald M. Cleveland, *Air Transport at War* (New York and London: Harper & Brothers, 1946), p. 36.

20. Craven and Cate (eds.), I, 332–33.

21. LaFarge, *op. cit.*, pp. 2, 4; Air Force planners estimated in 1941 a need for 160 four-engine transports. Craven and Cate (eds.), I, 147–50; VII, 6–7.

22. The Air Service Command procured materiel and operated the Air Force's domestic supply depots.

23. PAA operated for the Air Force in the Caribbean and South Atlantic, for the Navy in other areas. Cleveland, *op. cit.*, p. 154.

24. *Ibid.*, p. 76. It might be noted that under the press of war, the CAB became the equivalent of a full-time executive agency.

system of travel priorities (air travel was restricted to official business), and presided at the equity meetings held to decide which carriers would give up the aircraft to meet War Department requisitions.[25]

The arrangement made it possible for eleven airlines to abandon everything else they were doing and send all available aircraft to Alaska in 1942, when the Japanese threatened to advance too far along the Aleutian chain. Air Force historians describe this response of companies and crews as a "heroic" performance.[26] Civil-military integration even extended to the famous route across the "Hump" between India and China, the sole means of supply for American and Chinese forces in that area. This was the first high-density, high-utilization sustained airlift operation and, as such, was the precursor of the Berlin Airlift. The "Hump" ultimately became a largely military effort, but a PAA subsidiary did about 75 per cent of the first month's work.[27] As the ATC expanded, military and civilian pilots flew the same planes along the same routes, with both military and civilian ground crews stationed along the way.

The partnership produced some tensions due, at least in part, to misunderstandings on the military side. Relationships between civilian and military crews and mechanics were, for example, not wholly satisfactory. ATC had the lowest priority imaginable for getting pilots from Air Force flying schools, due to both the lure and priority of combat aviation. To find enough transport pilots, it was necessary to categorize some pilots as good enough to manage transports but unsuitable for combat. Known as "service pilots," they were private pilots appointed directly from civilian life and exempted from attendance at military flying school. Many ATC crews, however, were civilians flying with companies having leased aircraft, but wearing military officers' uniforms distinguishable only by rank insignia.[28] This dissatisfied military pilots,

25. *Ibid.*, pp. 58–59, 80–83; LaFarge, *op. cit.*, p. 32; Craven and Cate (eds.), I, 351.

26. *Ibid.*, VII, 156–57. There were many outstanding performances, especially in the first year or two of the war. Cleveland, *op. cit.*, p. 147; LaFarge, *op. cit.*, pp. 208–9. This entire subject forms a major part of a novel by a popular author who was one of the pilots involved. Ernest K. Gann, *Fate is the Hunter* (Greenwich, Conn.: Crest Reprint, 1962).

27. Craven and Cate (eds.), VII, 117.

28. There were even unfounded rumors that a policy decision restricted the use of civilian crews in combat zones. Years after the war, the Air Force still was trying to verify the existence of the policy. LaFarge *op. cit.*, p. 73; U.S.

largely because the pilots' union had negotiated substantial salary increases at the start of the war. The union was able to fix pilots' pay at the high rate of $1,000 per month ($1,150 in Alaska), a feat that established the union as one of the most successful in labor history.[29] At higher levels of organization, on the other hand, the tensions were more complicated.

As the war progressed, the military gradually assumed control of most of the en route base system, apparently intending to militarize the bases. When PAA objected to its bases being militarized, a bitter conflict ensued.[30] Another conflict also involved PAA and was the follow-on to the prewar decision to award AMEX an Atlantic route. PAA had secured landing rights in the Azores Islands from Portugal, but the U.S. could not establish a military base there because of Portuguese neutrality. These islands, two thousand miles east of Bermuda, were significant to international routes then, and have been ever since. During the war PAA apparently continued its commercial operation through the islands, and there is some evidence that the company refused to waive its landing rights. The British finally succeeded in making diplomatic arrangements for the use of the islands, thus making it possible for the U.S. to get its military base. The details of these transactions never have been made public, but the government apparently concluded that the airline had obstructed the national interest.[31]

Department of Defense, Military Air Transport Service, *Civilian Pilots in Combat Areas,* Historical Division (Mimeographed: February, 1959), p. 1.

29. These negotiations, helped along by military insistence that an agreement be quickly reached so as not to impede the war effort, established the salary scale that became so important after the war. Smith, *Airways Abroad,* pp. 312–15. Some other important factors probably were at work; the differences between civil and military pilot pay have, if anything, become greater and greater, but the difficulties virtually disappeared when civilian pilots resumed wearing company uniforms.

30. Craven and Cate (eds.), VII, 54. One version is that ATC exerted too much pressure on PAA civilian employees. Those who accepted direct military appointments found later that their colleagues had not been drafted but only diverted to other contract assignments. LaFarge, *op. cit.,* p. 92.

31. For maps of commercial routes in those years, see *CAB Annual Report, 1942,* pp. 35–36. During its hearings in 1957, the Celler Committee referred to classified Navy documents indicating that PAA had refused to waive its rights. Yet the testimony also revealed that PAA offered to waive its rights as early as 1939, if the CAB and the government considered it in the public interest. At the CAB's request, PAA executed a formal waiver in 1945. Arnold, *Global Mission,* p. 439; Craven and Cate (eds.), VII, 86–89; U.S. Congress, House, Committee on the Judiciary, *Monopoly Problems in Regulated Industries,* Hearings before Subcommittee, 84th Congress, 2nd Session, on Airlines

Because so much of this dispute is still classified, one can only speculate on the reasons for PAA's apparent obstinacy. The company might have concluded, for example, that the government intended only to use the Azores as a means of promoting transoceanic operation by PAA's competitors. Indeed, once the base was there TWA and American Airlines began using it, a factor that created yet other tensions within ATC itself. Military officers thought the airlines were much too interested in using ATC to establish their positions in the postwar market.[32]

There were other ways in which the ATC was quite different from traditional military organizations. The Army abandoned the historic principle, usually called *unity of command*, that the field commander must always have complete control of everything that enters his assigned area. During ATC's first year it took some doing to convince overseas commanders they should permit ATC aircraft to operate on schedule. The highest flying hour utilization and the greatest delivered tonnage could be achieved only by avoiding a "taxicab" operation in which individual commanders could divert aircraft to their own use.[33] After a few incidents in which theater commanders took over ATC transports, the War Department directed them to keep "hands off" except in the gravest of emergencies.[34] Given its new ground rules, the ATC predated SAC as the first truly centralized global command of the Air Force.[35] Indeed, ATC began as a delivery agency for the

---

(Washington: Government Printing Office, 1956), IV, 2819–25, 2901–3 (hereinafter *Celler Committee Hearings*); *Celler Report*, pp. 241–52. Without attempting to demonstrate that PAA did not engage in questionable business practices it may be noted that: (1) the Celler Committee majority was extremely hostile to PAA, and thoughtful minority members dissented strongly from the majority's evaluation, thus indicating that the majority's approach may have been biased; and (2) the policy in existence before the war, and until 1943, considered foreign landing rights as items suitable for private companies who then were free to build their own landing fields and facilities. It is understandable that any given company might have sought to retain its exclusive position in a foreign country and would look upon its own airfields as private property. On the government's postwar references to PAA, see the analysis of the policy of multiple competition in Chapter 4.

32. Craven and Cate (eds.), VII, 12. One of the ATC's wartime staff officers said later that the decision to militarize the routes was a compromise; the military had hoped to absorb the routes, the crews, and the mechanics. U.S. Industrial College of the Armed Forces, *Problems Involved in Mobilization of Civil Aviation*, Speech of Brigadier General Milton Arnold, now retired. (Mimeographed: February 19, 1953) p. 6.

33. Cleveland, *op. cit.*, p. 160.

34. Craven and Cate (eds.), VII, ix; LaFarge, *op. cit.*, pp. 63–66.

35. See note 1 above.

entire War Department, the British, and the State Department; this gave the command a task much broader than that of tending to the internal needs of the Air Force alone.[36] A War Department allocation agency managed the airlift capacity. Composed of individuals from all of the agencies using ATC, this War Department committee weighed the claims of overseas users and made the decisions.[37]

Individual ATC units were much more flexible than normally provided in military tables of organization. Military units generally are of a standard size, e.g., one heavy bombardment squadron might have twelve bombers and the ground units to support that number. This arrangement obtains at all levels of Air Force organization and is quite static; it depends on the combat notion that airplanes take off and land at the same base. Long-haul airlift, on the other hand, is a continuous flow operation which depends upon many bases, each suited to its particular task. One base, in an isolated area, might see only three or four transports each week; twenty transports might land daily at another base. The military unit at each location reflects these differences.[38]

Yet another unusual aspect of the ATC was its use of regional systems, each taking in an entire geographic area. It was obviously practical at the beginning to use domestic airlines in areas somewhat related to their normal operations. Northwest Airlines developed the Alaskan route from Minneapolis north, and Eastern Air Lines extended into the Caribbean and South America from its base at Miami.[39] From that start, ATC organized regional subsystems, e.g., South Atlantic Division, Pacific Division, each subsystem controlling the bases and transports within its area. Each division also was a link between the theater organization and agencies in Washington. Through constant negotiations with theater commands, the ATC subsystems kept strategic airlift as close to the forward areas as possible.[40]

It seems significant that the strategic airlift system, organized with the sole objective of efficiency, chose a form of regional or-

36. LaFarge, *op. cit.*, p. 56.
37. *Ibid.*, pp. 68–69; Craven and Cate (eds.), VII, 61.
38. *Ibid.*, pp. 48–49.
39. *Ibid.*, p. x; Cleveland, *op. cit.*, pp. 9–12; LaFarge, *op. cit.*, p. 7.
40. The negotiations took into account the difference between operating methodology in various areas. General MacArthur's theater, for example, was noted for its zealous protection of its prerogatives. One such agreement with

ganization that went against military tradition. Each regional subsystem had the characteristics of an airline, simply because that type of operation promised the greatest results. The success of the ATC was marred only by the internal conflicts mentioned earlier, none of which actually hindered operations; but the conflicts seem to have been traceable to the administration's determination to alter the international airline industry.

THE ATC AS AN INSTRUMENT OF POSTWAR POLICY

The administration quietly used the ATC to prepare for postwar changes in the international airline map and began using the military system itself to implement the new policies. The apparent objective was to end PAA's exclusive position by encouraging the domestic companies to enter the international market.[41] These companies, newly introduced to the international field, had their own hopes of expanding and continuing their wartime operations. The ATC civil-military partnership, which did not include PAA, thus became an instrument of postwar foreign economic policy. Several examples indicate that both the U.S. and its allies were aware that the development of four-engine transports had made domination of the international air travel market an important economic objective.

The U.S. proposed in 1944 that it establish a route across Norway to evacuate internees from Sweden. The British opposed the idea, arguing that it would set a precedent for U.S. commercial operation in that area. The U.S. persisted, however, and used unmarked airplanes operating under the label of "American Air Transport Service."[42] ATC then turned the route over to AMEX

ATC was described as "closely akin to an exchange of diplomatic notes between two sovereign states, if not to a formal treaty." Craven and Cate (eds.). VII, 191. These arrangements permitted the most efficient operation of the conveyor belt of strategic airlift. The airlines attained an average utilization rate of eight to ten hours, the military about six, although the four-engine C-54 (military version of the Douglas DC-4) achieved 11.6 hours on the Hump. U.S. Industrial College, *Problems Involved in Mobilization, op. cit.*, p. 17; Craven and Cate (eds.), VII, 117, 141.

41. The domestic airlines were convinced they had broken the "mystique" of PAA's international system. LaFarge, *op. cit.*, pp. 43–44.

42. Craven and Cate (eds.), VII, 110–13. There were other occasions on which the U.S. used unmarked planes manned by crews not in uniform. *Ibid.*, pp. 85–86.

in the summer of 1945, presumably to start off that company's permanent franchise in that region.[43]

A similar case involved Northwest Airlines and the Alaskan route, one that was not converted to military control. Northwest maintained throughout the war a force of about one thousand employees at bases between Minneapolis and Alaska.[44] The route, designed for delivery of supplies to the Soviets, declined in importance as the war progressed. Most of the later traffic carried along the route did nothing but support the personnel at the bases and the Army troops still protecting them, long after military danger had passed. Canada suspected U.S. and airline motives, and asked to no avail that the route be militarized like all the rest.[45]

Under Secretary of War Robert Patterson indicated U.S. interest in postwar aviation in a 1944 policy letter to General Arnold:

> It is important, in my opinion, that the Air Transport Command not carry traffic other than military traffic or traffic involved in the military effort if it can be reasonably handled by a United States civil air carrier, and I believe that any transportation of such non-military passengers by the Air Transport Command should be regarded as an interim or emergency measure until such time as the civil airlines are qualified and are operating over the various routes involved.
>
> ... It is in our interest to expedite the initiation of suitable air transportation by the Air Transport Command to all of the European points where our interests are involved. The service should operate with economy, but it should be so organized as to demonstrate the efficiency of this country in the air transportation field and should compare favorably with the air transport services operated by any of the other nations.[46]

The British also were interested in securing footholds in this important new industry. They began a scheduled military operation from Great Britain to Australia by way of Florida and Mexico, an action interpreted by the ATC staff as an attempt to inaugurate a permanent route.[47] The ATC also suspected the British of

43. The European War had ended, and there presumably was no immediate military need for the route.

44. Craven and Cate (eds.), VII, 156–57, 171.

45. During 1944, for example, supplies for the Soviets seldom exceeded 10 per cent of the traffic carried. Practically all the rest supported the airlift system and the Army. *Ibid.*, p. 157.

46. *Ibid.*, p. 107.

47. LaFarge, *op. cit.*, p. 181.

having convinced Eisenhower to forbid ATC contract companies from flying in the African area. The British operated between Tunis and Cairo, the ATC being limited to the western part of North Africa.[48]

The administration's activity generally was confined to the civil-military partnership of the ATC. Within the ATC structure, the domestic airlines appeared to be a unified whole that might serve as a useful counterweight to what the government believed to be PAA's drive for international supremacy. To defeat PAA the administration apparently was willing to forget, at least for a while, the long-term hostility associated with the 1934 air mail cancellations.

AIRLINES AND THE NATIONAL ECONOMY: WORLD WAR II AS A UNIQUE EXPERIENCE

Airline performance in World War II, while exemplary, led to some confusion about the relationship between transportation systems and national mobilization. The airlines were made to appear as a true national reserve, immediately available to support the military in any future emergency. This, in turn, buttressed the traditional view mentioned earlier that all transportation systems are part of the mobilization base.

The important study groups that studied aviation policy after World War II concluded that the airlines should continue in their quasi-military status. The Finletter Commission, appointed by the president, assumed that in future wars the military once again would "take over" the existing industry.[49] The Congressional Aviation Policy Board (Brewster Board) added that specific contractual arrangements should cover any future conversion to military operations. The board complained also that the lack of such arrangements had caused "serious dislocation of the commercial system" when the Secretary of War hastily commandeered airplanes.[50] In any case, said the board: "Such an operating fleet serves peacetime commerce and industry, while remaining avail-

48. Craven and Cate (eds.), VII, 82.

49. U.S. President's Air Policy Commission, *Survival in the Air Age* (Washington: Government Printing Office, 1948) p. 37 (hereinafter *Survival*).

50. U.S. Congress, Senate, Congressional Aviation Policy Board, *National Aviation Policy*, 80th Congress, 2nd Session, Senate Report 949 (Washington: Government Printing Office, 1948), p. 14 (hereinafter *National Policy*).

able for immediate conversion to military use in an emergency . . . as many transport aircraft as possible should be operated in commercial service . . . to provide a reasonable reserve."[51]

Because of these recommendations many government agencies have since put much effort into the attempt to make explicit arrangements for emergencies. The general statement that a commercially profitable industry is a reserve force has become standard rhetoric for both government and industry spokesmen. The idea is attractive, of course, because it promises a vast increase in military potential at no cost to the government. Upon closer analysis, however, industry's contribution to World War II seems to have been the result of its infancy, not its maturity.

The U.S. was able in World War II to refrain from imposing controls on public transportation in general. While the public was asked to minimize its travel, a vast expansion still was needed in national transportation capacity; fortunately, the railroads provided it with extra rolling stock that happened to be available. With airlines committed to a government-controlled system in which the airlines managed the priorities, the railroads quadrupled their capacity and handled more than 70 per cent of all domestic travel provided by common carriers. The domestic airlines, on the other hand, handled only about 3 per cent.[52]

Any future administration that seeks to prevent too great an increase in national frustration can be expected to do its utmost to avoid economic controls found unnecessary in World War II. The war may well have been a unique experience not to be repeated in the manner envisioned by the Finletter Commission and the Brewster Board, not to mention the executive agencies that have tried to implement their recommendations.

A Hoover Commission task force observed in 1949 that it might be dangerous to plan for military use of an industry needed for maintenance of the national economy: "It becomes clear, however the problem is approached, that military airlift requirements must in the main be provided by expanding the military air transport reserve. In view of the importance of military air transport, it would appear to be no less essential for the armed services to pro-

51. *Ibid.*, p. 15. The Finletter Commission agreed: "The soundest way to build up a pool for military use is to develop a plane that can be operated commercially on a profitable basis." *Survival*, p. 115.

52. George A. Lincoln, William S. Stone, and Thomas H. Harvey (eds.), *Economics of National Security* (New York: Prentice-Hall Inc., 1950), pp. 254–62, 282–90, especially chart on p. 255. For railroad statistics see U.S.

cure the transport equipment they need than to supply themselves with bombers or fighters."[53]

Another analyst who studied the wartime experience put it this way: "When air transport is relatively insignificant and its use can be readily dispensed with, then a peacetime air service is also a weapon of war. When, however, air transport becomes woven into the travel habits of individuals and the plans of merchants and manufacturers, it cannot be withdrawn from civil use without a great strain upon the economy."[54]

It can be added that the airlines, as of the 1960's, account for about half of all domestic travel by common carrier, and the importance of airlift to the national economy has changed in the same proportion.[55] The nature of aircraft and airlift systems, moreover, makes it unlikely that they suddenly can expand their total capacity the way the railroads did in the war.[56]

THE TACTICAL AIRLIFT QUESTION: SUPPLIES OR PARATROOPERS?

World War II's tactical airlift operations generated a conflict that has become a permanent part of the military airlift problem. In recent years the argument has been between the Army and the Air Force but, during the war itself, the disagreement was between low level Army commanders who thought their organizations neglected and higher commanders forced to take a broader strategic view. The entire problem has revolved around two principal functions of tactical airlift.

Annual Report of the Interstate Commerce Commission. 1942, p. 7; 1944, p. 7; 1945, p. 13; 1946, p. 8.

53. Charles L. Dearing and Wilfred Owen, *National Transportation Policy* (Washington: The Brookings Institution, 1949), pp. 145–47. The full commission did not consider these proposals, and the 1955 Hoover Commission followed the lead of the Brewster Board and the Finletter Commission. See Chapter 6.

54. Robert S. Einzig, "CAB Control over Domestic Routes and Competition 1938–1952" (unpublished doctoral dissertation, University of Michigan, 1953) pp. 450–51.

55. Exact percentages are difficult to determine, in part because the bus industry has adopted the practice of including charter traffic within a single total figure. Air Transport Association of America, *Facts and Figures About Air Transportation* (24th ed.; Washington: 1963), p. 8.

56. Two examples can be used to emphasize the point: (1) trains require relatively little maintenance while away from the railroad yards; and (2) railroads can expand their capacity a great deal without making substantial additions to important parts of the work force, especially engineers, but each additional airplane requires a full crew.

One function was the overseas corollary of the world-wide distribution systems of strategic airlift and sealift. Transports assigned to the overseas theater distributed supplies left at seaports and major airfields. This function was only an auxiliary to systems of surface transportation, e.g., Army truck convoys, but it became important in its own right as the Allies advanced across Europe. In performing the supply job, tactical airlift systems remained in constant operation (the conveyor belt again) and operated in the manner of "feeder" airlines.[57] The task, however, had not been contemplated in wartime planning; U.S. tactical transports came from the Troop Carrier Command, an Air Force organization dedicated to an entirely different function.[58]

The use of airlift to drop paratroopers in massive airborne assaults called for an entirely different type of organization. Airborne flight crews had to be expert in the specialized techniques of flying in close formation, picking out the proper target area for the airdrop, and the actual dropping itself. The entire operation vaguely resembled an old-fashioned bomber force in which the airplanes flew very slow, very close to the ground, with pilots picking out targets with the naked eye. In an assault, all airplanes had to be in the air at the same time. Training had to be conducted with the paratroopers themselves, and with some regularity. When a massive assault or training exercise was planned, several days had to be spent in "standing down" all of the airplanes to make repairs. It was, and is, difficult—virtually impossible—to use the same short-haul transports to accomplish both airborne assaults and conveyor-belt "feeder" operations. This incompatibility caused the wartime disagreement on airlift.

The dispute followed General Eisenhower's establishment of the first organization designed to blend ground and air forces into a single military command primed for airborne assault. This was the First Allied Airborne Army, composed of British and American troop carrier commands and airborne divisions, commanded by Air Force General Lewis Brereton.[59] Significantly, the U.S. division

57. A term commonly used for airlines which operate between major population centers and smaller outlying cities. One such system in World War II included British and French military transports, airplanes from the national airlines of those countries, and U.S. short-haul transports; this was the Mediterranean airlift system. Craven and Cate (eds.), VII, 81.

58. The command was formed as part of the 1942 reorganization which consolidated Air Force strategic airlift within ATC. *Ibid.*, p. 14.

59. Lewis H. Brereton, *The Brereton Diaries* (New York: William Morrow and Company, 1946), p. 308. Brereton's appointment was in accord with an old

commanders were Generals James Gavin, Matthew Ridgway, and Maxwell Taylor. These officers, and Brereton, disagreed with Eisenhower on how to use the airlift assigned to the Airborne Army.

Brereton's people and General Arnold in Washington argued that tactical airlift units should fly only in conjunction with the Airborne Army. Brereton criticized the requests of U.S. ground commanders for airlift supply assistance which, he said, wasted airlift on the performance of secondary tasks. Several outstanding opportunities for airborne assaults had to be passed up, he argued, because the airlift was busy delivering supplies in Europe.[60] Arnold agreed, complaining to Eisenhower that troop carrier aircraft were flying too few combat missions and too many supply trips. Arnold asked Eisenhower how he intended to use the airlift in the future, implying that he should terminate the supply job.[61]

Airborne commanders complained later that an airlift shortage was one of their major problems, but the evidence on this score is at best mixed. The most ambitious airborne assault of the war, for example, was an attack on Arnhem in September, 1944; it resulted in the "decimation" of a British airborne division that was one of the participating units.[62] The failure had many causes; an insufficiency of aircraft, the fact that the British division was dropped off target, bad weather that prevented the dropping of reinforcements, and failure of other ground units to connect up with the airborne units.[63] The British commander argued that the force of 1,100 aircraft and 500 gliders should have been at least twice that size.[64] Gavin asserted that a major defect in planning

---

military tradition that the agency charged with delivery should command the operation until after the delivery. In amphibious operations, for example, Navy officers usually are in charge until the troops are ashore.

60. Eighteen operations were planned and discarded, either because the front had advanced too fast or airplanes were unavailable. *Ibid.*, p. 343.

61. *Ibid.*, p. 333.

62. The evaluation is from Eisenhower, *op. cit.*, p. 349.

63. Wesley Frank Craven and James Lea Cate (eds.), *Europe: Argument to V-E Day, January 1944 to May 1945* (*The Army Air Forces in World War II* [Chicago: The University of Chicago Press, 1951]), III, 609–10; Major General J. F. C. Fuller, *The Second World War* (New York: Duell, Sloan and Pearce, 1949), p. 343; *The Memoirs of Field-Marshall the Viscount Montgomery of Alamein, K. G.* (Cleveland and New York: The World Publishing Company, 1958), pp. 266–67; Hart, *Strategy, op. cit.*, pp. 320–21; U.S. Department of Defense, Department of the Air Force, *USAF Airborne Operations, World War II and Korean War* (Mimeographed: USAF Historical Division Liaison Office, 1962), pp. 66–67.

64. Major General R. E. Urquhart with Wilfred Greatorex, *Arnhem* (New York: W. W. Norton & Company, Inc., 1958), p. 5.

caused the shortage of Douglas C-47's that so limited the paratroopers.[65] Simply stated, the airborne generals wanted many more aircraft (perhaps 5,000), all of them under the exclusive control of the airborne organization and never used for other purposes.[66]

Eisenhower's view was a bit different. The over-all needs of the Western front were more important than those of a small airborne force held in reserve, and he would decide when and how to use the paratroopers:

> As quickly as the defeat of the Germans on the Normandy front became certain, airborne forces were directed to prepare plans for drops in a number of successive positions, the appropriate spot to be selected when the developing situation should indicate the one of greatest promise. The mere paper planning of such operations was, while laborious, a simple matter. However, when actual preparation for a planned drop was undertaken, delicately balanced alternatives presented themselves. Preparation for airborne attack required the withdrawal of transport planes from supply purposes, and it was difficult, at times, to determine whether greater results would not be achieved by continuing the planes in supply activity.
>
> Unfortunately, this withdrawal of planes from other work had to precede an airborne operation by several days, to provide time for refitting equipment and for briefing and retraining of crews . . . with our supply situation growing constantly more desperate, and with all of us eagerly following combat progress in the search for another prospect of cutting great numbers of the enemy, the question of the Transport Command employment came up for daily discussion. On the average . . . our planes could deliver about 2000 tons a day to the front. While this was only a small percentage of our total deliveries, every ton was so valuable that the decision was a serious one.[67]

The argument, then, was between the theater commander and subordinate commanders eager to use their highly specialized skills. Commanders of an ostensibly elite organization understand-

65. This military equivalent of the commercial DC-3 carried between twenty and thirty paratroopers. James M. Gavin, *Airborne Warfare* (Washington: Infantry Journal Press, 1947), pp. 2, 36, 73.

66. The Arnhem operation involved four increments of aircraft and, says Fuller, two would have been desirable. Fuller, *op. cit.*, p. 343. There is evidence, however, that the airborne commanders really wanted enough aircraft to put all the airborne units (some 33,000 troops) in the air at the same time. *USAF Airborne Operations*, *op. cit.*, pp. 66, 69–70.

67. Eisenhower, *op. cit.*, pp. 341–42.

ably were disturbed at being bypassed in favor of the needs of ordinary infantry. The controversy is particularly significant to this book because of its later reappearance in a more distinctly political context; it thus requires further comment.

No matter how it has appeared in more recent years, the argument began as one between higher and lower levels of command, not between the Army and Air Force. To the extent that the wartime Air Force expressed itself, however, Brereton and Arnold sided with the paratrooper generals. Indeed, the Air Force conspicuously did not accept the role of airlift as a means of keeping the front lines supplied. The Air Force view was in the Mitchell tradition that airmen should fly airplanes in a combat environment. This was a most peculiar alignment of military interests; the Air Force supported a small elite group within the Army, in opposition to the supply requirements of the major part of the Army.

In a massive conventional war, the elite airborne divisions were an almost insignificant part of the total effort, Eisenhower managing a front that contained more than sixty American divisions alone. Yet these same airborne divisions became an important part of the much smaller U.S. Army of the Cold War years. The outspoken Army leaders of the 1950's, moreover, were the generals who had been so disappointed at Eisenhower's decisions during the war and perhaps even more disappointed in his actions as president. This is not to argue that any of the airborne generals had any but the highest of motives, but the fact remains that all of them looked upon airlift as something designed solely for paratroopers.

There is, it should be noted, some fundamental logic in one of the arguments advocated so strongly over the years by the paratrooper generals. The incompatibility of the two tactical airlift functions makes it unrealistic to assume that a single airlift organization can accomplish both tasks.[68] The generals were correct when they asserted that the World War II airborne divisions could

68. For a summary of the problems of the dual mission of tactical airlift, unsatisfactory attempts to use bombers as transports, and the improvements that occurred only after the end of airborne operations, see U.S. Department of the Army, Office of the Chief of Military History, *United States Army in World War II, Logistical Support of the Armies: May 1941–September 1944* (Washington: Government Printing Office, 1953), I, 572, 578, 583. *Ibid.* (1959), II, 42, 162. Airlift data from World War II is sketchy at best, so it is difficult to determine exactly what utilization rate was attainable. One study of the

not stay ready for combat unless airplanes remained constantly available to them, especially for periodic training. The separation of paratroopers from airplanes had the effect of splitting the airborne Army in two. To accept this part of the argument, however, raises a broader question.

As airlift has become progressively more necessary to the supply network for armies in the field, it has become virtually impossible to contemplate withdrawal of that airlift for any other purpose. But if the airplanes are left with the paratroopers, they must of necessity be superfluous to the supply function. Once airborne divisions are committed to combat, moreover, it takes far fewer airplanes to keep them supplied than it did to drop them in the first place.[69] The fundamental question of World War II, and since, is the marginal utility of an airborne organization which, given its own airlift, has an extraordinarily high cost. Have airborne assaults been effective enough to warrant that cost?

Airborne generals have sought over the years to increase the size of the airborne force and to establish military doctrine calling for assaults on an ever increasing scale. No standard, or limit, has been established, but airdrops of as many as 30,000 men have proved largely ineffective, and for various reasons. One of the obvious difficulties is the limitation on the size of the equipment that can be airlifted and airdropped. As in the Arnhem assault of 1944, if massive numbers of troopers are dropped a substantial distance behind enemy lines, it may not be possible for other ground forces to immediately provide a line of communications. If the line is not established, an elite force is cut off. Because such massive attacks cannot be concealed very well, troopers must be dropped with a

---

Burma campaign, however, concluded that when airlift (C-47's in that case) was used solely for air supply of ground troops, a six-hour rate was well within reach. U.S. Department of the Air Force, *Air Supply in the Burma Campaigns,* by Dr. Joe G. Taylor, USAF Historical Division, Research Studies Institute, Air University, 1957, p. 140.

69. The World War II division used 600 to 700 tons of supplies each day. To move all supplies in airplanes of the time required about 150 trips per day. Factors of distance, spare parts, etc., make it difficult to estimate exactly how many planes would have been needed to guarantee deliveries, but 250 probably is a high estimate. To deliver a combat element of 10,000 troops, however, would have called for 650 to 700 airplanes. Given the likelihood that even in 1964 there are practical reasons why a division cannot be supplied completely by air, it can be estimated that ten times as many transports must be used to drop a division than to supply it. Tonnage estimate from Eisenhower, *op. cit.,* p. 267.

high degree of precision, perhaps in the face of enemy ground fire and crosswinds which make accuracy impossible.

Much of the World War II history of airborne assaults consists of tragedy caused by attempts to be daring in the use of an exotic form of warfare.[70] While airborne operations gradually became more effective, they became more conservative as well until, in the final such attack of the war, troopers dropped only six miles behind enemy lines and well within the range of friendly artillery.[71] The results were little different than those associated with normal infantry attacks. Eisenhower's approach indicates that it was not important to his strategy to use the airborne divisions. Their position as the most glamorous of Army units and their prominence in all large-scale Army maneuvers seem out of proportion to their strategic significance.[72]

Small-scale airborne assaults, on the other hand, have proved reasonably effective, usually when 2,000 or fewer troopers have been involved.[73] It is easier to conceal such attacks and to drop the troopers in areas where a lack of immediate resistance gives them

70. Prime examples were the operations against Sicily; several planes were lost because of misdirected fire from friendly ships, and some 133 of 137 gliders were missing after being dropped over the ocean at night. Many Allied leaders concluded at the time that massive airborne assaults were too expensive, but troop carrier and airborne commanders were not convinced, emphasizing again that the problem is not one which pits the Army against the Air Force. *USAF Airborne Operations, op. cit.*, pp. 16–21. At Arnhem, there were 12,000 casualties in the eight days after the 33,000 troopers were committed. *Ibid.*, pp. 69–72. See also Fuller, *op. cit.*, pp. 263–65, 295.

71. Craven and Cate (eds.), III, 771; *USAF Airborne Operations, op. cit.*, p. 82.

72. These remarks apply to the Korean War also. The Army airdropped 2,860 and 3,447 troopers in two major assaults. Neither contributed anything of note to the total effort, contrary to press releases of the time. It was again necessary to withdraw airlift from other tasks at least two days before each assault, and both were delayed; one because of weather, the other because of a change in planning. This was not a problem, because of a relatively generous supply of tactical airlift in Japan. The Army complained during the war of too much airlift; it was forced to withdraw cargo from a reliable surface supply line just to fill the scheduled airlift service between Japan and Korea. Robert Frank Futrell, *The United States Air Force in Korea, 1950–1953* (New York: Duell, Sloan and Pearce, 1961), pp. 196, 325. U.S. Office of the Chief of Military History, Department of the Army, *United States Army in the Korean War, South to the Naktong, North to the Yalu (June–November 1950)*, by Roy E. Appleman (Washington: Government Printing Office, 1961) pp. 380, 657–58. Samuel Taylor Moore, *U.S. Airpower* (New York: Greenberg, 1958), p. 160.

73. Good examples were in New Guinea in 1943 and against Corregidor in 1945. *USAF Airborne Operations, op. cit.*, pp. 23–28, 73–78.

time to reorganize. Even in the event of failure, moreover, the total cost can only be much less than in the large scale disasters associated with massive airborne assaults.[74]

Under these circumstances, the postwar prominence of airborne strategy may be only coincidence. Had the Army's postwar leaders been those experienced in the Pacific, the apparently more successful strategy of amphibious assaults might have become more prominent.[75] The demands of the airborne generals always have been too expensive to meet, a factor that remains central to the military airlift dispute.

74. While Liddell Hart does not emphasize airborne operations, this description of how they have been most successful fits the definition of what he calls the "indirect approach." *Strategy, op. cit.*, pp. 15–22; 373–85.

75. Amphibious warfare generally has been quite effective. B. H. Liddell Hart, *Deterrent or Defense* (New York: Frederick A. Praeger, Inc., 1960), pp. 110–29.

[4]

The Postwar Policy Structure for the International Airlines

THE U.S. QUICKLY ADOPTED four new airlift policies as soon as World War II ended, and these have since governed the international airlines. The policies were adopted at different times and on different premises; this chapter is designed to demonstrate how this inconsistency led to economic difficulties for the carriers and domestic political conflict between the interest groups associated with them.

The administration established the first policy by quickly awarding permanent international routes to the domestic carriers so prominent in the wartime ATC. The decision produced multiple competition in the three geographical regions in which U.S. carriers have functioned since World War II (Europe, Pacific, Latin America). The administration thus attained its long sought victory over Pan American but, in a larger sense, the new policy epitomized the immediate postwar U.S. approach to world politics. The country seemed to visualize a world in which conflicts of national interest and power would be things of the past.[1] As nations turned to peaceful economic development, so went the argument, a vast expansion of both domestic and international air travel was certain to materialize. All that was necessary was for the government to provide a climate in which competing private enterprises could answer the demand.

The second policy, on the other hand, was forced upon the U.S. by other countries who felt unprepared for open competition between international airlines. After two years of disagreement, the U.S. had to accept the fact that international route concessions and landing rights would be decided by bilateral negotiation on a case-

1. See, for example, Hans J. Morgenthau, *Politics Among Nations* (2nd ed.; New York: Alfred A. Knopf, 1956), pp. 29–31; John W. Spanier, *American Foreign Policy Since World War II* (New York: Frederick A. Praeger, Inc., 1960), pp. 14–18; Gabriel A. Almond, *The American People and Foreign Policy* (New York: Frederick A. Praeger, Inc., 1960), pp. 87–115.

by-case basis. This was in accordance with the nationalistic traditions of European aviation law. Other countries understandably feared that U.S. economic strength, manifested in a group of international airlines equipped with modern transports and well established on ATC routes, quickly would overcome their own airlines.

As a corollary to the bilateral policy, the U.S. accepted the necessity for some sort of apparatus with which to avoid international price wars. The resultant *de facto* regulatory machinery has seemed on the surface only a private cartel, an appearance not wholly in accord with the evidence. In any case the combination of bilateral negotiation and *de facto* regulation was a more nationalistic approach than was then popular within the U.S., and thus contradicted the initial policy of open competition among all the world's airlines.[2] It is worth emphasizing, moreover, that the *de facto* nature of the price structure enabled the U.S. to perpetuate the fiction that international aviation was a matter for private enterprise.

The fourth policy was to promote the export sales of transport aircraft by granting financial concessions to countries promising to buy, a step recommended by both the Finletter Commission and the Brewster Board. Some congressmen proposed a domestic corollary, in which the government would have produced an "air merchant marine" by lending money to domestic carriers. The unsuccessful proposal, however, was closely tied to the broader question of what to do with the nonscheduled airlines.

REJECTION OF THE CHOSEN INSTRUMENT AND THE PERMANENT AIRLINE STRUCTURE: 1945-60

For the most part the routes awarded the domestic companies were the ones they had flown under the ATC banner throughout the war. The awards attracted little attention at the time, but they set a permanent pattern. The administration succeeded in turning back PAA's bid to be the "chosen instrument" after both had spent years preparing for the confrontation.

2. Like the term "liberal Democrats," the words "nationalist" and "internationalist" are used here in a noncommittal way. Europeans interpreted the U.S. approach as excessively nationalistic in that giant American companies would be sent overseas to smother foreign competition. To put it another way, only the very strong seemed able to afford free competition. Europeans considered their own approach internationalist in that each airline would be given some place in the market.

For its part the government announced in 1943 that the entire industry should apply for permission to service some twenty international routes.[3] This was part of a joint CAB-State Department announcement that the government henceforth would conduct all negotiations for landing rights and route concessions.[4] The joint nature of the announcement once again emphasized the CAB's position vis-à-vis international aviation, as did General Arnold's announcement that national security demanded that the U.S. maintain an "extensive system of air bases and airways outside the United States."[5]

PAA President Juan Trippe told his stockholders in 1943 that he would seek official approval for a single "community airline" to represent the U.S. in world aviation.[6] All the airlines would purchase stock in the company, but its operating nucleus quite clearly would be Trippe's company. The McCarran Bill, first introduced in 1944, followed Trippe's suggestions; it would have established the "All-American Flag Line, Inc."[7] The subsequent contest between the government and PAA was fought out during the Senate Commerce Committee's hearings on the bill, most of which were held in executive session. The committee came close to reporting out the bill in February, 1945, but administration officials, including General Arnold, reportedly provided additional information that was damaging to PAA.[8] This was enough to head off immediate congressional action.

In brief public hearings held during the spring of 1945, the government and the domestic airlines presented an almost united front against the bill. Acting Secretary of State Joseph C. Grew told the committee in writing, for example, that PAA's arrangements with foreign governments had been a "deterrent to the war effort."[9]

3. *CAB Annual Report, 1944*, pp. 12–14.
4. *Department of State Bulletin*, 9 (October 16, 1943), 265–66.
5. General Henry H. Arnold, "Air Power in the Future," from the Third Report to the Secretary of War by the Commanding General of the Army Air Forces, reproduced in Eugene M. Emme, *The Impact of Air Power* (Princeton: D. Van Nostrand Company, Inc., 1959), p. 312. Air Force officials reportedly met quietly with domestic airline officials to encourage their participation in international operation. Henry Ladd Smith, *Airways Abroad* (Madison: University of Wisconsin Press, 1950), p. 208.
6. *Ibid.*, p. 337.
7. *Ibid.*, pp. 205–6.
8. *Ibid.*, p. 223.
9. U.S. Congress, Senate, Committee on Commerce, *To Create the All-American Flag Line, Inc.*, Hearings before Subcommittee, 79th Congress, 1st Session (Washington: Government Printing Office, 1945), p. 8 (hereinafter

The domestic airlines organized a special lobby solely for defeat of the McCarran Bill.

The public arguments fell into the pattern that had become traditional.[10] PAA's opponents contended that overseas travel had an unlimited future, that more companies were needed to answer the demand, and that PAA had become overly secure in its belief that its high costs would continue to bring high subsidies. PAA replied that both the government and the other companies overestimated the future market and that, in any case, foreign airlines soon would furnish the most rigorous competition imaginable. PAA's only support came from the company that ultimately became the largest of the domestic carriers; President William A. Patterson of United Airlines, breaking away from the government-industry alliance, decided to stay out of the international market. Too many companies were in the market, he asserted, and all of them could not succeed. Monopoly was dangerous, argued Patterson, but some compromise with this abstract principle seemed necessary in the international field.[11]

The government held just enough support to prevent the Commerce Committee from reporting the bill, but the internal maneuvering was quite extensive.[12] The stalemate cleared the way for the administration to set up its own route system, and four important presidential decisions quickly established it. The first, made in July, 1945, designated TWA as the third international carrier in

Flag Line Hearings). The accusation did not prevent the Government from awarding PAA a Presidential citation for its wartime achievements. Reproduced in U.S. Congress, House, Committee on Armed Services, *Hearings before Special Subcommittee on National Military Airlift*, 86th Congress, 2nd Session (Washington: Government Printing Office, 1960), p. 4753.

10. The arguments are summarized in Lucien Zacharoff, *The World's Wings* (New York: Duell, Sloan and Pearce, 1946). TWA was leader of the group.

11. *Flag Line Hearings*, pp. 73–96.

12. McCarran wrote an interim report promising to submit a revised bill to meet the objections, but five committee members disagreed; they argued that even the hint of new legislation was an attempt to pressure the CAB. U.S. Congress, Senate, Committee on Commerce, *Creating the All-American Flag Line, Inc., and to Assure the United States World Leadership in the Field of Air Transportation*, 79th Congress, 1st Session Senate Report 805 (Washington: Government Printing Office, 1945) pp. 6, 7. McCarran lost important support when Senator Josiah Bailey, chairman of the Commerce Committee and an original backer of the bill, denounced Trippe's "too active" promotion of the legislation. Correspondence reproduced in U.S. Congress, Senate, Committee on Commerce, *Consolidation of International Air Carriers*, Hearings before Subcommittee, 80th Congress, 1st Session (Washington: Government Printing Office, 1947), pp. 90–94 (hereinafter *Consolidation Hearings*).

the Atlantic, joining PAA and AMEX.[13] Three additional decisions followed within a year and added competition in all of the major areas.[14]

The long-term alignment of companies should be briefly noted. In the Atlantic, AMEX became a subsidiary of American Airlines and was renamed American Overseas Airlines (AOA). When the parent company asked in 1950 to sell AOA to Pan American, the president agreed but awarded additional routes to TWA to make certain PAA had substantial competition.[15] In the Pacific, Northwest Airlines became PAA's principal competitor, with Eastern, Braniff, and (until it merged with Eastern) Colonial operating in Latin America. The most significant route decision since the 1945–46 period occurred in 1955 and illustrated once again the controlling position of foreign and defense policy in all international route awards.

The president permitted a small cargo airline (originally Seaboard and Western, now Seaboard World) to join the Atlantic route complex in 1955, principally because the Defense Department asserted a need for more commercial cargo operation in that area. Because the 1955 arguments were the logical follow-on to those used by General Arnold and represented the long-term view of the Defense Department, they are listed here:

> (1) The purchase and operation by the air transport industry of long-range aircraft in cargo configuration which can be converted expeditiously to military use is in the interest of national defense;
>
> (2) international developments emphasize the need for transatlantic air cargo service;
>
> (3) present air cargo service provided by United States-flag lines is considered inadequate to meet military and civil requirements;
>
> (4) it is not considered that a well-developed transatlantic air cargo industry is in existence at this time; and

13. 6 CAB 319 (1945) cited in John H. Frederick, *Commercial Air Transportation* (Homewood, Illinois: Richard D. Irwin, Inc., 1955), pp. 295–97.

14. Latin American Route Case 6 CAB 357 (1946); Hawaiian Route Case 7 CAB 43 (1946); Pacific Route Case, 7 CAB 209 (1946); all cited in *ibid.*

15. U.S. Congress, House, Committee on the Judiciary, *Report of the Antitrust Subcommittee (Subcommittee No. 5), Pursuant to H. Res. 107 on Airlines,* 85th Congress, 1st Session, Committee Print (Washington: Government Printing Office, 1957), pp. 241, 255–56 (hereinafter *Celler Report*). James M. Landis, acting as attorney for a group of AOA employees who sought to buy out the company, attacked the merger as a move toward monopoly. *New York Times,* August 30, 1950.

(5) the Department of Defense has an existing requirement for additional air cargo service to Europe.[16]

James M. Landis, CAB Chairman during part of the period in which the original awards were made, insists that the board worked out the over-all route structure in a series of internal staff studies.[17] The CAB apparently had not decided, however, which routes should be combined to form a sensible operation for any one company. If the route pattern resembled anything at all, it was the wartime ATC structure, but with the individual companies no longer unified within a single system. The president, moreover, made important changes to the CAB's original recommendations. He awarded two Latin American routes to a company that had not asked for them and he installed an additional carrier on the route to Bermuda.[18] It seems clear enough, then, that international route awards throughout the entire postwar period have been something more than the outcome of interplay between private companies and an independent regulatory agency. Nobody has ever defined, however, the precise nature of that "something more."

With the president awarding routes in the context of his broad authority in the areas of foreign policy and national defense, administrative confusion became the order of the day. The CAB, for example, has had to maintain some sort of vague liaison with the president. Nothing can be gained by submitting recommendations without finding out in advance what he is likely to approve. Any attempt to isolate the recommending agency from the approving authority could lead only to the submission of endless recommendations until the president discovered one he liked. This CAB-presidential relationship remains ambiguous, inexplicit, little discussed and little studied, but it cannot be lightly dismissed.[19]

16. Reproduced in U.S. Congress, Senate, Committee on Commerce, *Study of Military Air Transport Service*, Hearings before Subcommittee, 85th Congress, 2nd Session (Washington: Government Printing Office, 1958), p. 168.

17. U.S. Congress, Senate, Committee on the Judiciary, *Report on Regulatory Agencies to the President-elect*, 86th Congress, 2nd Session, Committee Print (Washington: Government Printing Office, 1960), p. 42 (hereinafter *Landis Report*).

18. In the Latin American case, the CAB had assembled 4,000 pages of evidence and 39 volumes of exhibits. Smith, *Airways Abroad*, pp. 268–73.

19. Because responsibilities are so ill-defined and because the president cannot devote much time to issues of this sort, too much room is left for excessive maneuvering by anonymous officials connected with the White House. On the effects of this type of administration without responsibility, see Henry J. Friendly, *The Federal Administrative Agencies* (Cambridge: Harvard University Press, 1962), pp. 147–62, especially 153–56.

Some scholars criticize interest groups for attempting to influence international decisions and the president for sometimes appearing to respond to such influences, and they suggest that the entire decision-making process should be removed from the reach of political influence.[20] When decisions are part of the political process and sometimes require that companies operate unprofitable routes in the national interest, this type of independence seems out of the question. Two contemporary examples can serve to clarify this point. As a part of Cold War maneuvering, the Soviets periodically threaten the continued operation into Berlin of scheduled PAA commercial flights. Newspaper accounts make it quite clear that State Department announcements on the subject are worked out in consultation with the airline itself.[21] Similarly, the U.S. and the Soviets periodically discuss the possible exchange of transport rights in which the Soviet national airline (Aeroflot) and PAA would begin scheduled operations between New York and Moscow.[22] All such decisions, in short, seem to require direct contact and discussion between the public decision-maker and the private interest at stake, each party attempting to clarify its position so as to avoid misunderstanding and either political or economic disaster.

A slightly different type of confusion also has survived within the executive branch. The Defense Department's constant support for airline expansion, epitomized in the 1955 decision mentioned earlier, became quite important to the controversy over the question of military traffic on commercial transports. Part Three of this book will explore this problem in somewhat more detail, but it is noted here that Defense Department arguments implied that it needed the carriers but did not feel responsible for their solvency. Approximately the same thing can be said of the Post Office, which has felt itself obliged to keep mail rates as low as possible.

The Post Office has, on more than one occasion, challenged in court the CAB's method of calculating air mail payments. In the case of companies operating both domestic and international routes, the Post Office argued that the CAB should not use domestic

20. Professor Schwartz, for example, proposes an Administrative Court, with lifetime tenure for its members, as a means of making the agencies more independent. Bernard Schwartz, *The Professor and the Commissions* (New York: Alfred A. Knopf, 1959), pp. 217–18, 272–75; James Landis argues that the president often has exceeded his authority. *Landis Report*, p. 14.

21. *Washington Post*, June 23, 1964.

22. *Aviation Daily*, Oct. 24, 1963.

profits to offset international losses.[23] In calculating the subsidies and mail payments due such companies, however, the CAB acts in its part-time capacity as a foreign policy agent of the executive branch. In this sense, the litigation seems roughly analogous to a situation in which the Defense Department might bring suit against the State Department over a foreign policy disagreement.

It might be concluded, then, that the principal result of the postwar presidential route decisions has been still more separation between theory and reality. Once the wartime partnership had served its purpose, the government retreated behind the abstract principle of regulation by independent agency. It seems unfortunate, in retrospect, that the 1945 confrontation occurred in an emotional atmosphere not conducive to compromise and at a time when wartime secrecy prevented full consideration of the issues. Assertions that PAA had hampered the war effort and that it sought only a global monopoly had moral overtones which demanded absolute victory for the administration. In such an atmosphere some of the more illogical outcomes went unnoticed. It seemed natural to the government that PAA should be the only airline denied any expansion, and the company thus has remained the only purely international airline. As the outsider in the case, United Air Lines admitted it did not need more competition along its own routes, but equity required that PAA be given a chance in the domestic market.[24]

Was any better alternative available? One was proposed, and it once again was the descendant of Postmaster General Brown and the 1935 Federal Aviation Commission. President Roosevelt indicated in 1943 a personal preference for assigning individual companies to different zones of the world. One company, in his view, would have the Mediterranean, another the east coast of South America, yet another the west coast. He implied that only one company would function in each zone and that foreign companies would provide sufficient competition.[25] Admiral Emory Land of the Maritime Commission said about the same thing during the public hearings on the McCarran Bill: "It would be the part of wisdom to divide the oceans in the world as you would a pie, and

23. The Post Office was successful and, if this pattern recurs, airlines will try to dispose of routes the president has directed them to operate. *CAB Annual Report, 1955*, p. 7; Frederick, *op. cit.*, pp. 232–33.

24. *Flag Line Hearings*, pp. 73–96.

25. Memorandum by Adolf A. Berle, one of those attending the White House meetings, cited in Smith, *Airways Abroad*, p. 151.

we would give to this American operator this segment of the ocean, we would give to another American operator this segment of the ocean, and so on, and we would not permit any cutthroat competition by American citizens in that segment."[26] In using the maritime analogy, Land described his proposal as the "multiple chosen instrument principle."[27] William Patterson, still not interested in overseas routes for United Air Lines, suggested to Congress in 1947 that only one U.S. company occupy each of the three major marketing regions.[28]

None of the appropriate government agencies seems to have given this idea serious attention, and it goes without saying that the president could not have expressed constant interest in the issue. During the 1945 hearings, the State Department official in charge of route negotiations insisted that any compromise with the principle of free competition would be a dangerous step toward monopoly.[29] There was no compromise, then, with the U.S. carriers, but the administration was singularly unsuccessful in imposing this arrangement upon other countries.

THE BILATERAL NEGOTIATION PROBLEM: 1946-57

The U.S. argument for international free competition may well have been only the extension of an old American economic principle.[30] The Europeans, however, saw it as incipient imperialism and an attempt to take undue advantage of wartime agreements. President Roosevelt and Prime Minister Churchill had agreed that the British should produce fighter planes while the Americans built

26. *Flag Line Hearings*, p. 314.
27. *Ibid.*, p. 320.
28. *Consolidation Hearings*, pp. 16–17.
29. *Flag Line Hearings*, p. 294.
30. The U.S. approach was known as the program of the "five freedoms." These can be summarized as the rights to (1) fly over another country; (2) land for technical reasons (refueling) but not to take on or let off passengers; (3) land and discharge passengers carried from the home country of the aircraft; (4) pick up passengers in the second country and carry them to the home country of the aircraft; and (5) pick up passengers in the second country and take them to a third country. These original five have been complicated by one now called the "sixth" freedom, i.e., picking up passengers in a second country and carrying them to a third country by way of the home country, as when a Dutch airliner takes on passengers in New York and delivers them to Paris via Amsterdam. See U.S. Congress, Senate, Committee on Commerce, *International Air Transportation Problems*, Hearings before Subcommittee, 87th Congress, 1st Session (Washington: Government Printing Office, 1961), pp. 5–7.

transports. Europeans thus faced the formidable threats of an aircraft industry with an almost insurmountable lead in transport development and global airlines already in business.[31] Well remembering the 1920's and 1930's, Europeans advocated some sort of regulation that would preserve a place in the market for each country's airline.

British Commonwealth countries took the lead in proposing regulatory schemes. Australia and New Zealand sought a supranational arrangement in which a world organization would own and operate all of the international airlines, presumably under the aegis of the United Nations. The British and Canadians asked for an international organization patterned after the CAB.[32]

At an international conference held in 1944 in Chicago, the U.S. attempted without success to convince fifty-one other countries in attendance to adopt its international free enterprise scheme. The British, while retreating somewhat from their initial position, held out for a minimal regulatory system of bilateral agreements to control the routes flown, the rates charged, and the frequency of schedules.[33] The conference deadlocked on the major issues and avoided complete failure only by agreeing on two technical matters. The U.S. and Britain spent the next year attempting to gain enough allies to break the deadlock, although evidence suggests the U.S. already had accepted the inevitability of the bilateral approach.[34]

The two countries reached agreement at a 1946 conference in Bermuda; the detailed provisions became known as the "Bermuda

31. Smith, *Airways Abroad,* p. 70; John C. Cooper, *The Right to Fly* (New York: Henry Holt and Co., 1947), pp. 171–72.

32. *Ibid.,* pp. 169–71. The evolution of agreement is traced in Virginia Little, "Control of International Air Transport," *International Organization,* 3 (February, 1949), 29–40. At one time the British Labor Party supported the Australia-New Zealand approach to global organization and control. *Ibid.,* p. 39. Regulatory proposals were not new in Europe; the French advanced one in 1932 which looked toward regional organizations. Oliver James Lissitzyn, *International Air Transport and National Policy* (New York: Council on Foreign Relations, 1942), pp. 91, 416–19.

33. "Routes" include ports of entry and airways. "Frequency" means the number of trips permitted across a given route within a given time period, e.g., two trips per week London–New York. The Soviets decided to attend the conference, then changed their decision and withdrew. Cooper, *op. cit.,* pp. 160–61.

34. The State Department began bilateral talks with Spain during the conference, and it was generally understood that CAB and industry opposed the multilateral approach. *CAB Annual Report, 1945,* p. 5; Smith, *Airways Abroad,* pp. 176–80.

Principles" and have since governed the route structuring of international operations. One aspect of the agreement left each country free to set allowable trip frequencies for its own carriers; this was a concession by the British who had sought international regulation.[35] A second provision specified that all future negotiations for route concessions and landing rights would follow the Bermuda pattern. This led to a furious round of negotiations and to further strains in the government-industry relationship.

The first order of business was to conduct all of the necessary negotiations while accommodating the policy of multiple competition.[36] The government-industry partnership was in full operation during the Bermuda conference, the country being represented by a combination of military experts, airline managers, and officials of the CAB and the State Department.[37] The airlines continued to participate in other negotiations until it became clear that the practice was unworkable when the interests of more than one U.S. company were at stake. The government next attempted to use the Air Transport Association to represent the entire industry, but this proved uncomfortable at best. During negotiations with Canada, for example, Stuart Tipton of ATA was a member of the U.S. delegation; the rules prohibited him from speaking with the airlines while conferences were in progress, and he could not make recommendations that might damage one company at the expense of another.[38] The ultimate solution was for the State Department to

35. The British and Europeans, of course, feared that the U.S., if left free to set its own trip frequencies, would turn its airlines loose and immediately dominate the world market. The U.S. was adamantly opposed to international regulation, and this was a substantial British concession. This agreement did not pose a problem for the U.S. so long as its airlines were the world's leaders. In more recent years, however, the steady decline in the U.S. position has forced something of a reversal in position; this country now asks others to show restraint. See Chapter 11.

36. Of 45 agreements in force in 1961, 35 had been negotiated before 1950. *CAB Annual Report, 1961*, p. 79.

37. Major General Laurence S. Kuter represented the Air Force. He became the first head of the Military Air Transport Service and would have been CAB Chairman had Congress agreed to let him retain his military seniority. *Ibid.*, p. 257; *New York Times*, January 10, 1948.

38. The negotiations turned on trade-offs involving concessions in eastern and western Canada. One of the airline officials forbidden to talk with Tipton was the ubiquitous James M. Landis, long in favor of multiple competition during his tour with the CAB, but now an executive of Colonial Airlines. U.S. Congress, Senate, Committee on Commerce, *Air-Line Industry Investigation*, Hearings, 81st Congress, 1st and 2nd Session (Washington: Government Printing Office, 1949 and 1950), Part 4, pp. 1513–16 (hereinafter *Air-Line Hearings*).

set its own position, negotiate the agreement in each case, then turn over to the CAB the remaining economic problems.

A theoretical approach to the problems of bilateral negotiation can serve to point out why the entire arrangement has proved so difficult to manage over the years. Many bilateral negotiations involving the U.S., for example, affect the interests of more than one American company, and one of four outcomes is likely to occur:

(1) The interests of both U.S. companies can be advanced at the expense of the airline of the other country;

(2) The interests of one U.S. company can be advanced, those of the second U.S. company not advanced, and those of the airline of the other country advanced;

(3) The interests of neither U.S. company can be advanced, while those of the airline of the other country are advanced; or

(4) The interests of all three companies can be advanced.

Of these possible outcomes, (1) and (4) seem unlikely to occur because the other country will feel it has given up too much; (2) seems unlikely because the CAB and the State Department are vulnerable to accusations they have favored one company over another. This leaves (3) as the likely outcome of any given negotiation, if only because it is easier to fashion a politically acceptable explanation. No matter the intentions of negotiators, damage to two private enterprises can be explained away as necessary to the national interest. Even the appearance of favoritism, on the other hand, promises only that the loser will seek political redress. At least in theory, then, the juxtaposition of direct competition and bilateral negotiations is an unfortunate arrangement.

Several examples can be used to demonstrate that these theoretical difficulties are indeed real. As other national economies gained in strength during the 1950's, the Eisenhower administration felt obliged to grant liberal route concessions. In one instance the State Department awarded the Australian national airline a route across the U.S. and on to Europe, in exchange for rights granted American companies to service two small cities in Australia. Some critics were annoyed at an apparently successful threat by the Dutch to withhold their contributions from the North Atlantic Treaty Organization until granted concessions for the Dutch national airline. Other complaints followed hard upon the award of significant routes to West Germany. The critics argued that in all cases the awards should be mutual and should be in proportion to the commercial traffic available within each of the negotiating countries.

Further, they asserted, the government should consult the affected airlines before and during negotiations.[39]

In replying to its critics, the Eisenhower administration produced the most comprehensive listing yet of the difficulties any administration encounters in implementing contradictory policies. The State Department spokesmen were Under Secretary Douglas Dillon and Thorsten Kalijarvi, then in charge of international aviation affairs. In two major speeches they argued that: (1) the U.S. often has no choice but to be liberal with other members of the free world alliance; (2) a *quid pro quo* is not always attainable; (3) the State Department permits the ATA to represent the industry as an observer but only when feasible; and (4) individual carriers cannot participate because they do not agree with each other.[40]

The administration thus implied that the inability of U.S. companies to agree left it with no choice but to favor foreign competitors, especially under the pressures of Cold War politics. Yet another administration, then, relied upon foreign policy arguments to defend its negotiations. The illogical outcome of all this is that the government still holds that it cannot assist U.S. airlines because they are private enterprises, but it must hurt them in the national interest. Given the limited range of possibilities outlined earlier, it is difficult to see how any better results can be expected so long as the policy framework remains unchanged.[41] The Eisenhower administration's arguments did not immediately silence the critics, and they tried to curb presidential authority.

Senator George Smathers of Florida introduced legislation in 1957 that would have distinguished between two categories of presidential decisions on international cases. One would contain cases in which foreign policy or national security interests were at

39. Critics included Senators Smathers, Mundt, and Capehart, and Representatives Jarman and Bow. *Congressional Record,* April 4, p. 5136; May 28 p. 7833; May 29, p. 7993; June 13, p. 9050; July 11, pp. 11426–29; all 1957. One estimate had it that the Dutch airline could earn $15.00 within the U.S. for each dollar earned in the Netherlands by U.S. carriers. *New York Times,* April 1, 1957.

40. *Department of State Bulletin,* 36 (June 24, 1957), 1012–13; (December 2, 1957), 877–89.

41. The *Project Horizon* task force, a generally nonpartisan group appointed by the Kennedy administration, concluded in 1961 that the U.S. should be more cautious in the future, especially when the richest markets are at stake. U.S. Federal Aviation Agency, Report of Task Force on National Aviation Goals, *Project Horizon* (Washington, Government Printing Office, 1961), pp. 26–27, 116–17 (hereinafter *Project Horizon*).

stake and could be decided only by the president. A second category would include cases in which only economic interests were involved and would be kept outside of presidential control. Smathers thought the distinction valid because the president had overturned CAB recommendations in some of the important decisions mentioned above.

In pressing their arguments for and against the bill, both sides avoided the principal issues. In reply to claims that the legislation would relieve the president of an unnecessary administrative burden, the administration said it would do no such thing; the president could not decide in advance if a given case involved foreign policy or national security and must, therefore, look at every case.[42] The Smathers bill, in sum, was an attempt to resolve a fundamental policy conflict by diminishing presidential authority in foreign policy. The bill postulated an imaginary world in which political, military, and economic interests are clearly distinguishable from each other, a view still widely held.[43]

THE DE FACTO REGULATORY APPARATUS: CARTEL OR LEGITIMATE INNOVATION?

In return for their acceptance of unilateral regulation of trip frequencies, the British succeeded at Bermuda in obtaining protection from the threat of global price-cutting wars. A curious arrangement evolved, one that satisfied the British demand for regulation but also made it appear that the U.S. had not retreated from its treatment of airlines as purely private enterprises not associated with any government. The mechanism chosen for international rate regulation was the International Air Transport Association (IATA), which forty-one of the world's international carriers formed in 1945. Britain's Director of Civil Aviation resigned his post to become Director-General of IATA.[44] The Ber-

42. The arguments are in U.S. Congress, Senate, Committee on Commerce, *Improvement of Procedures for the Development of Foreign Air Commerce,* 85th Congress, 1st Session Senate Report 119 (Washington: Government Printing Office, 1957), p. 119.

43. The *Project Horizon* group implied in 1961 that the distinction between economic and political decisions is valid and that the latter is an undesirable category. *Project Horizon,* p. 30. Others argue that making such a distinction can "remove temptation for the President to entertain economic appeals by the losing parties." "Notes," *Harvard Law Review,* 75 (June, 1962), 575–89.

44. Smith, *Airways Abroad,* p. 250.

muda Agreement stipulated that IATA would set international rates by unanimous agreement among members, each government retaining authority to approve or disapprove the rates. The U.S. intended that IATA regulation be only a temporary arrangement, but it has become the permanent one. It can be abandoned only if all concerned are willing to risk free competition or if the U.S. will accept a more formalized arrangement. Throughout almost two decades of IATA rate-making, however, the U.S. has remained philosophically committed to the removal of controls, other countries have disagreed, and the IATA's status has remained cloudy.

Statements issued at the time of the Bermuda Agreement indicated clearly that the U.S. had accepted for the first time a system of international regulation.[45] In practice, the government has kept its distance from IATA. U.S. companies usually discuss with the CAB the agenda of forthcoming IATA meetings, but this does not formally commit the board. The companies then carry back to the CAB the agreements reached at the meetings. On at least two important instances, the CAB has delayed taking action on proposed rate increases and then has tried to disapprove them at the last possible minute. In each case the State Department finally has instructed the CAB to accept the increase because the attempt to deny it had caused considerable tension between allies.[46] The CAB-IATA relationship thus remains one in which each side makes decisions on the basis of an uneasy estimate of what the other will accept.

The CAB continually has argued that the entire system is faulty and has asked for remedial legislation. Individual U.S. companies, says the CAB, occasionally are too adamant in the demands they advance at IATA meetings and use the principle of unanimity to force acceptance of their demands.[47] The CAB asks each year for

45. Cooper, *op. cit.*, p. 161.

46. One attempt in 1956 led to protests from the IATA, the companies, and other governments that the plans were made, the tourist season was at hand, and a reopening of the issue would cause only chaos. *Celler Report*, pp. 226–33. In 1963, the CAB delayed for five months, then announced that it would seek to overturn a 5 per cent rate increase. The *New York Times* called the ensuing events a "ludicrous joust" between allies; after the British and French threatened to withdraw landing rights from U.S. carriers, the State Department intervened. *New York Times*, May 15, 16, 17, 18, 1963; *Time*, May 3, 1963. See also Chapter 11, p. 291, n. 60.

47. *Celler Report*, p. 226; U.S. Congress, House, Committee on the Judiciary, *Monopoly Problems in Regulated Industries*, Hearings before Subcommittee, 84th Congress, 2nd Session (Washington: Government Printing Office, 1956), V, 2561–63.

authority to set the rates of all international airlines entering the country.[48] The State Department consistently takes the same position, especially during the 1963 rate dispute.[49] In the absence of a corollary proposal to establish an international regulatory apparatus, however, the logic of these CAB assertions is difficult to follow. The board appears to presuppose that other countries will accept unilateral U.S. rate regulation, a process that would require the CAB to examine the financial records of every international carrier whose routes touch the country. This seems unlikely, and it is equally doubtful that Congress would agree to any more explicit system of regulation.[50]

Important centers of opinion within and outside of Congress consistently have favored a unilateral approach. The Celler Committee of 1957, for example, described the IATA as only an international cartel organized by airline operators and, as such, offensive to American traditions.[51] During the 1963 dispute, several senators expressed dismay that the U.S. had "backed down" from the attempt to deny the rate increase.[52] The *New York Times* characterizes the IATA as antithetical to a U.S. tradition that emphasizes the relationship between efficiency and constantly decreasing rates.[53] The *New Republic* calls the IATA a "typical international cartel" which specializes in "price-fixing in restraint of trade."[54]

These descriptions demonstrate that the government may have been too successful in its approach to this international organization and to the international industry in general. Taking all of its

48. *CAB Annual Report, 1961,* pp. 10–11. The basic statute permits the CAB to disapprove only those rates considered unjust. 85th Congress, Public Law 726, Sec. 1002 (f).

49. *Denver Post,* May 14, 1963.

50. The 1963 dispute came closer than previous ones to producing new legislation. The administration asked for Congressional action, and the Senate passed S. 1540 (88th Congress) to give the CAB greater power over international rates. The bill did not pass beyond hearings in the House in 1964, but, in the meantime, the CAB was successful in applying some pressure on the IATA to reduce rates. The controlling factor, however, was the relatively low operating costs of the large jetliners of the 1960's; even with low passenger load factors, most airlines improved their revenue positions. *Congressional Quarterly Almanac,* XIX (1963), 83, 506; *Congressional Quarterly* (June 5, 1964), p. 1094; *The Aeroplane and Commercial Aviation News,* 100 (September 17, 1964), 10.

51. *Celler Report,* p. 218.

52. *New York Times,* May 16, 1963.

53. Editorial, *ibid.,* May 17, 1963.

54. *New Republic,* 148 (May 25, 1963), 5.

members into account, the IATA seems not a cartel at all, but an intergovernmental organization in which the U.S. is the only important member insisting that government and industry are completely separate. Practically all other member airlines are state-owned or state-controlled and, within each country, there is no real separation between airline and government.[55]

U.S. dissatisfaction with the IATA rate-making process seems traceable to two factors. First, those proposing unilateralism argue in effect that the U.S. should be able to veto any given decision. The argument follows from the difficult arrangement whereby U.S. members of the IATA cannot speak for the government and receive little or no effective guidance before negotiations begin. Second, the protests exemplify in yet another way the long-term confusion introduced by the removal of distinctions between domestic and international airline operation. Because the U.S. denies any connection between government and industry, it is a simple matter to assume that similar conditions obtain elsewhere.[56]

These factors have perpetuated a myth that obscures the larger political advantages of the IATA process. Free competition seems out of the question because no country can risk having its international airlines put out of business. Any regulatory apparatus more explicit than the IATA would, on the other hand, resemble the supranational agencies developed to manage the European economies. It is an open question whether all the countries having airlines, especially the U.S., would join a global organization of that type. The problem comes down to an American approach which refuses to consider the possibility that the IATA apparatus may be the best middle ground available. IATA's severest critics might recognize that: (1) the veto is not too appropriate for international airline regulation; (2) if the Soviet bloc continues to expand its air transport activity, it may have to be drawn into IATA; and (3) it may yet be necessary to reduce the administra-

55. IATA authority is greater than any ever exercised in the field of shipping and is the first such instance of direct co-ordination between an international industry and the representative governments. Daniel Marx, Jr., *International Shipping Cartels* (Princeton: Princeton University Press, 1953), pp. 282–84.

56. In attempting to deny IATA rate increases, the CAB consistently poses as the champion of lower rates. That CAB history does not justify such a pose will be demonstrated in Chapter 5. It might be added also that this book is written at a time when airline subsidies are less of an issue than heretofore, due both to substantial amounts of military traffic and to the lower operating costs of jets.

tive and philosophical distance between government, industry, and the IATA.

The major policies analyzed thus far are the only ones concerned with direct management of the airlines themselves. In each case, the government rejected partnership in favor of fictitious separation. There was yet another postwar policy, however, and it contemplated a partnership between the government and the aircraft manufacturing industry.

EXPORT SALES AND THE AIR MERCHANT MARINE PROPOSALS OF 1949-50

Long-term government promotion of export sales of transport aircraft was a major policy contribution of the Brewster Board and the Finletter Commission. Significantly, no important study group explored the policies outlined above; these emerged quietly at the end of the war, and the Brewster Board and the Finletter Commission had no choice but to accept them as givens. Both groups thus concentrated more upon airplanes than upon the airlines.[57]

Both emphasized export sales, but with some important differences in accent. The Finletter Commission wanted to sustain an aircraft manufacturing industry to meet the mobilization requirements of a future emergency.[58] The government, said the commission, should use the industry's required level of operation to calculate the number of aircraft to be sold abroad, and the Export-Import Bank should assume a larger share of the credit risk in such sales.[59] The Brewster Board entered a warning; export sales should not reach any level that might threaten the competitive position of U.S. carriers.[60] Both groups also recommended that the government participate in developing new cargo transports, apparently out of fear that the British might otherwise take the lead. The Finletter Commission wanted civilian control through the Department

57. These groups have been criticized for having concentrated upon strategic bombardment, especially since the Finletter Commission emphasized the need for a 70-group Air Force. Walter Millis with Harvey C. Mansfield and Harold Stein, *Arms and the State* (New York: Twentieth Century Fund, 1958), pp. 205–10. Both, however, seem to have tried quite hard to cover the entire range of aviation issues.

58. U.S. President's Air Policy Commission, *Survival in the Air Age* (Washington: Government Printing Office, 1948), p. 69 (hereinafter *Survival*).

59. *Ibid.*, pp. 45–59.

60. U.S. Congress, Congressional Aviation Policy Board, *National Aviation Policy*, 80th Congress, 2nd Session, Senate Report 949 (Washington: Government Printing Office, 1948), pp. 39–40 (hereinafter *National Policy*).

of Commerce, with military representation on the board of directors of an Aircraft Development Corporation.[61] The Brewster Board sought more direct military participation through use of the military budget and Air Force chairmanship of the board of directors. Moreover, said the board, the government should sell transports only to airlines promising not to modify them without military approval.[62]

Government policy over the years has tended to follow the lead of the Finletter Commission, presumably because it was the creature of the executive branch. Despite some similar proposals made in Congress, however, it proved impossible to reach agreement as to how the government should develop and sell transports within the U.S. Senator Edwin Johnson of Colorado and Representative John Kennedy of Massachusetts introduced bills to create an air merchant marine. Both would have set up government corporations, capitalized at $100 million, to develop and purchase new transports suitable for commercial and military use.[63] Private operators then would lease the airplanes, applying the rentals and any excess profits to the purchase price. The Kennedy bill provided that 75 per cent of the airline personnel involved in operating such aircraft be members of the military reserve. The proposals differed mainly on the question of administrative control; Johnson favored Air Force management, but Kennedy wanted a five-man board not under military control. With Air Force support, Brewster almost succeeded in having his own proposal passed in 1948, but neither the Kennedy nor Johnson bill advanced to the hearing stage.[64] The

61. *Survival*, pp. 137–39.

62. *National Policy*, p. 13.

63. H. R. 448, S. 237, 81st Congress.

64. The Truman administration opposed on budgetary grounds Brewster's 1948 bill to finance transport development under Air Force control. A companion bill passed the House but died in the Senate. Some Congressmen objected to military control, but Brewster said he could not determine the source of opposition. U.S. Congress, Senate, Committee on Commerce, *Development of Civil Transport Aircraft Suitable for Auxiliary Military Service*, Hearings Before Subcommittee, 80th Congress, 2nd Session (Washington: Government Printing Office, 1948), pp. 44–45; *Congressional Record*, July 29, 1948, p. 9466; *Air-Line Hearings*, I, 219.

Trade publications reported that the Kennedy and Johnson bills failed because (1) Congress opposed concealed expenditures on commercial aircraft; (2) no matter the administrative arrangements, one type of aviation might dominate others; and (3) tentative designs examined within the government did not seem worth the money. *Aviation Week*, January 10, p. 14; March 21, p. 12; August 8, p. 14; all in 1949.

significance of all these notions, including the export policy that was adopted, lay in the new partnerships they contemplated.

The only limitations on government development, purchase, and sale were to be the needs of the manufacturers and the calculated defense requirements. Limitations of this sort could not be measured with precision, nor were they related at all to the commercial market expected to support the aircraft and produce the revenues to repay government loans. Oddly enough, those advancing the proposals were adamant in demanding more stringent airline regulation.

The Brewster Board and the Finletter Commission advocated restricting the entry of too many new airlines into the market, wanted disinterested agencies concerned only with efficiency to redraw the airline map, and recommended that the CAB suspend or reduce mail payments as a means of forcing airlines into self-sufficient operation.[65] Kennedy led a congressional group determined to end what it considered unjustified subsidies.[66] Aircraft apparently were to be built and sold within the U.S. and abroad on the basis of one set of calculations, but airlines were to be regulated according to other rules.

The export promotion policy and the over-all rationale of the Export-Import Bank have combined to confuse the relationship between U.S. international carriers and their foreign competitors. The Bank, created by the Roosevelt administration to stimulate American exports, has had the equivalent of an open-end license to sell as many airplanes as it can and to guarantee the purchase loans needed to buy them. The standard administrative requirement for such loans is that the borrower be unable to get funds elsewhere, a provision that has tended to relieve foreign carriers from proving

65. *Survival*, pp. 111–14; *National Policy*, p. 25.

66. He accused the airlines of "getting away with murder" and introduced legislation to separate subsidy from mail pay, crediting the proposal to James M. Landis, the author of parallel proposals by the Hoover Commission. Representative Carl Hinshaw, co-Chairman of the Brewster Board, disagreed with Landis and Kennedy and argued that the CAB could not allocate costs between the statutory objectives of air mail, national defense, commerce, and public convenience. The Brewster Board and Finletter Commission opposed separation. *Congressional Record*, December 6, 1960, p. 16215; *The Hoover Commission Report* (New York, London, and Toronto: McGraw-Hill Book Company, Inc., 1949), p. 230; *Survival*, pp. 100–4; *National Policy*, p. 23. For an analysis of why subsidies had indeed gone up, see Chapter 5.

that the commercial market is sufficient to protect the loans. If market estimates indicate that aircraft can be gainfully employed, money is available through normal commercial channels. The policy never has been stated in such terms, but it amounts to subsidizing foreign carriers while insisting that U.S. carriers be self-reliant.

The U.S. thus acquired through its export promotion policy a vested interest in the financial success of the foreign companies purchasing American aircraft. The Eisenhower administration, stung by the criticisms of its apparent generosity to foreign companies, added to the argument the assertion that the Dutch were the largest single foreign purchaser of U.S. transports and that the route awards were only protection for the interests of American manufacturers.[67] This entire problem will be mentioned again in Chapter 11, but it is important to note here that both the export program and the air merchant marine proposals implied the acceptance of other political partnerships.

One partnership would guarantee to keep the aircraft manufacturers in business. There were, of course, sound political reasons for such a guarantee; the industry always has represented a significant geographic concentration of economic and political strength, and organized labor always has been quick to support programs designed to keep the factories operating. The U.S., moreover, has tended over the years to define mobilization requirements and national defense needs in terms of factories and production. The airlines, by contrast, suffer from geographic dispersion that makes them a relatively ineffective political force.[68]

A second partnership was much more complex. It involved something of a traditional liberal view carried forward from the Roosevelt years, to the effect that the airlines were relying upon permanent subsidies. Content with their lot, so went the argument, the airlines simply had overlooked the vast possibilities for expansion of the domestic and international travel markets. Once the markets expanded, it would easily be possible to keep busy in peacetime all the aircraft that would be needed in war. The air merchant marine advocates did not contemplate the established carriers as members of the new organization, because (1) the old-line

67. *Department of State Bulletin*, 36 (June 24, 1957) 1012–13.

68. It can be added that several Congressmen identified as airline supporters of the airlines were from areas where neither the airlines nor the manufacturers were significant factors, e.g., McCarran of Nevada, Brewster of Maine.

companies were too conservative and unimaginative, and (2) the old-line companies were to be strictly regulated and not given any more government assistance. The most likely candidates for the air merchant marine, then, were airlines not tied to specific routes designated by the CAB and most anxious to find loans they could not otherwise obtain. There was a group of such airlines in the postwar era and, in the eyes of those holding the views outlined here, this group seemed most attractive. The air merchant marine idea apparently was a bid to connect permanently the government and the nonscheduled airlines that emerged after the war.[69]

Within the proposed scheme the government would have financed airlines completely outside the established regulatory apparatus and completely inside the military reserve, an arrangement the military was ready to accept.[70] The nonskeds failed to get guaranteed financing for new aircraft on the virtually unlimited basis foreseen in the air merchant marine program. They managed, however, to ally themselves with congressmen who favored small business in general, in addition to those extremely critical of the scheduled carriers. These nonscheduled lines (some still in operation in the 1960's) added a new and complex dimension to the airlift problem, for their very existence cut across all of the policies outlined in this chapter. In effecting important political alliances, these companies managed to become one of the most potent political forces in U.S. airline history.

It is important to add a footnote designed to put in proper perspective these postwar policies of multiple competition, *de facto* regulation, and export promotion. They have governed U.S. airline operation for almost two decades and, in retrospect, are not the best that could have been developed. When they were adopted,

69. One of Hinshaw's objections to the Kennedy bill for separation of air mail and subsidy was that it would have permitted air mail applications from the non-scheduled carriers. *Congressional Record*, December 6, 1950, p. 16227. A representative of the nonskeds said later he had helped Kennedy in his attempts to pass the air merchant marine bill. U.S. Congress, Senate, Select Committee on Small Business, *Role of Irregular Airlines in United States Air Transportation Industry*, Hearings before Subcommittee, 82nd Congress, 1st Session (Washington: Government Printing Office, 1951), p. 177.

70. Military airlift officials hoped to organize reserve units around the structure of the nonscheduled companies. U.S. Archives, *Air Force Special File #2*, Letter from Major General Laurence S. Kuter to Secretary of the Air Force, February 26, 1948.

however, the Cold War was still in the future, international airline travel was more of a dream than a reality, the domestic carriers were busily trying to introduce untested new equipment, and the administration was insistent upon what it thought a long-overdue victory over Pan American. One of this book's major purposes, taken up again in Chapter 11, is to demonstrate that it may well be time to take another look at the policy structure.

[5]

Supplemental Carriers: A Political Force Outside the Policy Framework

THE NONSCHEDULED AIRLINES became an important political force because they seemed the embodiment of cherished American traditions, and they did not begin to lose their political strength for more than fifteen years after their beginnings in the aftermath of World War II. The sources of that strength, its apparently erroneous premises, and the reasons why this industry became so important to airlift history are the subjects of this chapter.

Wartime pilots built these new airlines from war surplus stocks, seeing an opportunity to use an exciting skill acquired in the armed services. Along the way these companies found needed allies among all those who had been suspicious of the major companies since the 1934 debacle. Additional support came from sources who favored small business of any sort, looked kindly upon an apparent eagerness to provide cheap transportation, viewed the CAB as the ally of a monopolistically inclined industry, and advocated a closer relationship between airlines and the military. The new companies responded so quickly to the needs of the Berlin Airlift and the Korean War that they acquired a permanent reputation as an important military reserve. For its part, the military provided an impressive series of testimonials to the continuing need for such an industry.

The nonskeds had so much appeal that they took on a political strength totally out of proportion to their economic contribution. For reasons that seem logical enough in retrospect, they never have commanded more than a 5 to 8 per cent share of the market, and the Defense Department has provided most of that. Their political supporters long have been reluctant to explore the possibility that the policy framework outlined in the previous chapter cannot accommodate such a group of carriers. Given such strong support, the nonskeds kept their reputations intact until a new series of tragic aircraft accidents in 1960 and 1961 made continued support inadvisable. By that time the industry had stayed alive for a longer period than its predecessors.[1]

THE SOURCES OF POLITICAL STRENGTH AND THE IMAGE OF THE INDUSTRY

The supplemental industry (the most recent official designation of the nonskeds) has become one of four categories of airlines which operate long-haul aircraft capable of transcontinental or transoceanic operation. Of the nine categories shown in Table I

Table I • Airlines as of April 1, 1963

Classification	Number	Fixed Routes	Authorized Subsidy	ATA Members
Domestic trunklines	11	Yes	Yes	Yes
International Carriers[a]	17	Yes	Yes	Yes
All-Cargo[b]	6	Yes	No	2 Yes, 4 No
Supplemental	15	No	No	No
Local Service[c]	13	Yes	Yes	Yes
Intra-Hawaii[c]	2	Yes	Yes	Yes
Intra-Alaska[c]	7	Yes	Yes	4 Yes, 3 No
Intra-State[d]	—	Authorized by individual state		
Helicopter	3	Area	Yes	Yes

[a]Includes 10 domestic trunklines, and 5 companies with very short routes, e.g., Florida to Bahamas.

[b]Important companies are Flying Tiger, Slick, Riddle (since renamed Airlift International), Seaboard World.

[c]These are essentially the same types of operation, i.e., "feeder" service.

[d]Exact number not obtainable. One example is an airline that operates between San Francisco and Los Angeles.

Compiled from ATA *Facts and Figures, 24th ed., 1963,* p. 35. U.S. Civil Aeronautics Board, *Subsidy for United States Certificated Air Carriers* (Mimeographed: Washington, 1962).

above, the CAB has awarded fixed long-haul routes to only the first two (domestic and international trunklines); these are the only classifications permanently eligible for subsidy. In policy terms,

1. The period before Postmaster General Brown drew his airline map in 1930 can be described as a period of unregulated competition, as can the policy vacuum of 1934–38. See Chapter 1.

then, only the scheduled airlines operate in the national interest.

A third classification, the all-cargo airlines, is a much smaller group that has CAB permission to service long-haul routes, mostly within the U.S. The most important of these companies (Slick, Flying Tiger, Airlift International, Seaboard World) are graduates of the supplemental group. Because the cargo airlines are not eligible for subsidy, they have had no choice but to suspend operations whenever the market is insufficient to keep them going, a relatively frequent occurrence. One of these companies is an ATA member but retains most of the characteristics of the supplementals. The fifteen supplementals are the residue of a much larger group.

When the war ended, thousands of trained pilots, surplus military transports that hardly had been flown at all, a widespread belief that air transportation would undergo unlimited expansion, and government desire to get a partial return on expenditures all combined to form the new industry. The CAB forecasted a tremendous increase by 1955 in the air cargo market, one that materialized only by 1962.[2] The Reconstruction Finance Corporation and the War Assets Administration arranged government loans for the purchase of transports.[3]

There were indeed some phenomenal bargains available. An ex-ATC pilot, also the heir to an oil fortune, began Slick Airways when he bought nine twin-engine C-46's for the original price of one of them.[4] Those unable to buy planes could lease them; four-engine C-54's rented for $2,000 per month, C-46's for only $300.[5]

2. The CAB predicted a cargo market of one billion ton miles by 1955. Richard Malkin, *Boxcars in the Sky* (New York: Import Publications, Inc., 1951), p. 234. A chart showing the trends is in U.S. Congress, House, Committee on the Judiciary, *Monopoly Problems in Regulated Industries,* Hearings before Subcommittee, 84th Congress, 2nd Session (Washington: Government Printing Office, 1956), I, 183, 189 (hereinafter *Celler Committee Hearings*). For more recent trends see U.S. Civil Aeronautics Board, *Supplement to the Monthly Report of Air Carrier Statistics* (Mimeographed: Washington, 1961), pp. 3–4.

3. U.S. Congress, Senate, Select Committee on Small Business, *Report on Role of Irregular Air Carriers in United States Air Transportation Industry,* 82nd Congress, 1st Session, Senate Report 540 (Washington: Government Printing Office, 1951), p. 4 (hereinafter *Small Business Report, 1951*).

4. Malkin, *op. cit.,* p. 98. New C-46's had cost the government $221,550 each. Alfred Goldberg (ed.), *A History of the United States Air Force 1907–1957* (Princeton: D. Van Nostrand Company, Inc.), p. 8.

5. U.S. Congress, Senate, Select Committee on Small Business, *Role of Irregular Airlines in United States Air Transportation Industry,* Hearings before Subcommittee, 82nd Congress, 1st Session (Washington: Government Printing Office, 1951), p. 8 (hereinafter *Small Business Hearings, 1951*).

Those using this government assistance fit within a cherished image.

One pilot had flown in World War I, had been a stunt flyer between wars and a PAA ferry pilot in World War II; he started an airline that by 1951 included two C-46's and two C-47's. During the first six postwar years, his personal investment amounted to $2,000, with government loans providing the rest.[6] Richard Malkin tells of the young veteran who purchased a single plane and then asked Malkin how to begin an air freight business. The pilot had not studied rates and did not know that more cargo moved from east to west than in the other direction, a factor making it difficult to operate profitable round trips.[7]

The number of companies formed during those years never has been determined with accuracy. When the CAB tried to register them all in 1946, there had been so many sales, leases, trades, and managerial turnovers that a complete accounting was impossible.[8] The Civil Aeronautics Administration estimated later that 3,600 new airlines tried to establish themselves.[9] Many of the long-term difficulties might be traced to the CAB's hesitancy in deciding what to do.

The board acted only very gradually to cope with the new industry, apparently for several reasons. For one thing, the CAB itself still was quite new; it had regulated airlines for little more than two years before war intervened. For another, both the board and the executive branch were deeply involved in building the policies outlined in the previous chapter. For yet another, the sale of surplus war stocks was a broad government policy, and it was easy to overlook the fact that military transports then were identical with those in commercial service. A fleet of Army buses, in contrast, hardly would have been suitable for a municipal bus system. There was, finally, enough of an immediate demand for travel to make the CAB hesitate to withdraw capacity from the market, especially given its own optimistic estimates of the future. Nevertheless a CAB staff examiner suggested in 1945 that the board not permit new companies to carry passengers on routes assigned

6. *Ibid.*, p. 54.

7. Malkin, *op. cit.*, p. 195.

8. John H. Frederick, "American Air Cargo Development," *Air Affairs*, 2 (Autumn, 1947), 93.

9. U.S. Department of Commerce, Civil Aeronautics Administration, *Statistical Handbook of Civil Aviation, 1948* (Washington: Government Printing Office, 1949), p. 37.

to the domestic trunklines.[10] The board, however, stayed with its policy of gradualism.

Between 1946 and 1948 the CAB asked companies to register, but allowed them to continue using the statutory exemption from regulation. Those who registered were divided into three categories known as the "large irregular carriers" (109), the "small irregular carriers" (1,701), and the all-cargo airlines (12).[11] During the subsequent two years the CAB tightened regulation of the large irregulars, some of which operated transcontinental routes.[12] The CAB prohibited them from operating with "more than a reasonable degree of regularity"; while not spelled out, this meant that irregulars could not advertise themselves as scheduled airlines.[13] The companies were for the most part operating between the country's major population centers and, of necessity, trying to fly regular schedules without appearing to do so.[14] The board also required all the companies to apply for individual exemptions or for permanent licenses to fly specific routes.

Processing and hearings promised to take several more years, so the CAB abruptly announced in March, 1951, that a new rule would become effective in April. Known as the "3 and 8 provision," the rule would have restricted the amount of flying between major cities.[15] When companies protested the rule to Senator John Sparkman of the Small Business Committee, he asked the board to delay implementation.[16] The carriers began simultaneous court action, and the Federal District Court in Washington thereupon enjoined the CAB from enforcing its rule. The court held

10. *CAB Annual Report, 1945*, p. 16.

11. *CAB Annual Report, 1946*, p. 23.

12. The largest of the companies concentrated on routes such as New York–San Juan, Seattle–Alaska, New York–Pacific coast. *CAB Annual Report, 1948*, p. 24.

13. U.S. Congress, Senate, Select Committee on Small Business, *Future of Irregular Airlines in United States Air Transport Industry*, Hearings before Subcommittee, 83rd Congress, 1st Session (Washington: Government Printing Office, 1953), p. 304 (hereinafter *Small Business Hearings, 1953*).

14. Testimony of Ramsay Potts, at the time a representative of a group of small carriers, *ibid.*, p. 397. For his later role as representative of the trunklines, see Chapter 7.

15. There was no limit to charter flights, but the irregulars could operate only three flights in the same direction between any one of eleven specified pairs of cities, e.g., New York–Miami, in any four consecutive weeks. Any company could make three round trips between New York and Miami, for example, but it could not make six one-way trips from New York to Miami. *CAB Annual Report, 1951*, pp. 27–29.

16. *Small Business Report, 1951*, p. 2.

that the irregulars had become established companies which the CAB could not put out of business without observing the Administrative Procedures Act.[17]

The Berlin Airlift and the Korean War gave the nonskeds a much stronger claim to survival as a matter of national interest. Impressive political support appeared almost immediately when the Small Business Committee issued its first report on the subject:

> Beginning in 1948 the Board apparently commenced a program of strict enforcement... the certificated carriers took an active interest in these efforts to eliminate the nonscheduled air carriers....
>
> To the extent that the "nonskeds" still exist today, they have managed to survive since 1948 in spite of constant harassment....
>
> ...In view of the Board's responsibility for the route pattern and the determination of mail pay allowances, it is only reasonable to expect that the five Board members would generally view air transportation problems in very much the same way as they are viewed by members of the industry.
>
> ...The nonscheduled industry has kept alive and developed technical skill and administrative know-how which has proved to be of great value in defense emergencies....
>
> These carriers offer a flexible airlift capacity that can be brought into military use far more rapidly than the certificated airlines whose equipment is obligated by commitments and schedules to service regularly appointed routes....[18]

The committee cited with approval the statement made by an industry witness that the irregulars were part of the "reserve pool for military use" the Finletter Commission had recommended.[19] While the Senate and House Commerce Committees had shown a mild interest in the subject in 1949, this 1951 confrontation cemented the important political alliance of parts of Congress and the nonscheduled operators.[20]

17. *Ibid.*, pp. 2–3.
18. *Ibid.*, pp. 5, 8, 17.
19. *Ibid.*, p. 8.
20. The chairmen of the two Commerce Committees, Senator Edwin Johnson and Representative Robert Crosser, had sent a letter to CAB Chairman McConnell, asking that the board find some way to preserve the irregulars as part of the "air reserve" until Congress could make an orderly determination. The CAB reply left little doubt that it felt it had little choice but to apply strict regulation and eliminate some of the companies. U.S. Congress, Senate, Committee on Commerce, *Air-Line Industry Investigation*, Hearings, 81st Con-

The combined effect of the Small Business Committee's intervention, the Cold War, and the court decision was to transfer a part of the regulatory process into a new environment. While the CAB investigated each of the companies, the Small Business Committee held more hearings and issued a new report praising the irregulars and asking the board to preserve them. The committee asked also that the CAB permit the new industry to enter the international market.[21] As its contribution, the Defense Department reported that the Berlin and Korean crises had demonstrated both a present and future need for such carriers.[22]

The CAB issued its first formal decision in 1955, stating that continued operation of the nonskeds was in the national interest; the board praised them for contributions to the Cold War and to domestic transportation. In words that echoed the Small Business Committee's first report, the CAB declared that: "An assessment of the importance of the irregular air carrier industry must include reference to the vital services rendered by these carriers in the interests of national defense. . . . It is evident that the irregular air carriers have the necessary flexibility to meet the demands of the military, while . . . the certificated carriers, due to their commitments to render adequate service to certificated points, are not as flexible in meeting emergencies."[23]

Furthermore, said the CAB, the irregulars had developed the charter market to the point where it had become important; they had provided specialized services, such as the shipment of lobsters, cattle, and turkey poults; they were a necessary adjunct to the trunklines during peak travel seasons; and they were the innovators who had pioneered in air coach travel.[24] In giving such weight to the defense argument and in using words supplied by the Small Business Committee, the CAB implied that regulation of this new industry was at least partly the responsibility of outside agencies.

Since 1955 the CAB and various congressional agencies have

gress, 1st and 2nd Session (Washington: Government Printing Office, 1949 and 1950), Part 1, p. 512 (hereinafter *Air-Line Hearings*). See also *Celler Committee Hearings*, I, 300.

21. U.S. Congress, Senate, Select Committee on Small Business, *Future of Irregular Airlines*, 83rd Congress, 1st Session, Senate Report 822 (Washington: Government Printing Office, 1953), reproduced in *Celler Committee Hearings*, II, 762.

22. Reproduced in *ibid.*, I, 580–81.

23. CAB Docket 5132, *Large Irregular Carrier Investigation*, reproduced in *ibid.*, I, 542–63, esp. 547–48.

24. *Ibid.*, p. 547.

quoted each other in support of the general proposition that the industry is worth preserving. The Celler Committee's 1957 report, for example, cited the 1955 CAB decision as authority for the conclusion that the military needed the irregulars.[25] The Senate Commerce Committee referred in 1960 to the same CAB decision and, a year later, CAB Chairman Alan Boyd referred to the committee's quotation of the 1955 decision as his authority for yet another statement supporting the supplemental industry.[26] The Defense Department consistently has issued its own supporting statements, as in 1961: "The continued existence of the irregular air carrier fleet is of real value in terms of national defense and it is evident that the future of the irregular air carriers to service the military, as they are doing now and have done so ably in the past, depends upon their ability to operate their planes in commercial activities when not engaged in service for the military."[27] The Defense Department, moreover, often has told the industry itself that it is indispensable to defense needs: "The preservation of those airlines comprising your association is of genuine concern to us. In any future military action, your group would unquestionably make a vital contribution."[28]

The supplementals thus drew their political support from the CAB itself, from the Defense Department, and from Congress. It was essential to the political strength of the carriers, however, that some supporters deny that the CAB was anything but unfriendly to fledgling airlines.

One rather traditional view has it that the CAB, like all other regulatory agencies, is the prisoner of the industry it regulates.

25. U.S. Congress, House, Committee on the Judiciary, *Report of the Antitrust Subcommittee (Subcommittee No. 5) Pursuant to H. Res. 107 on Airlines*, 85th Congress, 1st Session, Committee Print (Washington: Government Printing Office, 1957), pp. 95–96 (hereinafter *Celler Report*).

26. U.S. Congress, Senate, Committee on Commerce, *Supplemental Air Carriers*, Hearings before Subcommittee, 87th Congress, 1st Session (Washington: Government Printing Office, 1961), p. 6 (hereinafter *Supplemental Carrier Hearings, 1961*).

27. Letter from Assistant Secretary of the Air Force Joseph Imirie, reproduced in U.S. Congress, House, Committee on Commerce, *Supplemental Air Carriers*, 87th Congress, 1st Session, House Report 1177 (Washington: Government Printing Office, 1961), p. 59.

28. U.S. Congress, Senate, Committee on Commerce, *Study of Military Air Transport Service*, Hearings before Subcommittee, 85th Congress, 2nd Session (Washington: Government Printing Office, 1958), p. 166 (hereinafter *MATS Hearings*); Richard E. Caves, *Air Transport and its Regulators* (Cambridge: Harvard University Press, 1962), p. 278.

The nonskeds have, over the years, seemed the small and defenseless victims of this alliance between the regulators and the regulated. Economists have argued that the real task of any regulatory agency is to introduce new competition, not to stifle it, and political scientists have agreed.[29]

Echoing somewhat the views of the Small Business Committee and the majority of the Celler Committee, Senator Wayne Morse put it this way in 1962:

> We have the picture of the aviation industry arriving at a state of total monopoly, coming to this state through the machinations of the major trunk carriers. A pliant Civil Aeronautics Board has historically responded to their bidding... and has functioned as their . . . devout champion over the years . . . .
>
> It is one of the brightest memories in the history of all aviation to recall what these independent and nonsubsidized airlines did at the time of the Berlin and Korean airlifts.... They were still a very young group at the time of the Berlin Airlift. They borrowed money to acquire equipment, and carried over 25 per cent of the commercial tonnage into that beleaguered city, flying right alongside the American and allied military craft.[30]

AN ANALYSIS OF THE POLITICAL IMAGE

Simply stated, the facts of the matter never have supported the political image of the supplemental airlines. Their economic history demonstrates they had become largely superfluous by 1947, were sustained only by the Berlin and Korean emergencies, both of which occurred before the U.S. had rebuilt its military strength, and are legal and economic anachronisms almost solely dependent upon Defense Department largess. And, it must quickly be added, that department has done its utmost to deny a responsibility it has brought upon itself. The CAB, moreover, has made over the years a reasonable attempt to find a place for these airlines within the framework of national policy objectives. These conclusions no

29. See testimony of Professor Horace M. Gray in *Celler Committee Hearings*, I, 76–88; Professor Marver H. Bernstein in *ibid.*, pp. 59–86; Professor Bernard Schwartz has implied that any regulatory agency violates the public interest any time it increases rates, Bernard Schwartz, *The Professor and the Commissions* (New York: Alfred A. Knopf, 1959), pp. 118–21. Also Walter Adams and Horace M. Gray, *Monopoly in America* (New York: The Macmillan Company, 1955), pp. 65–66; *Celler Report*, pp. 74–110.

30. Reprint of speech from *Congressional Record*, April 11, 1962.

doubt appear to be an abrupt dismissal of an attractive image, but an analysis of postwar airlift history eliminates any other conclusion.

The supplementals began operation when travel demands were at a peak. Until 1947 the optimistic forecasts seemed borne out by a sudden tripling in the number of domestic passengers.[31] The passenger load factor, a statistic which can be used to measure both traveler convenience and the general level of demand, was extremely high, reaching 89 per cent in 1945 and 71 per cent in 1947.[32] The scheduled companies added to public frustration by an apparent inability or unwillingness to develop adequate machinery for dealing with impatient travelers.[33] Convinced by the evidence that the travel increase would continue, the CAB quickly lowered both passenger fares and mail payments.[34] Unfortunately, the predictions did not stand up.

The market suddenly leveled off in 1947, just as the airlines were introducing new aircraft ordered at the end of the war. These transports combined a great expansion in capacity with an increased need for high passenger load factors to cover the added costs. Financial reverses began immediately; by the time of the Finletter Commission's 1948 report, trunklines were losing money at the rate of $22 million per year.[35] There were numerous strikes, some dramatic accidents involving the new aircraft, and temporary groundings until mechanical defects could be overcome. Such a drastic reversal confronted the CAB with a major policy crisis.

The board had little choice but to reverse all of its recent actions and to admit to miscalculation. Where the CAB had lowered mail payments and passenger fares, it now raised both, and it set higher mail rates than had existed in the first place. Airline losses were so

31. *CAB Annual Report, 1948*, Chart, p. 50.

32. The load factor is the percentage of revenue space actually sold. If an airplane has 100 seats and 50 of them are occupied by paying passengers, the load factor is 50 per cent. As a rule of thumb, anything above 65 or 70 is thought too high for the convenience of the general public; given the fluctuations in day-to-day demand and the need to assist customers having last minute changes in plans, this is about the best that can be done. John H. Frederick, *Commercial Air Transportation* (Homewood, Illinois: Richard D. Irwin, Inc., 1955), pp. 383–84.

33. "What's Wrong with the Airlines?" *Fortune*, 32 (August, 1945), 159–64.

34. *CAB Annual Report, 1945*, pp. 7–8.

35. U.S. President's Air Policy Commission, *Survival in the Air Age* (Washington: Government Printing Office, 1948), p. 99.

great that the CAB paid retroactive subsidies to spread those losses over a three-year period.[36] In some cases the board forced the industry to raise passenger rates when it did not want to and, even more importantly, denied the trunklines permission to experiment with air coach operation.[37]

The situation was not conducive to friendly relationships between the CAB and the major companies, each side accusing the other of having caused the confusion.[38] The CAB disliked explaining to Congress why mail payments and subsidies had to be raised, and the carriers did not relish accusations that they were the beneficiaries of unjustified expenditures.[39] One casualty of the skirmish was James M. Landis; President Truman quietly decided in 1948 not to reappoint him to the CAB.[40] Landis admitted later he had been too optimistic, but he added that arguments with the major companies had led to his departure.[41]

Once the trunklines had their new aircraft and the market had leveled off, there was no real market for the supplementals. During the period of peak demand, they had flown the same types of aircraft and had offered lower rates between the major cities.[42] They had enjoyed this prosperity, however, only because they were free of regulation and because of the scarcity of seats. Because the supplementals could experiment at will, set any rate, and choose any route, they concentrated on the big cities; even so, they depended upon their ability to fill every seat on every flight. Once

36. *CAB Annual Report, 1948*, pp. 1–2. If the financial charts in the CAB Reports of those years are compared, the additional payments appear to have come to about $30 million.

37. Specific instances are indicated in *Celler Committee Hearings*, III, 2192–93, 2212–16.

38. The airline case was made by Ralph Damon, "Airline Outlook," *Air Affairs*, 3 (October 1950), 459–68; Carleton Putnam, "Airline Outlook," *Air Affairs*, 2 (Winter, 1949) 491–504. Damon was then president of TWA, Putnam president of Chicago and Southern (later merged into Delta Airlines). W. A. Patterson of United Air Lines said that the CAB had awarded too many competitive routes, thus preventing sound development. Chairman McConnell of the CAB blamed the carriers for having asked for the routes. *Aviation Week*, November 8, 1948, p. 4.

39. See Chapter 4, note 66.

40. *New York Times*, January 1, 1948. January 3, 1948.

41. *Air-Line Hearings*, I, 213.

42. There is no way to prove or disprove the old argument over whether the nonskeds actually offered low-cost transportation. When travel demand was at its immediate postwar peak, these companies may well have been charging premium rates—there was little in the way of records and, of course, the CAB had no control.

the public could get reservations on newer aircraft without a long delay, the supplementals lost most of their attractiveness. To those observing the industry, it was clear even in 1946 that a great "weeding out" lay ahead.[43]

Faced with the necessity to make up the losses of the major companies, the CAB understandably decided to take a closer look at carriers duplicating the routes of those companies. As the board processed each new application for permanent exemption from regulation, it began to deny requests made by companies operating on regular schedules.[44] By 1947, then, the nonskeds were trying only to survive until something yet unforeseen would improve their lot.[45]

The industry's staunchest supporters have since admitted that survival would have been impossible without the help afforded by international crises.[46] During the Small Business Committee's 1951 hearings, one of the irregulars admitted that "if the Russian timetable in Korea had been delayed as much as 3 months, Overseas National Airways would have been out of business."[47] In this typical case, the company's entire 1951 operation was five airplanes and 119 employees, all engaged in military contract work between California and Japan.[48] Most of the companies that survived became more and more dependent upon these military contracts.

Both the Small Business Committee and the CAB used the phrase "flexible airlift capacity" to argue that the nonskeds could develop a specialized domestic market, yet remain constantly ready to help out in military emergencies. This has been a superficially attractive argument, but it overlooks the nature of airlift systems, commercial airlift in general, and normal business practices as well. Even industry representatives have agreed, for example, that they can attain efficiency only by a scheduled operation across a

43. "Freight by Air," *Fortune*, 35 (May, 1946), 134–39.

44. *Small Business Hearings, 1951*, p. 42. The litigation growing out of these decisions continued until 1956. Robert B. Bendiner, "The Rise and Fall of the Nonskeds," *Reporter*, 16 (May 30, 1957), 29–34. These were anything but partisan disputes. The largest of the nonskeds employed Joseph O'Mahoney while he was between terms as a Wyoming Senator. The same company's counsel was Murray Chotiner (a prominent California Republican), whose letter to Sherman Adams, then a ranking Assistant to the President, caused Adams to go over the case with the CAB in some detail. Schwartz, *op. cit.*, pp. 233–34.

45. Frederick, "American Air Cargo Development," p. 93.

46. *Celler Report*, p. 107.

47. *Small Business Hearings, 1951*, p. 65.

48. *Ibid.*

stable route structure.[49] The specialized market for charter services, moreover, always has been too sporadic and unreliable to enable any company to stay in business only on that basis.[50] The unregulated carriers thus protested strongly whenever the CAB limited the extent of their scheduled operations.

If these companies, on the other hand, ever had developed a market that depended upon their services, they would have been much less anxious to accept military business. Normal business practice demands that customers once cultivated be kept satisfied. "Flexible airlift capacity" has been only a euphemism for "unemployed airlift capacity," more than eager to accept any paying assignment. During the Berlin and Korean emergencies, the established companies followed the same practice and made available to the military their currently unemployed transports.

The supplementals, finally, exaggerated their Cold War contributions to an extent that approached misrepresentation. The greatest exaggeration was the new industry's description of its participation in the Berlin Airlift. In presenting the case, former Senator O'Mahoney put it this way: ". . . I am sure every American thrills with pride over the fact that American carriers were able for more than a year to feed and fuel and supply the beleaguered city of Berlin in the face of the Communist threat. . . .

. . . The Kremlin was sure that we would not have the courage to go through, but we went through, and in the period from July 1, 1948 to December, 1, 1949, irregular carriers flew 25 per cent of the cargo that was flown."[51]

Congressional supporters used this statement for several years. The Small Business Committee said about the same thing in its 1951 and 1953 reports, and Senator Morse adopted it.[52] Commercial airlift actually operated only between the U.S. and airfields in the Western Zone of Germany; between those airfields and Berlin,

49. Testimony in *Small Business Hearings, 1953*, p. 397.

50. There has been some past evidence that tourist season travel demand sometimes is high enough to force the trunklines to temporarily charter additional aircraft from the supplementals. Generally speaking, however, the scheduled companies often have some of their own aircraft temporarily unemployed at various times of the year. The supplementals have, for the most part, been tied to military traffic all along. CAB Docket 5132 in *Cellar Committee Hearings*, I, 349; Examiner's report in *ibid., p.* 605.

51. *Small Business Hearings, 1953*, pp. 267–68; also testimony of industry witness at pp. 115–16.

52. *Small Business Report, 1951*, p. 8; 1953 Report reproduced in *Celler Committee Hearings*, II, 755. Quote from Senator Morse is on p. 98.

the operation was entirely military and for very practical reasons.[53]

Statements about the Korean War were less exaggerated but produced an equally misleading impression. The supplementals provided some necessary help, as did the established companies. The Pacific Ocean, however, was at the very limit of the ability of the four-engine C-54's then the mainstay of both the military and supplemental fleets. Strategic airlift still was in its infancy and, taken together, military and commercial fleets took six months to move the equivalent of a single Army division.[54] Sealift provided most of the strategic support for the Korean War, and it was misleading to imply that war surplus transports barely able to traverse the distance from California to Hawaii could answer future needs for strategic airlift, especially cargo.[55] The Defense Department praised the supplementals in general terms but, in so doing, lent credence to the exaggerations. In retrospect, then, the defense aspects of the 1955 CAB decision are open to question; so are the domestic aspects.

The CAB apparently overlooked important evidence collected

53. In the first six months of the Berlin crisis, commercial airlift made 610 transatlantic trips, divided about equally between scheduled and nonscheduled operators. *Aviation Week*, January 31, 1949, p. 16. Within the corridors the operation was controlled by an Anglo-American Combined Airlift Task Force, headed by U.S. Air Force General William H. Tunner. W. Phillips Davison, *The Berlin Blockade, A Study in Cold War Politics* (Princeton: Princeton University Press, 1958), p. 195. The British used some commercial operators in their portion of the corridor. Lowell Bennett, *Berlin Bastion* (Frankfurt am Main, Germany: Fred Rudl, 1951), pp. 102–3. Also *The Berlin Airlift* (London: His Majesty's Stationery Office, Air Ministry and Central Office of Information, 1949), pp. 25, 50. British commercial participation was unified and managed as such. Seven U.S. companies were involved, however, and it would have been almost impossible to manage them within the corridors. An airlift of this sort requires complete interchangeability of flight crews, ground crews, and all other facilities. Had the seven commercial companies been one, the Air Force might have been able to integrate it within the corridors.

54. Computed in Hiram C. Caroom, "Management Responsibilities and the National Interest: An Analysis of the Practicability of the U.S. Flag-International Passenger Airlines' Transporting more Military Traffic in Peace and War" (unpublished doctoral dissertation, Harvard University, 1958), p. 39.

55. Three routes were in use: one was through Alaska and the Aleutians to Japan, often difficult to fly because of weather conditions; the others were through Hawaii and other mid-Pacific islands to Japan. The distance between California to Hawaii is greater than that on any part of the transatlantic route, and the whole problem was magnified by the prevailing headwinds. Discussion of routes is in U.S. Office of the Chief of Military History, Department of the Army, *United States Army in the Korean War: South to the Naktong, North to the Yalu* (June–November, 1950), by Roy E. Appleman (Washington: Government Printing Office, 1961), p. 115.

during its investigation of the supplementals, thus following a practice common to regulatory agencies. The 1955 decision said nothing of hundreds of complaints submitted by the Better Business Bureaus of many large cities. The complaints mentioned unsavory practices of the sort that might be expected in an industry struggling for survival.[56] What is the nature of a decision-making process that chooses to overlook such evidence?

The quasi-judicial appearance of the CAB seems to mask a process that is anything but judicial in nature. Like other agencies of its type, the CAB relies for the writing of its reports on a special staff office whose members seldom attend the relevant hearings. In many cases the opinion writers receive only one-sentence directives from board members who, in turn, do not review the opinions before publication. Each opinion must only conform to the decision and does not have to connect logically with any other opinion. From his detailed study of many CAB decisions, Samuel B. Richmond concludes that the precise intent of the board never can be ascertained with accuracy.[57] Because the 1955 decision did not comment on important evidence and used language apparently borrowed from the Small Business Committee, it must be considered only a direct response to congressional pressure.[58] The CAB produced this key decision during the single year of Ross Rizley's

56. Chicago, San Francisco, New York, and Los Angeles submitted complaints, as did several other cities. There were reports that two-engine airplanes had been used instead of the four-engine planes advertised, that return tickets had not been honored, that refunds had been refused, and that advertised meal services had not been provided. *Celler Committee Hearings*, I, 583. The seamier aspects of the story, including the manner in which Puerto Ricans were crowded aboard old airplanes far beyond the number of seats available, are well described in Ernest K. Gann, *Fate is the Hunter* (Greenwich, Connecticut: Fawcett Publications, Inc., Crest Reprint, 1962), pp. 334–46.

57. Samuel B. Richmond, *Regulation and Competition in Air Transportation* (New York and London: Columbia University Press, 1961), pp. 132–35. A parallel description of the Federal Communications Commission is in Schwartz, *op. cit.*, pp. 190–92. Nor can this problem be overcome simply by insisting that board members do all of their own work. Speaking of all such agencies, Anthony Lewis observes that "commission members can be shown statistically to have no time to read the record themselves." Anthony Lewis, "To Regulate the Regulators," in Samuel Krislov and Lloyd D. Musolf, *The Politics of Regulation* (Boston: Houghton Mifflin Company, 1964), p. 10. This suggests that only a broad change in the entire theory of regulation will suffice. See Chapter 11.

58. This is not to argue that Congress has no legitimate place in the process, just the opposite; the entire process is quite political, involves military, foreign, and economic policy, and everyone concerned with these matters is a logical participant in the policy process.

tenure as Chairman, and by a three-two vote; he is credited with having generated many such decisions in a very short period.[59]

The Defense Department and the CAB both have supported the supplementals, then, but the diffusion of responsibility has caused a type of confusion somewhat similar to the situation of the scheduled international companies. Each agency has held the other responsible for the supplementals and neither has been willing to accept its own share of responsibility. The CAB decision relied heavily upon the Cold War contributions of the supplementals. The Defense Department has insisted that its praise of the industry does not require it to use those companies when they are not needed. It is not a military function, said a Defense Department spokesman in 1958, to subsidize commercial enterprise.[60]

To an important degree, the political image of the supplementals has depended upon a single major assumption; that an increase in unregulated competition is the answer to most, if not all, airline problems. The purpose of this brief analysis of the image is not to dispose of the assumption, but rather to demonstrate that postwar airline history and the nature of the supplementals' contribution to national defense are not precisely what their supporters have claimed. It is just as important to note what appears as a major contradiction; those championing the supplementals have insisted they be turned loose to compete against the scheduled carriers and that subsidies simultaneously be reduced. Indeed, the very existence of the unregulated carriers stood out in such bold relief against the national policy framework outlined in previous chapters that the courts, beginning with the 1951 injunction against the CAB, struggled for a full decade trying to make legal sense out of the problem and figure out what to do with airlines seemingly not contemplated in the basic legislation.

THE UNUSUAL LONG-TERM STATUS OF THE SUPPLEMENTALS: 1955-63

A final solution has yet to be found to the problem represented by these carriers. Tables II, III, and IV indicate the dependency

59. See exchange between Representative Celler and Chairman Rizley in *Celler Committee Hearings*, I, 123–25, with Celler praising the CAB's action. Also Richmond, *op. cit.*, p. 99; Caves, *op. cit.*, pp. 283–84; 296–97. Conversely, the same period has been called the "wildest spree in CAB history." Henry J. Friendly, *The Federal Administrative Agencies* (Cambridge: Harvard University Press, 1962), pp. 154–55.

60. Testimony of Air Force Secretary Dudley Sharp, in *MATS Hearings*, p. 27.

Table II • Activity of Supplemental Air Carrier Industry

	1960 (000)	1961 (000)
Total passenger-miles	2,421,669	1,571,771
Military passenger-miles	1,762,243	821,188
Total cargo ton-miles	95,097	172,862
Military cargo ton-miles	88,395	165,979
Over-all military percentage	75	68

Source: U.S. Civil Aeronautics Board, *Supplement to the Monthly Report of Air Carrier Traffic Statistics,* Quarter Ended December 31, 1961 (Mimeographed: Washington, 1962).

Table III • Activities of Four Selected Supplemental Carriers[a]

	1960 (000)	1961 (000)
Total passenger-miles	1,761,741	681,329
Military passenger-miles	1,583,450	457,627

[a]Capitol Airways, U.S. Overseas Airlines, World Airways, Overseas National Airways.
Source: Same as Table II.

upon military business of both the supplementals and the cargo airlines. What remains of the nonskeds has survived a great deal of litigation; immediately following the 1955 CAB decision, it was the scheduled industry's turn to challenge the CAB in court.

The Circuit Court of Appeals invalidated the CAB decision in 1956, on grounds that important parts of the decision violated the basic statute.[61] In allowing the supplementals to continue flying, the CAB had restricted each of them to a maximum of ten flights per month between any pair of cities. The total number of flights permitted each carrier was unrestricted, but the limitation prevented any company from setting up a permanent schedule. The Court held that the board had relieved the supplementals of the normal licensing process without making the administrative finding that the process was too great a burden on those companies. After three more years of CAB investigation and research, another board decision produced inconclusive results.

When, in 1959, the CAB issued certificates to twenty-five supplementals, the scheduled carriers challenged once again. The same

61. This summary is from the testimony of CAB Chairman Alan Boyd, in *Supplemental Carrier Hearings, 1961,* pp. 4–6.

Table IV • Activities of Three Selected Supplemental Carriers[a]

	1960 (000)	1961 (000)
Total cargo ton-miles	89,826	155,703
Military cargo ton-miles	87,974	152,857

[a]Capitol Airways, U.S. Overseas Airlines, World Airways.
Source: Same as Table II.

Table V • Revenues of Three Selected Cargo Carriers[a] (millions of dollars)

	1960, 1961	Fiscal Year 1961, 1962
Total Revenue	119	
Charter Revenue	72	
Revenue from MATS Contracts		80

[a]Flying Tiger, Slick, Seaboard World.
Sources: U.S. Civil Aeronautics Board, *Quarterly Report of Air Carrier Financial Statistics, December, 1961* (Mimeographed: Washington, 1962) pp. 30, 37. Also U.S. Department of Defense, Military Air Transport Service, *Airlift Service Management Report,* July, 1961–June, 1962, pp. 20–21 (Discrepancy is because CAB reports by calendar year and MATS by fiscal year. Since revenues are distributed fairly evenly throughout year, military revenue constitutes approximately two-thirds of total revenues for two years).

court invalidated the new certificates because (1) as blanket authorizations to operate anywhere, they violated the statutory requirement to assign fixed routes; (2) they imposed a limitation of ten trips per route per month, but the statute permitted carriers to set their own trip frequencies on assigned domestic routes; and (3) they authorized nationwide operation for small carriers which did not seem ready for such a broad grant of authority.

Taken together, the 1951, 1956, and 1959 court decisions made their own contributions to policy confusion.[62] The first prevented the CAB from imposing strict regulations on the supplementals, while the others threw out CAB attempts to license the same carriers. A decade of furious activity thus produced only a legal stalemate which forced the CAB to go to Congress for assistance.

Congress decided in 1960 not to permanently resolve the position of the supplementals. A temporary statute authorized the CAB to issue two-year certificates, thus giving Congress time to review

62. This is the author's view, not Mr. Boyd's.

the subject in detail.[63] When they took up permanent legislation the following year, the Senate and House deadlocked on the question of how broad the grant of authority should be. The Senate passed a bill generally favorable to the supplementals, but the House proved more cautious. When Congress recessed in 1961, these disputed points were still at issue:

(1) The Senate would allow ten trips per month on any route. The House would allow only temporary thirty-day certificates for use during the tourist season.

(2) The Senate would encourage the Post Office to ship air mail at freight rates, generally lower than mail rates. The House would prohibit the shipment of mail on the supplementals.

(3) The Senate would authorize charter flights for groups of tourists having no other common association. The House would continue existing restrictions which limited overseas charters to groups having an established identity of their own.[64]

The House-Senate disagreement reflected the contrasting views of the chairmen of the Aviation Subcommittees: Representative John Williams of Mississippi and Senator A. S. Monroney of Oklahoma. The deadlock was broken by an unfortunate set of circumstances.

As had occurred after the 1934 air mail cancellations, during the policy vacuum of 1934–38, and during the 1950's when a series of mid-air collisions demonstrated the need for more sophisticated control of the airways, tragedy produced legislation. The supplementals quickly lost a substantial amount of their political support when seventy-four Army recruits died in a crash near Richmond, Virginia, in 1961. This was one of several accidents involving the supplementals which led Najeeb E. Halaby, head of the Federal Aviation Agency, to voice concern at a dangerous trend.[65] Investigations by the CAB and the House Armed Services Committee disclosed that the supplemental industry was in a questionable condition. The CAB concluded, for example, that the flight crew in the Richmond crash was "not capable of performing the function

63. Public Law 86–661.

64. Summarized in *Congressional Quarterly Almanac*, 17 (1961), 502.

65. U.S. Congress, House, Committee on Armed Services, *Richmond Air Crash, November 8, 1961*, Hearings before Subcommittee, 87th Congress, 2nd Session (Washington: Government Printing Office, 1962), p. 3047 (hereinafter *Richmond Crash Hearings*).

assigned to it" and that the company's maintenance practices were "substandard."[66] The CAB admitted to Congress that it had licensed all such companies on the basis of evidence gathered before 1955, in some instances as long ago as 1951.[67] The companies had changed ownership and management so often that the board's limited resources had prevented continuous supervision.[68]

This particular accident prompted Secretary of Defense McNamara to instruct the Military Air Transport Service (MATS) to make its own independent investigation of these companies, instead of relying upon CAB certifications.[69] McNamara's action brought a mixed reaction from the press and the airline industry, but the Armed Services Committee agreed with him that something had to be done.[70] This view was reinforced by Congressman Walter of Pennsylvania, a very senior Democrat and the Representative for many of the recruits killed in the crash. Reporting on his own investigation of the supplementals, he concluded that many of them were marginal and nearly bankrupt and, he added, congressional action was long overdue.[71]

The result of all this was that Congress, in legislation passed in 1962, installed more stringent provisions for regulating the supplementals. Those still in business could continue airline-type operations for two more years, but they could not exceed their average level of operations between 1959 and 1961. Congress directed the CAB to re-examine each company and to issue certificates when this was in the public interest. By April, 1963, the board had certified fifteen such carriers, generally those still kept busy on mili-

66. Extract of CAB Report reproduced in U.S. Congress, House, Committee on Armed Services, *Report of Special Subcommittee to Investigate Crash of Imperial Airlines, Inc., Richmond, Va., Nov. 8, 1961*, 87th Congress, 2nd Session, Committee Print (Washington: Government Printing Office, 1962), p. 3109 (hereinafter *Richmond Crash Report*).

67. Testimony of CAB in *Richmond Crash Hearings*, p. 2979.

68. *Ibid.*, p. 2975.

69. Policy statement reproduced in *Richmond Crash Report*, p. 3096.

70. One example of support was an editorial in the *St. Louis Post-Dispatch*, December 3, 1961. Some industry members feared MATS would expand its international operations into the domestic field, since the crashed aircraft had been flying recruits from one camp to another within the U.S. *Aviation Daily*, December 4, 1961. See also *Richmond Crash Report*, pp. 3094–95.

71. *Richmond Crash Hearings*, pp. 2882–86. Congressional supporters of the supplementals almost immediately became much more cautious in their support. See U.S. Congress, Senate, Committee on Commerce, *Regulation of Supplemental Air Carriers*, Hearings before Subcommittee, 87th Congress, 2nd Session (Washington: Government Printing Office, 1962), pp. 28, 34.

tary contracts; the number apparently will remain quite stable.[72]

What remains of the industry is no longer a group of small entrepreneurs. Some of them have become entangled in complex financial arrangements which, if widely publicized, might open broader questions of public policy. In an arrangement generated by pressure from some segments of Congress and the CAB, the Defense Department has issued tacit guarantees; if these companies purchase modern jet transports suitable for military use, the Department will provide enough business to keep the aircraft busy throughout an operating lifetime of seven to ten years. The Department now awards contracts for three-year periods on grounds that it has a stake in the promotion of a "sound commercial industry."[73] The companies use the tacit guarantees as collateral for the loans needed to purchase transports at about $5 million each.

One small airline contracted in June, 1962, for $6.2 million in military business for the following year, with the likelihood that more would be forthcoming during that year.[74] The company ordered two new Douglas jets immediately after signing the contract.[75] Four months later the Studebaker Corporation purchased the same airline as one of a group of profitable businesses that would offset Studebaker's losses in other enterprises.[76] Similarly, the Greyhound Corporation has been operating a profitable subsidiary which leases new jet transports to companies having military contracts.[77] The military contracts, moreover, include rates set by the CAB and, because the rates take into account the higher operating costs of established international airlines having fixed routes and associated facilities, they promise substantial profits for small operators with low overheads.[78]

72. Assuming, of course, there is no substantial fall-off in military contract business. A summary of airline categories is in Air Transport Association of America, *Facts and Figures About Air Transportation,* (24th ed.; Washington: 1963), p. 35.

73. Testimony of Edward J. Driscoll, Office of the assistant Secretary of the Air Force for Installations and Logistics, in U.S. Congress, Senate, Committee on Commerce, *Aircraft Guarantee Loan Program,* Hearings before Subcommittee, 87th Congress, 2nd Session (Washington: Government Printing Office, 1962), p. 152. This office is responsible for passing out the military contracts so important to the supplemental industry.

74. *Aviation Daily*. June 4, 1962.

75. Douglas Aircraft Company Press Release (Mimeographed: June 11, 1962).

76. *Time,* October 19, 1962.

77. *Ibid.*, April 26, 1963.

78. Small companies are able to realize profits as high as 45 per cent of total revenues before taxes. U.S. Congress, House, Committee on Govern-

The constantly increasing expense of airlift operations, the strict new regulatory system for the supplementals, and the virtual lack of a commercial market for nonscheduled carriers has completely transformed the nature of those companies. As in the case of the small New York–Washington airline that Postmaster General Brown rejected in 1930, there probably was little hope all along that the entrepreneurs produced after World War II could find a market secure enough and profitable enough to permit constant modernization of their operating fleets. The most marginal of the operators have by this time been largely swept away, but at least some of the survivors seem to violate important tenets of public policy. They still look like small enterprises but they are, in reality, organizations investigated and approved by the military, almost completely dependent upon military business; most significant of all, they are only avenues along which government subsidies travel to corporations not usually thought eligible for such assistance.

THE SUPPLEMENTALS AND INTERNATIONAL AVIATION

Given the circumstances, the supplementals inevitably became by the 1950's a pressure group dedicated to securing increased government expenditures for commercial airlift, not only during military crises but between them. The industry's very existence, however, was so contradictory to the national and international regulatory apparatus that the long-term political support of the industry can be explained only as a commitment to one particular economic theory, i.e., as many businesses as possible, and in any case the airline industry, should be composed of as many companies as possible—all of them quite small. The attempts to apply the theory ran head-on into some difficult practical problems, especially in the international field.

The most important one was the vague notion that the supplementals should enter international aviation. Some critics have

---

ment Operations, *Military Air Transportation*, Hearings before Subcommittee, 87th Congress, 1st Session (Washington: Government Printing Office, 1961), pp. 96–98, 156–57. The CAB admitted in 1963 that the gross profit margin of some of these companies was about 25 per cent, with a return on investment of 27.9 per cent. *Ibid.; Military Air Transportation, 1963*, 88th Congress, 1st Session (Washington: Government Printing Office, 1963), table on p. 82. The regulated companies have been kept at a profit margin of around 10 per cent.

reproached the Eisenhower administration for having quietly intervened to make sure the 1955 CAB decision was not applied to international aviation.[79] The administration hardly could have done otherwise. If applied across the board, the logic of permitting the supplementals to operate outside the country would require that every airline in the world be allowed to select any route it wanted, including domestic routes within other countries. This seems an extreme proposition, but the political support mobilized behind the thought that regulation should go no further than enforcement of the antitrust laws. How this might have worked out in international aviation can only be imagined.

In both domestic and international operations, reduced activity in the certificated industry is one of the surest roads to an increase in subsidy payments. The subsidy problem can be avoided, of course, by prohibiting it altogether while permitting airlines to abandon unprofitable routes. The supplementals' advocates did not explicitly demand this type of action, perhaps because of an unwillingness to face the implications. They were left to ask simultaneously for increased competition and less subsidy, while saying very little about the substantial indirect subsidies which had put the supplementals in business. Nor were those advocates prepared to examine in any detail the relationship between the CAB and the established carriers; far from being one of mutual friendship, it was but another form of government-industry hostility. Neither the trunklines nor the CAB wanted to be held responsible for causing increases in subsidy payments.

These contradictions indicate that the fundamental problem has been an intellectual one; an economic theory espousing a steady increase in the number of companies and a progressive removal of regulation does not help policy-makers who must face hard questions of subsidy and international negotiation. There will be much more to say about the question of economic theory, particularly in Chapter 11.

It is necessary to point out here that the nonskeds who first came to political attention in a purely domestic context inevitably became an issue of defense policy. They could find no commercial market and the all-cargo carriers could not find a substantial one, and so both turned largely to military contract operation.

79. Professor Schwartz argues that the president exceeded his authority in international cases, implying either that the CAB has or should have authority to grant foreign rights on its own. Schwartz, *op. cit.*, p. 218.

The practical limit to the expansion of these carriers, and the creation of still more of them, became the amount of military traffic they could encourage the government to release to them. It hardly is surprising that by the mid-1950's, lobbying activity had reached a fairly high level of intensity. By that time, however, the lobbying collided with yet another phenomenon. Strategic airlift, virtually unknown to the military before World War II, had become a full-fledged task of military aviation and, like everything else connected with the armed forces, had greatly expanded in size under the impact of the Cold War.

[6]

Military Airlift as the Source of Two Controversies

MILITARY AIRLIFT first became a source of dispute because its strategic aspects bore some resemblance to private enterprise and, later, because everyone in the armed services who had an interest in airlift, especially its tactical aspects, advanced contradictory notions of how to organize and operate it.

The first dispute evolved from the Defense Department decision to maintain strategic airlift as a permanent part of the military establishment. In the Cold War environment this permanent airlift organization, the Military Air Transport Service (MATS), posed an unusual problem. When bombers or fighters engage in peacetime training flights, they perform functions completely identifiable by their uniquely military characteristics. When long-haul military transports do the same thing, a new question presents itself: Should these airplanes carry cargo and passengers as part of their normal training routine? If the transports fly empty the training is unrealistic, at least in part, and usable airlift capacity goes to waste. Yet the global network of strategic airlift routes is similar to the route system of the international carriers which, in turn, evolved from the wartime ATC. The most prominent spokesman for the airline industry, if not for the supplementals, turned out to be the 1955 Hoover Commission which protested vehemently against the organization and operation of MATS and forced major changes in policy.

The second dispute, on the other hand, was more narrowly military in origin and was an updated version of the World War II disagreement between the airborne generals and their superiors. The former, now leaders of the Cold War Army, continued to press for an airlift organization devoted exclusively to delivery of paratroopers. Other military leaders felt that airlift should support all military forces and that it was both impractical and too expensive to provide a part of the Army with a separate airlift force. Impressive support for the Army's leaders came from scholars, in the main political scientists, now deeply interested in defense policy. The military dispute became increasingly complex as improved technology made possible the design of long-haul transports capable of both crossing the ocean and dropping paratroopers.

The organization and operation of MATS thus was opposed by the scheduled airlines, the supplementals, the Army, and supporters of all three; this was a curious alliance of military and economic interests, but it had a substantial political impact. Alliance members probably could not have agreed on anything but general opposition to the existing airlift situation, because some of them argued there was too much military airlift while others were insisting there was too little. Nevertheless the agreement served to thrust airlift into the Congressional arena where it attracted considerable attention up to and including the presidential campaign of 1960. Because the two disputes matured within the initially separate dimensions of strategic and tactical airlift, this chapter is organized in that fashion.

THE POSTWAR EXPANSION OF STRATEGIC AIRLIFT: 1948–58

In a decade of postwar activity, strategic airlift progressed from the position of a remnant of World War II to that of a large and comprehensive organization attracting much opposition. Indeed, airlift became a matter of dispute almost as soon as the war ended; it was the first important function subjected to the emphasis on unification. As the Finletter Commission had recommended, long-haul airlift was partially unified in 1948 by merging the three regional systems of ATC with parts of the Navy's airlift organization. The new MATS was a compromise with the Navy which had protested that the principle of *unity of command* required it to keep its own airlift.[1] At the time MATS had only 234 transoceanic

1. U.S. National Archives, Air Force Special File #2, Correspondence on Formation of MATS. MATS never has become a "unified command" in the sense in which the term is used, i.e., commands made up of forces from all the services and usually responsible for a given geographical area, as with the U.S. commands in the NATO and Pacific areas. SAC, on the other hand, is a "specified command," comprised of units from a single service. MATS

transports flying at the low utilization rate of less than three hours per day.[2]

The ATA protested in 1949 that even this level of activity was depriving the airlines of badly needed business, but General Laurence S. Kuter of MATS retorted that his aircraft carried only nonessential items; if MATS did not take them, they would only be sent by sealift which remained the primary means of strategic supply.[3] Three aspects of the continued expansion of MATS are significant here: (1) The evolution of airlift from a secondary to a primary means of strategic supply for forces overseas but with important differences in degree between services; (2) the extent of MATS' expansion; and (3) its organizational position within the Air Force and the way this affected the political dispute.

All during World War II and until the Korean War, airlift was a secondary means of strategic supply used to deliver high priority and emergency shipments. When supply lines were properly managed, there was never a massive demand for airlift. Under these circumstances, it was inevitable that some relatively nonessential items would be carried on airplanes being flown largely to keep open the route system.

As Samuel P. Huntington points out, Cold War rearmament combined the two concepts of mobilization and deterrence.[4] The MATS role included parts of both concepts. The officially approved war plans of the 1950's visualized a general war in which SAC bombers would be the first military forces employed against

---

has remained a command within the Air Force, but with some membership from all services. Definitions from U.S. Congress, House, *United States Defense Policies in 1960,* 87th Congress, 1st Session, H. Document 207 (Washington: Government Printing Office, 1961), p. 29.

2. The initial idea was that only enough flying be done to keep open the route system. Alfred Goldberg (ed.), *A History of the United States Air Force 1907–1957* (Princeton: D. Van Nostrand Company, Inc., 1957), pp. 147–48.

3. U.S. Congress, Senate, Committee on Commerce, *Air-Line Industry Investigation,* Hearings, 81st Congress, 1st and 2nd Session (Washington: Government Printing Office, 1949 and 1950), V, 2098. The scheduled airlines also protested against the government's use of nonscheduled airlines. *New York Times,* February 7, 1948.

4. The Korean War was the watershed. The government adopted a partial mobilization because total mobilization would have brought severe domestic strain. The intent was to deter the massive attack in Europe thought to be the likely Soviet objective. Samuel P. Huntington, *The Common Defense* (New York and London: Columbia University Press, 1961), pp. 57–58, 344; Bernard Brodie, *Strategy in the Missile Age* (Princeton: Princeton University Press, 1959), p. 317.

the Soviets.[5] At the time SAC included mostly short-range bombers which would move to overseas bases before launching their attacks. The MATS task was to airlift SAC's supporting equipment and personnel after the decision to deploy the bombers.[6] The SAC war plans thus determined the number of transports needed by MATS and also set the utilization rate at which they would fly during the emergency. This was the first instance in which airlift became the primary means of deploying an important military force, and it was directly related to the SAC deterrent.[7]

The theory was that MATS would do enough training in peacetime to enable it to meet the wartime need. Like other military organizations, MATS would then be augmented in wartime by reserve personnel, the utilization rate and the airlift capacity would expand, and this would make possible the deployment of other forces and supplies. This mobilization concept was somewhat vague, because there were no specific plans to deploy any forces other than SAC. Neither mobilization nor deterrence, however, was of much help in deciding what should be done with airlift in the absence of hostilities.

The initial decision was to permit each military service to use for its own purposes a portion of the airlift capacity generated by the MATS training program. The Joint Military Transportation Committee (JMTC), an agency of the Joint Chiefs of Staff (JCS), allocated the monthly capacity until 1958. There always were more requests than airlift, but the decision to seek commercial assistance required additional appropriations.[8] The Army made lit-

5. The official argument was that limited war situations were too difficult to predict with accuracy and that the available airlift could meet most situations. Summary in *U.S. Defense Policies in 1960,* p. 25. For later developments in war planning, see Chapter 9.

6. This mission gradually changed in the 1950's as SAC converted to longer range bombers and missiles. The plan was that intercontinental bombers would carry out their combat missions, then would return to "poststrike" bases for rendezvous with MATS transports. A MATS aircraft would meet each bomber and would have on board a crew of technicians to prepare the bomber for another mission. This task required instantaneous response by MATS but less total airlift capacity, because there was no need to move entire wings.

7. See testimony in U.S. Congress, House, Committee on Appropriations, *Department of the Air Force Appropriations for 1957,* Hearings before Subcommittee, 84th Congress, 2nd Session (Washington: Government Printing Office, 1956), pp. 1414–17.

8. U.S. Commission on Organization of the Executive Branch of the Government, *Subcommittee Report on Transportation* (Washington: Government Printing Office, 1955), pp. 249–52 (hereinafter *Subcommittee on Transportation*).

tle or no attempt before 1959 to use airlift as the primary means of strategic supply for its overseas forces, nor did the Navy.[9] The Air Force, on the other hand, quickly shifted to airlift as the primary means of such supply and reorganized its entire supply system.

The Air Force concept dates from 1952, when General William H. Tunner, graduate of the wartime ATC and commander of the Hump, Berlin, and Korean airlift operations, described the logistics potential of strategic airlift:

> Consider a hypothetical case of an aircraft engine in current use in Korea or Europe; for 5 months it powers an airplane and for the next 7 months it is in the maintenance pipeline, spending a good deal more time in transit and in various receiving and shipping depots than in the maintenance shops. By conservative calculations, air transport of the engine from the plane to the maintenance depot and back to the plane would get it out of the pipeline 3 months faster than if it moved by surface transportation.
>
> Suppose there were 10,000 engines in this year-round operational maintenance cycle. Airlift could eliminate at least 1 month of the cycle; hence one-twelfth of 10,000 or 833 engines would be saved. At a cost of $50,000 each, the value of the engines saved would be $41,650,000.[10]

Moreover, argued Tunner, the Army could get much better utilization of its troops if it were to use airlift to rotate soldiers overseas in either peace or war.[11]

The Air Force has pointed ever since to the vast savings realized by using airlift to ship engines, spare parts, electronic components, and similar items.[12] The initial conversion to this method, however,

9. The Army submitted its first major request to use airlift for overseas rotation of troops in late 1959. *Department of Defense Press Release*, 1191–59, October 14, 1959. As recently as 1964, the Army has been giving cash awards to supervisors who devise methods of sending less mail by air, thus cutting down on expenditures. *Washington Star*, October 24, 1964.

10. Major General William H. Tunner, "Technology or Manpower," *Air University Quarterly Review*, 5 (Fall, 1952), 13.

11. *Ibid.*, p. 14.

12. Douglas N. Jones, "Economics of Logistics Support," *Air University Quarterly Review*, 13 (Fall, 1961), 15–16. Before using the new system, the Air Force found that it used 150 days to send messages, return an engine to the U.S., overhaul it, then return it overseas. Testimony in U.S. Congress, Senate, Committee on Armed Services, *Study of Air Power*, Hearings before Subcommittee, 84th Congress, 2nd Session (Washington: Government Printing Office, 1956), Part 1, p. 415 (hereinafter *Air Power Hearings*).

occurred in 1955 before Congress had become directly interested in commercial airlift procurement. The Air Force had some difficulty in getting the funds for commercial airlift needed to move such items from factories and domestic depots to the coastal bases operated by MATS. Military authorities admitted that they could not prove the new system actually was a savings; they could not accurately assign costs to an airlift organization provided to meet wartime needs.[13] Savings were computed, in short, on the basis of airlift that represented no initial cost. In theory, the airplanes had to fly anyway, and it was only common sense to use the space.

Many observers in Congress and on the Hoover Commission looked upon airlift as a very expensive item in the business world, one that never should be used to carry trivia. These differences in approach generated much of the confusion that has surrounded the entire question; the Air Force treated airlift as a military necessity in both peace and war but tended to exaggerate the savings involved, the Army and Navy considered airlift as only a bonus to the regular supply system, all military services understood that the airlift was "free," and the Hoover Commission looked only at the similarity between strategic airlift and commercial enterprise.

Strategic airlift also took on other Cold War tasks not contemplated in war plans but quite important to the national interest. MATS evacuated refugees from Hungary in 1956, deployed a United Nations Emergency Force to the Suez that same year, and performed numerous mercy missions, all within the framework of its peacetime training program.[14]

13. The Air Force added that the Soviet submarine menace was an additional reason for using airlift. U.S. Congress, House, Committee on Appropriations, *Department of the Air Force Appropriations for 1956*, Hearings before Subcommittee, 84th Congress, 1st Session (Washington: Government Printing Office, 1955), pp. 1439–50 (hereinafter *HCA DAF Appropriations for 1956*).

14. The most important nonmilitary operation was the Berlin Airlift; it had an "extraordinary impact" upon the Germans. W. Phillips Davison, *The Berlin Blockade: A Study in Cold War Politics* (Princeton: Princeton University Press, 1958), pp. 328–31, 355–65. MATS evacuated 14,263 refugees from Hungary, most of them in December of 1956. Commercial airlift handled about one-fourth of the total; the civil carriers had one substantive advantage, in that they could land at Vienna instead of Munich. U. S. Congress, House, Committee on Appropriations, *Department of Defense Appropriations for 1958*, Hearings before Subcommittee, 85th Congress, 1st Session (Washington: Government Printing Office, 1957), Part 2, pp. 1693–94, 1799–50 (hereinafter *HCA DOD Appropriations for 1958*). The U.N. Suez Force included troops from Colombia, India, and Brazil. *Ibid.*, pp. 1775–76. For a list of other unusual operations, see *Congressional Record*, June 1, 1960, p. 10747.

The second aspect of MATS' expansion that is significant here is the precise physical dimensions of that expansion during the 1950's. The growth is not measurable solely in numbers of aircraft; the nature of airlift systems makes growth a product of aircraft utilization rates, speed, size, and the numbers of each type. The JCS decided during the Korean War that MATS should be able to attain an eight-hour utilization rate in emergencies, a figure that virtually averaged out the experience of commercial and military operations during the Berlin Airlift and the Korean War.[15] This wartime objective has remained relatively stable in the intervening years, but peacetime (Cold War) utilization has fluctuated. MATS gradually increased from its pre-Korea level of three hours to four hours by 1955, five hours by 1957, and had planned an increase to six hours by 1959 before being headed off by Congress and the Hoover Commission.[16] By 1957 the MATS fleet included 451 four-engine transports, only 80 of which were the C-54's of World War II vintage.[17] Airlift capacity doubled between 1954 and 1957 alone, most of it used on a scheduled basis between fixed bases in the U.S. and overseas areas.

Because of the nature of this scheduled operation and because many of the transports were only modifications of commercial types, MATS seemed something other than purely military. The Air Force heightened this appearance by giving MATS a peculiar organizational position in its heirarchy.

MATS remains, even in the 1960's, largely outside the central organization and philosophy of the Air Force. To some extent this has been the case with every Air Force command other than SAC,

15. In the Berlin Airlift, peak operation was about eight hundred trips per day, with a utilization of nine hours on the C-54. Charles J. V. Murphy, "The Berlin Airlift," *Fortune*, 38 (November, 1949), 89–93. *The Berlin Airlift*, Air Ministry and the Central Office of Information, His Majesty's Stationery Office (London: 1949), p. 37. Commercial airlift operating between the U.S. and Japan during the Korean War attained about a ten-hour utilization. MATS said it used eleven months in raising military utilization from three to six hours. U.S. Congress, House, Committee on Armed Services, *Hearings Before Special Subcommittee on National Military Airlift*, 86th Congress, 2nd Session (Washington: Government Printing Office, 1960), p. 4174 (hereinafter *Rivers Committee Hearings*).

16. *Subcommittee on Transportation*, p. 289; *HCA DAF Appropriations for 1957*, pp. 1414–17; *HCA DOD Appropriations for 1958*, Part 2, p. 1637.

17. U.S. Congress, House, Committee on Government Operations, *Military Air Transportation*, Hearings before Subcommittee, 85th Congress, 2nd Session (Washington: Government Printing Office, 1958), p. 833 (hereinafter *Military Air Transportation, 1958*). The fleet included 318 C-124's (cargo), 81 C-118's (military version of the DC-6), and several miscellaneous types.

but the MATS case has been more peculiar than others.[18] The postwar Air Force, for example, consistently has been described and measured in terms of the number of wings assigned at any given time; during Cold War rearmament, the number changed from 48 to 70 to 143 and, with the advent of the Eisenhower administration, to 137. The specific number always has included only the combat wings maintained by SAC, the Tactical Air Command (TAC), and the Air Defense Command (ADC). MATS airlift wings always have been listed in an auxiliary category defined as supporting forces and, as a consequence, were virtually unknown before airlift became an important issue. In making budgetary presentations to Congress, Air Force leaders have referred to combat wings and, when individual field commanders have been brought in to testify, only the chiefs of SAC, TAC, and ADC have appeared.[19] Strategic airlift thus underwent its greatest expansion with hardly any attention from people who concentrate on defense matters. One important example underscores the point.

The Eisenhower administration reduced the Air Force structure to 137 wings by transferring six troop carrier wings (counted within the combat forces) to the active reserve, each of them equipped with the twin-engine short-haul C-119, an aircraft designed for paratrooper use. The action appeared to reduce airlift capacity, and prominent scholars concerned with defense policy mistakenly used it as evidence to demonstrate administration neglect of strategic airlift.[20] During the air power hearings of 1956,

18. SAC has been a relatively new type of military command. Officers have been able to spend an entire career in the single command, and with a special promotion system. Huntington, *op. cit.*, p. 311; Morris Janowitz, *The Professional Soldier* (Glencoe, Illinois: The Free Press, 1960), p. 306.

19. See, for example, *HCA DAF Appropriations for 1956*, pp. 238–40. There has been one major exception to this practice; beginning with 1959, the House Appropriations Committee asked that the MATS Commander report on the status of his organization. See Chapter 8.

20. In his study of the Eisenhower "New Look," for example, Glenn Snyder dealt only with the Air Force combat wing structure, and he reported that the reduction in troop carrier wings was a reduction in strategic airlift. Glenn H. Snyder, "The 'New Look' of 1953," in Warner R. Schilling, Paul Y. Hammond, and Snyder, *Strategy, Politics, and Defense Budgets* (New York and London: Columbia University Press, 1962), pp. 459, 483. *International Security: The Military Aspect*, Report of Panel II of the Special Studies Project, Rockefeller Fund (Garden City: Doubleday & Company, Inc., 1958), p. 29; Henry A. Kissinger, *Nuclear Weapons and Foreign Policy* (New York: published for the Council on Foreign Relations by Harper and Brothers, Inc., 1957), pp. 163–65, 408–9; Huntington, *op. cit.*, p. 83. No analyst of Eisenhower defense policy made the distinction between strategic and tactical airlift.

Senators Stuart Symington and Henry Jackson, both fairly close to Air Force affairs, exhibited almost total unawareness of any form of airlift other than the C-119 wings.[21] The Air Force organizational structure and the method of presenting budgets to Congress must be held responsible, at least in part, for misinforming prominent legislators and scholars about the precise status of military airlift.

MOBILIZATION PLANNING AND THE CIVIL RESERVE AIR FLEET: 1948–63

While MATS was growing apace in the 1950's the government simultaneously was trying to formalize the contractual framework within which the commercial industry would respond to any new war emergency. The activity was traceable to the Brewster Board and the Finletter Commission. First the Air Coordinating Committee (ACC) and later the National Security Resources Board (NSRB) assumed the planning responsibilities which led to the organization of a Civil Reserve Air Fleet (CRAF) within the airlines themselves.[22] Since 1952 CRAF theoretically has remained available for government use, but it never has been activated.

In its attempts to design a national mobilization policy the NSRB commissioned a number of important airlift studies, most of them undertaken by volunteers from the industry. The *Douglas Report* of 1950 was the most important one; it was named for James A. Douglas, former airline official and ATC officer, later Secretary of the Air Force and, in 1960, Deputy Secretary of Defense.[23]

The *Douglas Report* argued that industry should designate 350 of its four-engine transports, about half the fleet then in service, to

21. Jackson did not understand the testimony of a Navy witness who spoke of the transport utilization rate. Symington was unaware of the types of planes flown by MATS. Because MATS did not make a detailed presentation to the Symington Air Power Subcommittee, committee members got their airlift testimony from Army witnesses. *Air Power Hearings*, pp. 515, 583, 992.

22. The ACC was an interdepartmental committee formed at the end of World War II to co-ordinate aviation policy matters. It was the first agency to implement the recommendations of the Finletter Commission. U.S. Congress, House, *Report of the Air Coordinating Committee, Calendar Year, 1949*, 81st Congress, 2nd Session House Document 476 (Washington: Government Printing Office, 1950), p. 5; *Calendar Year, 1950*, 82nd Congress, 1st Session, House Document 55 (Washington: Government Printing Office, 1951), p. 5.

23. U.S. Executive Office of the President, National Security Resources Board, *Report on Utilization of Airlines for Wartime Airlift and Proposals to Aid Expansion of the Civil Air Fleet* (Mimeographed: Washington, 1950) (hereinafter *Douglas Report*).

begin military operations within 48 hours after a declaration of emergency. The government would supply the added communications facilities needed for overseas operation, as well as the spare parts required for a ten-hour utilization rate around the world.[24] The study was completed during the lull between the Berlin and Korean crises, and the supplementals had relatively few four-engine planes in service at the time; but the *Douglas Report* suggested that all thirty-one of them be assigned to CRAF.[25] The report explicitly rejected any air merchant marine program that might require the government to purchase or lease aircraft, arguing that only the travel market should determine the number of aircraft in commercial service. [26]

The Defense Air Transportation Administration (DATA) took on the management of the new reserve fleet in 1952; it was a Commerce Department agency initially organized in 1950 to handle the insurance of commercial aircraft on military contract work against the risks of war, loans made under the Defense Production Act of 1950, tax amortization certificates for Korean War purchases, and preparatory planning for CRAF.[27] Ray W. Ireland, an ex-ATC general and one of the airline officials involved in the 1934 cancellations, was called in to head DATA. The agency's rationale was quite simple; upon receiving from the Defense Department the latest approved airlift needs for a hypothetical general war, DATA distributed them as equitably as possible among companies pledged to CRAF. As the Secretary of Commerce put it to the airline presidents when he announced the formation of CRAF, the auxiliary fleet was to move the supporting equipment needed by SAC.[28]

In subsequent years the Defense Department and DATA constantly updated both the official requirements and the CRAF fleet itself, including after 1960 the proposed needs for various sorts of

24. *Ibid.*, pp. 2, 10.

25. It did not seem at the time that there might be a future connection between membership in the CRAF and eligibility for peacetime military traffic. Had the scheduled airlines foreseen the future, they might not have suggested that the supplementals join CRAF. *Ibid.*, p. 19.

26. *Ibid.*, pp. 27–36.

27. U.S. Department of Commerce, Defense Air Transportation Administration, *History of the Defense Transportation Administration to June 30, 1953* (Mimeographed: Washington, 1953), p. 9.

28. U.S. Department of Commerce and Department of Defense, *Report to the Airline Presidents on Civil Aviation Mobilization* (Mimeographed: Washington, 1952), p. 5. Detailed arrangements were in U.S., Department of Defense, *The Department of Defense Plan for the Civil Reserve Air Fleet* (Mimeographed: Washington, 1952).

limited war. The 1963 CRAF fleet, for example, contained some 342 aircraft, most of them the most modern available to the industry.[29] Even so, CRAF gradually has declined in relative importance, reflecting the long-term buildup in military airlift capacity. In 1957 the Defense Department counted on CRAF to lift 40 per cent of its strategic airlift requirements for the first ten days of a general war but, by 1962, the MATS commander estimated that reliance upon CRAF had decreased by 25 per cent.[30]

The CRAF plan installs airline managers in regional committees located within the MATS subcommands based on the east and west coasts of the U.S.; each company is supposed to manage its own aircraft and crews within an over-all government-industry partnership at the operating level. Cost estimates have varied, but it appears that the Defense Department has spent something over $30 million in providing parts and facilities for the fleet itself.[31]

CRAF remains, of course, an ambitious attempt to institutionalize World War II experience. The CRAF theory has seemed attractive, principally because the government has seemed able to guarantee itself a sort of merchant marine without making any substantial expenditures.[32] As mentioned in Chapter 3, however,

29. List in U.S. Congress, House, Committee on Government Operations, *Military Air Transportation, 1963*, Hearings before Subcommittee, 88th Congress, 1st Session (Washington: Government Printing Office, 1963), p. 101 (hereinafter *Military Air Transportation, 1963*).

30. The 1957 figures are cited in Hiram C. Caroom, "Management Responsibilities and National Interests: An Analysis of the Practicability of the U.S. Flag International Passenger Airlines' Transporting More Military Traffic in Peace and War," (unpublished doctoral dissertation, Harvard University, 1959), p. 44. The 1962 position is from U.S. Congress, House, Committee on Appropriations, *Department of Defense Appropriations for 1963*, Hearings before Subcommittee, 87th Congress, 2nd Session (Washington: Government Printing Office, 1962), Part 3, p. 539 (hereinafter *HCA DOD Appropriations for 1963*).

31. The Air Force reported $34 million had been spent by 1960. *Rivers Committee Hearings*, p. 4286. In 1963, however, the Air Force told another committee that expenditures had totalled $31.6 million as of 1962. *Military Air Transportation, 1963*, p. 23.

32. One estimate was that the Government had saved capital costs of $550 million and yearly upkeep of $350 million to $400 million. William H. Hessler, "America's Merchant Marine of the Air," *U.S. Naval Institute Proceedings*, 85 (May, 1959), 48–63, especially 53; "Civil Airlift and the National Defense: How the Scheduled Airlines Contribute to the Nation's Security," Public Relations Service, Air Transport Association of America, 1962 (booklet). By way of contrast, Congressman Rivers produced statistics to show that the merchant marine had not been sufficient in either World Wars I or II, but had been forced to expand by a factor of three or four. *Congressional Record*, June 24, 1959, p. 10753.

World War II experience may well have been too unique to permit repetition.

Beyond that, there are economic, political, and strategic reasons for doubting the workability of the CRAF concept. Industry, for example, hesitated for many years before signing contracts committing it to perform unknown tasks under unforeseeable conditions. The military has indicated from the beginning a preference for the assignment to CRAF of the most modern transports available, a factor that immediately made it impossible to spread the hypothetical workload on a completely equitable basis.[33] An initial hope that CRAF might become a project of the entire Free World never has been realized, although one European airline once asked to join.[34] The prevailing fear, and an understandable one in view of the direct competition between U.S. airlines, has been that the first company to completely commit itself to CRAF would be the first to lose its competitive position should an emergency arise. No single company has felt itself able to go first, and this has made all of them reluctant. Indeed, when contracts ultimately were signed they included only a revised CRAF that is not a reserve fleet at all; companies have agreed to make available airplanes already committed to military contract work on a day-to-day basis.[35]

In the complex nuclear age there are important strategic difficulties to consider before putting too much reliance upon CRAF. The government cannot quietly withdraw very many transports from commercial service without raising the level of tension in any given crisis situation. If, on the other hand, any crisis has matured to the point where the entire CRAF fleet must be pressed into service, it can be argued that domestic economic needs also will have increased by some substantial amount. Unfortunately, military leaders have tended to overlook these fundamental difficulties and have instead attacked some other aspects of CRAF.

33. U.S. Department of Commerce, Defense Air Transportation Administration, *Report of the Clemson Committee* (Mimeographed: Washington, 1952), p. 4.

34. The initial objective was to enroll in CRAF the 126 U.S. transports belonging to foreign airlines. *Douglas Report*, p. 19. There was an informal agreement at one time that U.S. and European airlines would attempt to maintain after any emergency the competitive balance in existence before that emergency. Interview with Theodore Hardeen, Administrator of DATA, 1962. One prominent European airline asked permission in 1959 to join CRAF, but nothing came of the request. Should the CRAF ever be activated in a NATO context, however, someone will have to face this question.

35. See Chapter 10.

Military officers have criticized the industry for its alleged indifference to military needs in times of incipient crisis. During the 1958 Lebanon crisis, for instance, the MATS commander announced in a speech that no airline had responded to ten specific invitations to bid for military contracts, out of a total of 205 invitations issued at the time.[36] The Air Force's Chief of Staff later agreed with a House Subcommittee that the Lebanon crisis demonstrated the deficiencies of CRAF.[37] Yet there is no evidence that the government was ready to declare an emergency and take over the aircraft allocated to CRAF.

To avoid imposing direct controls, the government often has relied upon informal contacts between high level officials and various segments of the airline industry. This occurred at the outset of the Korean War and during the Berlin crisis of 1961, but the officials generally have confined themselves to asking for the maximum contribution that would not severely damage any company.[38] This did not happen during the Lebanon crisis and the ATA, speaking only for the scheduled airlines, implied that more airlift might have been found if appropriate officials had contacted the industry in some way other than through the normal procurement procedures.[39]

Military leaders thus have criticized the industry for failing to respond while, at the same time, the administration was deciding not to take the drastic step of commandeering transports. The same officers also asked for legislation to make certain that commercial flight crews would obey the instructions given them during emergency.[40] One scholar concludes that the flight crew problem is the principal issue remaining unsettled in the CRAF program, and at least one prominent military analyst agrees.[41] But the Ken-

36. Speech by General William H. Tunner, quoted by Brigadier Gen. T. R. Phillips in *St. Louis Post-Dispatch*, November 9, 1959.

37. *Rivers Committee Hearings*, p. 4897.

38. At the outset of the Korean War, the government called an emergency meeting of all airlines in Washington. Robert E. Gross, "The Transport Plane in Peace and War," *Air Affairs*, 3 (December, 1950), 471. In the 1961 Berlin crisis, the Secretary of the Air Force quietly contacted the ATA, a practice reportedly not uncommon. Interviews.

39. "Outline of U.S. Airline Responsiveness in Emergencies," prepared by Air Transport Association of America (Mimeographed: Mar. 31, 1960).

40. Testimony of General Thomas D. White, Air Force, and General Lyman L. Lemnitzer, Army, in *Rivers Committee Hearings*, pp. 4086, 4157–58, 4887.

41. Caroom, *op. cit.*, pp. 51–54; Brigadier General Phillips in *St. Louis Post-Dispatch*, March 27, 1960.

nedy administration rejected MATS proposals to put the CRAF fleet under direct military command in time of war and sent to Congress draft legislation that would give the president authority to take control of all or any part of a transportation system in time of war or other emergency.[42] The bill was explained as necessary to "fill a gap in statutory authority" by making it possible to take over transportation systems in situations short of war.[43] To implications that its members may not be too responsive, the pilots' union protested that pilots always have been ready to meet the needs of war.[44]

Government and airlines reached some sort of accommodation in 1963 when they agreed to a revised contractual arrangement for CRAF participation in emergency operations. The companies agreed to negotiate agreements with participating unions, the latter agreeing not to stop work during crisis. The complex contractual provisions permitted the government to activate CRAF in stages, depending upon the nature of the emergency and the precise airlift requirements. On this basis the administration decided that it needed no other legislation and withdrew its request.[45] The result was to make it appear that the airlines would be more readily available in crisis conditions, even though the agreement rests upon the premise that the government can quickly withdraw civil aircraft from commercial operations. The real issue remains whether any administration can afford to commandeer a major transportation system or impose a form of economic control not found necessary in World War II.

The 1963 agreements had at least the salutary effect of dissipating the acrimonious argument about the trustworthiness of civilian crews. The real problem in situations such as the 1958 Lebanon crisis has been the tourist season; commercial transports were unavailable at that time because of summertime demands for travel.

42. H. R. 12111, S. 2799, 87th Congress, introduced by Congressman Rivers and Senator Magnuson.

43. Correspondence reproduced in *HCA DOD Appropriations for 1963*, Part 3, pp. 565–73.

44. Letter reproduced in *Rivers Committee Hearings*, p. 4939.

45. Testimony in U.S. Congress, House, Committee on Armed Services, *Hearings Before the Special Subcommittee on National Military Airlift*, 88th Congress, 1st Session (Washington: Government Printing Office, 1963), pp. 5990–91, 5999–6000; *Ibid., Report of Special Subcommittee on National Military Airlift*, 88th Congress, 1st Session, Committee Print (Washington: Government Printing Office, 1963), pp. 4–5; the subject is covered in great detail in *Military Air Transportation, 1963*.

There has been no substantial evidence that civilian flight crews are undependable in time of stress.[46]

Since its inception, then, the unanswered questions of the CRAF program have been political and economic. Especially in the 1950's, however, the program was attractive to those who sought to combine a low government budget, a high level of preparedness, and support for private enterprise.

THE TWO-PRONGED ATTACK OF THE 1955 HOOVER COMMISSION

The 1955 Hoover Commission reports made two broad proposals with respect to MATS, and both had considerable effect. One was that MATS, as a military airline, should stop competing with its civil counterparts; the other was that MATS should be managed as a business enterprise. The sweeping nature of the proposals was typical of the 1955 Commission which took a broader view of its mandate than had its 1949 predecessor. It sought wherever possible to turn over government functions to private business.[47] Given this approach, the theme of ending government competition against private enterprise took on moral connotations.

The Commission demanded that the government apply the same

46. Commercial crews long have operated in combat zones and on unusual assignments. The best example has been in the Far East, beginning with CAT, Incorporated; this was an American subsidiary of a Chinese Nationalist airline based on Taiwan. Formed by General Clair Chennault, CAT operated 15,000 trips to Korea during the Korean War, went into every airstrip that could take its twin-engine C-46's, and lost several crew members in the process. Captain Annis G. Thompson, *The Greatest Airlift* (Tokyo: Dai-Nippon Printing Company, 1954), pp. 449–54. Either CAT or some similar companies have operated under U.S. aegis in the Far East area since that time, especially in Southeast Asia. There have been numerous reports that some of these companies function under supervision of U.S. intelligence agencies, individual budgets running to $10 million per year. Yet other small airlines operate U.S. contract schedules between Japan, Korea, Okinawa, Taiwan, and the Philippines, in support of U.S. military operations. All of these operations are in an area that would be within the combat zone of an expanded conflict in the area. *Washington Post, Chicago Tribune,* January 22, 1961; *Aviation Week,* May 21, 1962; *New York Times,* August 27, 1964.

47. The commission's tendency to make broad recommendations brought protests from a member who later became a leader in the Congressional phase of airlift policy-making. Congressman Holifield of California complained that the commission should not attempt to advise Congress on such "disparate matters as Alaska railway tariffs, parcel post rates, naval shipyards, military sea and air transportation, production of helium and nickel, and the TVA fertilizer program." U.S. Commission on Organization of the Executive Branch of the Government, *Business Enterprises* (Washington: Government Printing Office, 1955), p. 109 (hereinafter *Business Enterprises*).

test as any businessman to the use of airlift. It was an expensive and unusual method of shipment, said the Commission, and must be used with prudence. The Commission criticized the airlift of ping-pong balls to Berlin, 25,000 pounds of cement to Bermuda, and dog food to Okinawa.[48] It had been similarly wasteful, said the Commission, to send an empty airplane from Bermuda to Massachusetts merely to airlift an Air Force band to Bermuda; the job should have been turned over to the scheduled airlines. If the government would make such decisions more often, continued the Commission, it might be possible to reduce subsidies.[49]

The Commission was equally disturbed at a post-Korea decrease in government spending for commercial airlift. To remedy the decline, the Commission asked that MATS be prohibited from airlifting anything that the industry could handle and that MATS withdraw from all commercial routes.[50] Furthermore, said the Commission, the government should implement a recommendation made by its own Air Coordinating Committee: "The Government should, to the greatest extent practicable, adjust its use of air transportation so as to use existing unutilized capacity of United States air carriers."[51]

The ACC statement became a favorite quotation for those petitioning for more government business, as did this paragraph from the report of the Hoover Commission's Subcommittee on Transportation:

> The Committee recommends that the level of MATS's peacetime operations be limited to that necessary to maintain the minimum wartime readiness of the Command. The peacetime operations of the integrated service should be restricted, and realistically limited, to air transportation of persons and cargo carefully evaluated as to the necessity for such transportation, and only after all forms of commercial carriers have handled traffic appropriate and properly assignable to their service. Failure to accomplish this means a continuing and ex-

48. *Subcommittee on Transportation*, p. 19.

49. *Ibid.*, p. 286.

50. U.S. Commission on Organization of the Executive Branch of the Government, *Transportation* (Washington: Government Printing Office, 1955), pp. 58–59 (hereinafter *Transportation*); *Ibid.*, *Business Organization of the Department of Defense* (Washington: Government Printing Office, 1955), p. 136 (hereinafter *Business Organization*).

51. U.S., Air Coordinating Committee, *Civil Air Policy* (Washington: Government Printing Office, 1954), p. 17 (hereinafter *Civil Air Policy*). Quoted in *Transportation*, p. 57.

> panding airlift socialism over all air transportation and extending down into other forms of commercial transportation.[52]

If the government would adopt its recommendations, argued the Commission, the industry would develop a transport fleet large enough to meet the needs of any future emergency.[53]

Some of the reasons why all of the Commission's recommendations did not immediately take hold should be reviewed at this point, because they are indicative of the complexity of the total problem. (1) In pointing to the post-Korea reduction in spending for commercial airlift, the Commission established as a norm a wartime period (Korea) in which the government had no substantial airlift of its own. (2) The first sentence of the Hoover Commission's long paragraph quoted above was somewhat at odds with the remainder of the paragraph. The first sentence conceded the necessity to prepare for war, but the Commission then added that MATS should airlift only what was left after the industry had taken all it could handle. (3) A careful reading of the ACC's 1954 *Air Policy Report* makes it clear that the ACC statement was quoted at least partly out of context. The particular section of the report referred to the domestic travel of individual government employees. Indeed, the same section praised the Defense Department and the Post Office for their already extensive use of commercial airlift.

The Hoover Commission's use of the phrase "unutilized capacity" was somewhat ambiguous. Outside groups could interpret it as a call for increased military purchases of charter airlift, and it became a limitless proposition when applied to the supplementals. They stood ready to buy or lease as many transports as possible and, no matter how much business they secured, the next aircraft in line was "unutilized." Even had the term been more explicitly limited to the scheduled companies, moreover, it also would have had limitless connotations. For reasons mentioned earlier, commercial carriers maintain a gap between the number of seats available and the number sold. Had the military filled one airplane, another would have been added to the schedule and this one, in turn, merely would add still more "unutilized" capacity.

The Commission made two rather specific proposals in connection with its observation that MATS should be operated and organ-

52. *Subcommittee on Transportation*, p. 295.
53. *Transportation*, p. 58.

ized along business lines; one concerned methodology, the other personnel. The first was that MATS should operate with a revolving fund; the Defense Department should give it an initial allocation of cash, and its customers thereafter should reimburse it for airlift services performed.[54] At the urging of the House Appropriations Committee and over objections from the Air Force and MATS, the Defense Department decided in 1955 to put the recommendation into effect in 1957.[55] In theory at least, MATS became the "single manager" for airlift services needed by all Defense Department agencies.[56] This was the methodology used to operate miscellaneous military "industrial fund" activities in which agencies "sell" their products to military customers, e.g., laundries and printing plants.[57]

The industrial fund device never has proved completely compatible with airlift or sealift, where it also is in effect. The size of the airlift force, the number of crews, the level of operation, and the necessity for modernization, all remain a function of JCS war plans and the normal budgetary process. The basic mission of MATS thus makes it impossible to include within the industrial fund such things as flight crew salaries and aircraft costs. This has made the fund something less than satisfactory to critics of MATS who looked to the fund to demonstrate once and for all that commercial enterprise is more efficient and economical than its military counterparts.[58] The fund has proved surprisingly workable, however, if for reasons apparently unforeseen in 1955.

The Air Force and MATS list these benefits as having resulted from industrial fund experience:

> MATS is now able to meet all airlife requirements. . . . The only limitations to satisfying peacetime airlift requirements . . . are the amounts of money the services have available to spend for airlift and the civilian airlift capacity.
>
> . . . The Airlift Service as well as the users are forced into

54. *Business Organization*, pp. 78–79.

55. W. J. McNeil, Comptroller of the Defense Department during most of the Eisenhower years, was credited with having pressed for adoption of the industrial fund. Dialogue between General Tunner and Congressman Rivers in *Rivers Committee Hearings*, pp. 1621–26, 1701–3.

56. *Department of Defense News Release* 1286–56, Dec. 13, 1956.

57. Herbert L. Fowle, Jr., "A Critical Analysis of the Airlift Service Industrial Fund" (unpublished master's thesis, University of Texas, 1962), pp. 16–17.

58. See remarks of Congressmen in *HCA DOD Appropriations for 1958*, Part 2, pp. 1764–66; also testimony of Stuart Tipton in *ibid.*, pp. 2276–78.

> better forecasting techniques. Dollar controls have imposed a requirement to estimate much closer than was done when airlift was paid for solely by the Air Force.
>
> . . . only air-eligible cargo is now moved by the Airlift Service.
>
> . . . this method of financing has allowed long-range planning. MATS can now enter into long-term contracts with the airlines. . . .
>
> Several years ago, MATS was criticized for carrying for the services and for itself noneligible air cargo. We had the case of the dogfood and the ping-pong balls. . . . With the industrial fund, the services are much more careful not to send those things that should not be going by air in the first place.[59]

To put it simply, the industrial fund has reversed the budgetary process. MATS is able to buy commercial airlift from its revolving fund without a specific appropriation; this presents the Army, Navy, and Air Force with a *fait accompli* in the form of the rates they pay during the subsequent year. One example cited by the MATS commander shows how the system works: ". . . when we moved the National Guard to Europe it took all of our capability, so we had to go out and buy commercial augmentation to replace MATS capability. Without this industrial fund, none of the services would want to come up with that money on the line right then for that service. We can buy it, contract for it, and then collect."[60]

The military services have, in addition, not been hesitant to use MATS transports on an individually chartered basis, even though the charter price is quite expensive.[61] This has tended gradually to convert MATS into more and more of an "air-taxi" organization. By making it possible to avoid the budgetary cycle, the industrial fund may have encouraged greatly increased expenditures, despite the apparently businesslike methodology.

In the personnel area, the Hoover Commission asked that the administration remove from direct military control those Defense Department functions most receptive to business-type manage-

59. U.S., Department of the Air Force, unclassified briefing sheet prepared for House Armed Services Committee (Mimeographed: Washington, 1962).

60. *HCA DOD Appropriations for 1963*, Part 3, p. 529.

61. The charter rates are between $500 and $1,000 per flying hour per airplane. MATS carries passengers and cargo on regularly scheduled flights at rates computed on the basis of weight and distance. Explanation in U.S., Department of Defense, Military Air Transport Service, *Airlift Service Management Report, July 1958–June 1959*, p. 5.

ment. The Commission suggested, for example, that a single supply agency buy everything needed for all the military services; at its head should be a presidential appointee coequal with the military chiefs of service.[62] Similarly, said the Commission, there should be a Director of Transportation subject only to instructions from the Assistant Secretary of Defense for Supply and Logistics; this new official could centralize all military traffic management functions.[63] The Commission added that the Defense Department should upgrade in both status and salary its civil service traffic managers, and should make them independent of the "military mind."[64]

This approach affected MATS almost immediately. The Assistant Secretary for Supply and Logistics for each of the services became a member of the administrative committee which oversees MATS operation under the general direction of the corresponding Assistant Secretary of Defense.[65] As the civilian traffic managers praised by the Hoover Commission have taken over an increasing share of the management of MATS, the resulting two-headed system of control has perpetuated the appearance that MATS is essentially nonmilitary in character.

The ATA, representing the scheduled airlines, worked closely with the Hoover Commission while it was studying MATS and the question of commercial airlift.[66] The general tone of Commission reports, and of the 1954 ACC report, was favorable to the scheduled industry and largely overlooked the supplementals. The Commission's language, however, lent itself to quotation by any airline looking for business. When the discussions moved into the congressional arena, the supplementals had a good deal more built-in support than did the old-line companies.

The Hoover Commission thus brought into focus one major dimension of the airlift controversy that occupied so much congressional attention in the late 1950's. The military services generated the other dimension of the controversy by engaging in some rather heated intramural disputes on the subject.

62. *Business Organization*, pp. 47–50.

63. *Transportation*, p. 63.

64. *Ibid.*, pp. 64–67. For the similarity between these Hoover recommendations and Defense Department trends in the 1960's, see Chapter 10.

65. *DOD Directives* 5160.12, January 31, 1956; 5160.2, December 7, 1956.

66. Description of ATA role is in U.S. Congress, House, Committee on the Judiciary, *Report of the Antitrust Subcommittee (Subcommittee No. 5), Pursuant to H. Res. 107 on Airlines*, 85th Congress, 1st Session, Committee Print (Washington: Government Printing Office, 1957), p. 133.

THE POSTWAR EVOLUTION OF TACTICAL AIRLIFT: 1948–58

The in-house military airlift dispute grew out of the old airborne problem, largely because the Army's airborne divisions were the most important part of the strategic reserve kept within the U.S. The Army's object remained consistent throughout the decade, if not completely understood. It made several requests for more airlift capacity, all of them qualified by the corollary assertion that any increase was acceptable only if the airlift remained constantly available to the airborne divisions. Army leaders declined to include MATS within any of their airlift calculations, a stance that was traceable to the Eisenhower administration's actions with respect to two particular types of transport aircraft.

The twin-engine C-119 was a new aircraft during the Korean War but the administration, as mentioned earlier, transferred it to the reserve forces.[67] These reserve wings (fifteen of them by 1959) were prepared to airlift paratroopers in the event of emergency and were among the most efficient organizations in either the active or reserve Air Force.[68] General Maxwell Taylor, then the Army's Chief of Staff, admitted in 1958 that if the C-119 wings were included, the Air Force had come "pretty close" to meeting the Army's needs.[69]

The four-engine C-124 also appeared during the Korean War, apparently as the outcome of one of General Arnold's conclusions at the end of World War II: ". . . the key to success lies in the seizing of airfields or landing strips in the initial phases of operations. . . . I feel quite strongly that future airborne operations must be focused about airbases if the air transported force is to remain active and effective."[70] The aircraft itself could carry a massive load over a short range; its loading ramps permitted easy access for large vehicles, and it could airlift some 225 troops. It was not very useful, on the other hand, for the airdropping of

67. See above, p. 121.

68. By 1960 the Air Force rated the Reserve C-119 units more nearly ready for combat than wings having the new C-130's, *Rivers Committee Hearings*, pp. 4241–42, 4251; number of wings from U.S. Congress, House, Committee on Appropriations, *Department of Defense Appropriations for 1959*, Hearings before Subcommittee, 85th Congress, 2nd Session (Washington: Government Printing Office, 1958), p. 893.

69. U.S. Congress, Senate, Committee on Appropriations, *Department of Defense Appropriations for 1959*, Hearings before Subcommittee, 85th Congress, 2nd Session (Washington: Government Printing Office, 1958), pp. 81–82.

70. Quoted in Lewis H. Brereton, *The Brereton Diaries*, (New York: William Morrow and Company, 1946), p. 365.

either paratroopers or heavy equipment.[71] Within the U.S. the C-124's became part of TAC, still charged with the old troop-carrier task. Two of TAC's wings flew the C-124 and belonged to the 18th Air Force, a joint planning command formed after the Korean War to work with the airborne divisions.[72]

The C-124's had other miscellaneous tasks to perform in addition to training with the Army. They delivered supplies to Navy Antarctic exploration teams and to Arctic radar sites.[73] Strategic airlift remained the MATS task, but since the JCS had not yet approved strategic airlift of the Army, MATS initially expressed little interest in the C-124.[74] By 1956, however, a reluctant MATS had more of them than did the two troop carrier wings (163 to 136), and the C-124 had become an important, if expensive, strategic transport; it was especially good for airlifting Air Force engines.[75] For its part, the Army looked upon the two TAC wings as its own, a feeling intensified by a 1956 rotational exercise.

In "Operation Gyroscope," eighty of the TAC C-124's moved seven thousand troops to Japan and brought back the same number, in a maneuver designed only to keep units intact for reasons of morale. Because of the basic differences between strategic and tactical airlift systems, the maneuver stretched over a considerable period of time.[76] This was the only instance until 1960 in which the Army used airlift for the strategic movement of troops.

When MATS became the single manager for airlift services in 1957, a corollary action transferred the two TAC C-124 wings to the new industrial fund organization. While this was in line with the premise that as many long-haul transports as possible should be parts of the command responsible for strategic airlift, the Army generals felt they had lost the only transports constantly available to them. With the C-119's in the reserve forces and the C-124's

71. Army testimony in *Rivers Committee Hearings*, p. 4224.

72. Edward H. Nigro, "Early Troop Carrier Operations in Korea," *Air University Quarterly Review*, 7 (Spring, 1954), 86–89.

73. *Air Power Hearings*, p. 474.

74. Because it was designed for short-haul airlift, the C-124 appeared much less efficient than other types for long-haul operation; this was the MATS conclusion, based upon economic calculations involving payload, distance, fuel, etc. *HCA DAF Appropriations for 1957*, p. 1547.

75. Expansion of the C-124 fleet from *Military Air Transportation, 1958*, pp. 833–34.

76. Forty-four extra aircraft carried spare parts and mechanics. Because the TAC wings had only one aircrew per aircraft, the fleet had to stop at several places along the way for purposes of resting the crew. *Air Power Hearings*, p. 522.

subject to business-type management and, it might be added, equally available to all customers, Army criticisms of Eisenhower administration defense policies became more strident in tone.

THE ARMY VERSUS THE JCS: THE DENIAL OF THE PRINCIPLE OF ALLOCATION, 1951–60

According to General James Gavin, the Army submitted in 1951

Table VI • Army Requests for Airlift

Date and Source	Nature of Request
Army, 1951:[a]	2⅔ divisions tactical, 1 division strategic
Gavin-Wheeler, 1956:[b]	2 divisions strategic (5,000 tons per division to established area, 11,000 tons to unestablished area)
Kissinger, 1957:[c]	5,000 tons in division combat element, 14,300 tons in total division (airlift needed for at least the former—272 C-133 aircraft)
Congressman Flood, 1957:[d]	Move 3 divisions within 15 days, not in the same direction
Army, 1958:[e]	Move spearhead force of 2-division force (5,840 troops, 7,438 tons of supporting equipment)
Congressman Flood, 1958:[f]	1 division within 7 days, one month's supplies for that division within the next 7 days (division needed 1,000 tons per day)
Lemnitzer, 1960:[g]	Combat element of division within 14 days, 2 divisions within 4 weeks (size apparently same as Army request of 1958)
Army-Air Force agreement, 1960:[h]	Size of force is 4,082 tons to established area, 8,504 tons to unestablished area (1 division combat element)
Congressmen Flood, 1960:[i]	2 divisions within 14 days, 1 in each direction

Sources:

[a] James M. Gavin, *War and Peace in the Space Age* (New York: Harper & Brothers, 1958), p. 172.

[b] Testimony of Generals Gavin and Wheeler in *Air Power Hearings*, pp. 833–46. At the time, Taylor and Army Secretary Brucker felt that airlift was adequate, and Brucker said he "took exception to any statement to the contrary." *Ibid.*, Part 2, pp. 1281, 1327. Gavin said later that Taylor had succumbed to pressure from Secretary of Defense Wilson and JCS Chairman Radford. Gavin, *op. cit.*, pp. 283–84. Taylor said only that he had a slight disagreement with Gavin. Maxwell D. Taylor, *The Uncertain Trumpet* (New York: Harper & Brothers, 1959, 1961), p. 98.

[c] Henry A. Kissinger, *Nuclear Weapons and Foreign Policy* (New York: Published for the Council on Foreign Relations by Harper & Brothers, 1957),

p. 163. The book was written after participation in a seminar which included Gavin.

[d] Congressman Daniel Flood of Pennsylvania, in his own words, "stumbled into" the airlift problem as outlined by Gavin to the House Appropriations Committee. Flood was sympathetic to Gavin because the latter came from the same part of Pennsylvania, and Flood said later that he received much of his information from sources within the Army. U.S. Congress, House, Committee on Appropriations, *Department of Defense Appropriations for 1957*, Hearings before Subcommittee, 84th Congress, 2nd Session (Washington: Government Printing Office, 1956), p. 2094; *Ibid., Department of the Air Force Appropriations for 1957*, Hearings before Subcommittee, 84th Congress, 2nd Session (Washington: Government Printing Office, 1956), pp. 1542–44; *Ibid., Department of Defense Appropriations for 1959*, Hearings before Subcommittee, 85th Congress, 2nd Session (Washington: Government Printing Office, 1958), Part 1, p. 765.

[e] Testimony of Army Secretary Wilber Brucker, in *ibid., Department of Defense Appropriations for 1961*, Hearings before Subcommittee, 86th Congress, 2nd Session (Washington: Government Printing Office, 1960), Part 2, pp. 424–25.

[f] *Ibid., Department of Defense Appropriations for 1959*, Hearings before Subcommittee, on Air Force, 85th Congress, 2nd Session (Washington: Government Printing Office, 1959), p. 375.

[g] *Rivers Committee Hearings, op. cit.*, pp. 4075–76, note 16.

[h] Independent agreement between Air Force Chief of Staff Thomas D. White and Army Chief of Staff Lyman L. Lemnitzer, *St. Louis Post-Dispatch*, March 7, 1960.

[i] U.S. Congress, House, Committee on Appropriations, *Department of Defense Appropriations for 1961*, Hearings before Subcommittee, 86th Congress, 2nd Session (Washington: Government Printing Office, 1960), Part 2, p. 390.

its first formal request for airlift to the Air Force.[77] The Army sought two packages of airlift; one would be able to airlift two and two-thirds divisions in a tactical airborne assault, while the other would be large enough to move one division from the U.S. to any area of the world.[78] The Army submitted periodic requests from that time on, but the separate proposals tended to merge into one that combined strategic and tactical aspects. The general nature of the requests is shown in Table VI.

There are several reasons for the wide variations in Army estimates of its needs, all of them generally attributable to the fact that these were the first attempts to plan for strategic deployment by air. Airborne operations within combat zones never have required the accurate weight computations needed for strategic overwater routes. Wind conditions vary a great deal from month to month, and the difference in the number of flights required can be

77. James M. Gavin, *War and Peace in the Space Age* (New York: Harper & Brothers, 1958), p. 172.

78. This was designed to make the Army ready to mount an airborne assault in two theaters at the same time. *Ibid.*

as much as 30 per cent. The Army usually has agreed, moreover, that any strategic deployment will call for some combination of airlift and sealift, in proportion to the time available and the usable airfields and seaports.[79] Army generals have advocated also that some equipment should be redesigned to make it weigh less and that other items be stockpiled at selected overseas locations.[80]

The Army decided in the 1950's that it could not depend upon the JCS to allocate MATS airlift for the deployment of Army divisions, a decision that interacted with the transfer of the TAC C-124's to MATS. Before the transfer, for example, one Army general advised the House Appropriations Committee that neither strategic nor tactical airlift had increased in capacity during the previous year.[81] He apparently referred only to the TAC wings, since MATS had added 77 aircraft during the year, 56 of them C-124's.[82] Also before the transfer, General Taylor announced that the available airlift was adequate, while agreeing with the House Appropriations Committee that "Operation Gyroscope" had used practically all the Air Force's airlift capacity.[83]

After the TAC wings joined MATS, Taylor said he hoped the Army would not "lose any of the tonnage capacity we have counted on in the past."[84] The commander of the Army forces in the Pacific provided an even better example in 1959; he defined his problem as uncertainty about airlift. The 25th Infantry Division in Hawaii was ready to move on four hours' notice, he said, but there was no airlift; with no troop carrier units in Hawaii, some would have to fly in from the mainland.[85] Yet an entire wing of MATS C-124's had been based in Hawaii all along, and numerous

79. Testimony of General Wheeler in *Air Power Hearings,* pp. 836–37; General Taylor in U.S. Congress, Senate, Committee on Appropriations, *Department of Defense Appropriations for 1958,* Hearings before Subcommittee, 85th Congress, 1st Session (Washington: Government Printing Office, 1957), p. 135 (hereinafter *SCA DOD Appropriations for 1958*); also Maxwell D. Taylor, *The Uncertain Trumpet* (New York: Harper & Brothers, 1959, 1960), p. 151.

80. Taylor, *op. cit.,* p. 98; Gavin, *War and Peace, op. cit.,* pp. 283–84.

81. Testimony of General Paul D. Adams in U.S. Congress, House, Committee on Appropriations, *Department of Defense Appropriations for 1956,* Hearings before Subcommittee, 84th Congress, 1st Session (Washington: Government Printing Office, 1955), p. 648a.

82. Tables in *Military Air Transportation, 1958,* pp. 833–36.

83. U.S. Congress, House, Committee on Appropriations, *Department of Defense Appropriations for 1957,* Hearings before Subcommittee, 84th Congress, 2nd Session (Washington: Government Printing Office, 1956), p. 474.

84. *SCA DOD Appropriations for 1958,* p. 135.

85. *Army-Navy-Air Force Journal,* December 5, 1959.

other MATS transports traveled through Hawaii each day. Henry Kissinger's strategic airlift calculations followed the same pattern; he included only the airlift belonging to TAC.[86]

Army generals based their case on the likelihood of U.S. involvement in peripheral, or limited, wars, as opposed to general war.[87] They added that SAC bombardment units probably would move overseas in the event of any crisis, so as to deter expansion of the peripheral conflict. The airlift could not serve two masters, said General Earle Wheeler (later Army Chief of Staff and JCS Chairman); the Army must have its airlift available to it "100 per cent of the time."[88] Gavin added that he looked with "considerable doubt upon the wisdom" of planning to use any part of MATS under any circumstances.[89]

Kissinger adopted the same argument, that the equally necessary general and limited war airlift packages should not augment each other.[90] This was something of a direct challenge to the JCS mechanism, for the argument implied that the JCS were powerless to allocate airlift in accordance with the precise needs of national strategy. Admiral Arthur Radford, then JCS Chairman, responded that the Army's conditions were physically impossible to meet. There was no system of airfields, argued Radford, that could accommodate the simultaneous movement of SAC, the Army, and supplies and reinforcements for units already overseas.[91]

86. Kissinger used as evidence the testimony given the Symington Air Power Subcommittee by General O. P. Weyland, commander of TAC. Weyland said his airlift could move one regimental combat team to Laos. Later in the same testimony, however, Weyland explained carefully that he had not included MATS and that the job really belonged to that command. Kissinger, *Nuclear Weapons and Foreign Policy*, p. 163; *Air Power Hearings*, pp. 520–22. Kissinger said later that the Lebanon crisis had "tied up most" of the U.S.'s strategic airlift. Henry A. Kissinger, *The Necessity for Choice* (Garden City: Doubleday & Company, Inc., Anchor Edition, 1962), p. 99. Yet while MATS provided 36 aircraft for Lebanon, it operated 120 other trips from the U.S. to Lebanon. The 1958 Taiwan crisis erupted at about the same time, and MATS had to respond to those requirements. *Rivers Committee Hearings*, pp. 4170–72. Taylor concluded that the problem in 1958 was the shortage of airfields and seaports in the Middle East. Taylor, *op. cit.*, p. 289.

87. See, e.g., General Gavin's testimony in *Air Power Hearings*, p. 740.

88. *Ibid.*, p. 841.

89. *Ibid.*, p. 845.

90. Kissinger argued for a separate airlift package (272 C-133 transports) that could move the combat element of a single division. Kissinger, *Nuclear Weapons and Foreign Policy*, pp. 163–65.

91. Radford added that the Army's requests convinced him that its spokesman did not comprehend the logistics problems. *HCA DOD Appropriations for 1958*, Part 2, pp. 2049–97, especially p. 2085.

THE ARMY'S SUBSTITUTE FOR ALLOCATION: AIRLIFT PREDESIGNATION AND ITS ALTERNATIVES, 1955–60

Army leaders took the argument yet another step and suggested that airlift intended for its use be categorized specifically as such or assigned outright to the Army. The most important proposal was that of airlift predesignation, a concept that seems simple enough at first glance.

Army spokesmen did not ask for the separation of MATS into separate airlift packages, but they tried to explain how centralized airlift management could meet their needs. General Taylor conceded in his book that the "principle of pooling the common-use strategic airlift in the Military Air Transport Service is sound and should be continued as it is."[92] The term "common-use," however, signified a sharing of airlift capacity, and this was not enough. Taylor explained it in more detail:

> Since it would be plainly uneconomical to tie down aircraft while they await a possible Army mission, I have never suggested any such action. . . . It would be most helpful, however, if the Air Force would predesignate certain squadrons of strategic airlift for joint planning with Army commanders . . . In an emergency, their entire units would immediately be ordered to specific airfields to move specific Army units. . . .
>
> . . . such predesignation would improve joint planning without immobilizing any of the predesignated aircraft.[93]

Army Secretary Wilber Brucker agreed on the necessity to "earmark" specific aircraft without forcing them to "sit on airports and wait for the gong to ring," an eventuality he described as "stupid."[94] When General Lyman Lemnitzer succeeded Taylor as Chief of Staff, he cautioned against any labeling of aircraft "for Army use only."[95]

Yet the conditions set by Army spokesmen could have led only to separate airlift organizations. Taylor was ready to permit predesignated aircraft "to go about their business in the United States unhampered by this requirement," and Gavin agreed.[96]

92. Taylor, *op. cit.*, p. 172.
93. *Ibid.*, pp. 99, 143.
94. U.S. Congress, Senate, Committee on Appropriations, *Department of Defense Appropriations for 1960,* Hearings before Subcommittee, 86th Congress, 1st Session (Washington: Government Printing Office, 1959), p. 104. (hereinafter *SCA DOD Appropriations for 1960*).
95. *Rivers Committee Hearings,* p. 4075.
96. Taylor, *op. cit.*, p. 99; *Air Power Hearings,* p. 841.

Brucker wanted Army units in the air within fifteen minutes after notification, with entire divisions underway in two hours.[97] Lemnitzer specified that units be airborne within one hour.[98] MATS did not operate within the U.S., however, and time limits of this sort are attainable only by keeping aircraft and crews on constant alert at airfields quite close to the Army posts.

On other occasions, Army spokesmen asked for something more than predesignation. Before he retired, Taylor proposed a new joint command to be responsible for "organization, movement, and dispatch" of the joint task forces needed in limited war.[99] General Paul Adams, later the commander of such an organization (STRIKE Command, formed in 1961), put the argument about as strongly as possible in 1955:

> We think that the Army should have at its disposal the instruments that are necessary for fighting the battle. If that includes aircraft to move into the assault zone, even if they come from a thousand miles to the rear, we should command them. . . .
>
> . . . It is universally recognized that a military commander must have control of the means with which he is to fight the battle . . . he must be given the necessary forces to fight the battle. This is based on the principle of unity of command and unity of effort. . . .
>
> . . . The Army has a use for transport support rendered by tactical transport aircraft. This type . . . provides . . . logistic support for Army units from rear supply bases forward to Army supply points. This means of transportation releases the Army commander from dependence upon ground lines of communication. . . .
>
> Transport support aircraft is also essential to the air mobility of large assault forces. Such operations encompass airborne assaults by parachutists, up to formations as large as an Army Corps, and movement of corps and Army reserve divisions in exploit and pursuit of operations. Obviously, the use of transport aircraft in the situations I have just mentioned are part

97. *SCA DOD Appropriations for 1960*, p. 104.
98. *Rivers Committee Hearings*, p. 4075.
99. Speech reprinted in U.S. Congress, House, Committee on Appropriations, *Department of Defense Appropriations for 1960*, Hearings before Subcommittee, 86th Congress, 1st Session (Washington: Government Printing Office, 1959), Part 1, pp. 97–101; also Taylor, *op. cit.*, p. 98. One general criticism of Eisenhower defense policy was the lack of a command charged with fighting limited wars. Huntington, *op. cit.*, p. 352.

and parcel of the land battle for which the Army commander is responsible.[100]

On still other occasions, Army spokesmen tried to accommodate their views on airlift to the emerging problem of tactical nuclear war. The best known expression of the Army's first attempts to deal conceptually with such warfare is Kissinger's first book on defense policy, written after he participated in a study group which included Gavin. Tactical nuclear war, said Kissinger, would call for small, self-contained, highly mobile units using air transport for communications within combat zone.[101] Taylor added that nuclear battlefields would be extremely large, Army units being widely dispersed within them.[102] Gavin coined the phrase "sky cavalry" to describe the new transport and observation forces needed for survival on the battlefield.[103]

The Air Force had failed the Army, argued Gavin, because the smallest Air Force transports weighed about 30 tons each, and had to use runways 5,000 feet in length.[104] Secretary of Defense Wilson, he added, had unjustly prohibited the Army from equipping itself with aircraft weighing more than 5,000 pounds or helicopters exceeding 20,000 pounds in weight.[105]

To cope with nuclear war and dispersed units, the Army thus asked for aircraft weighing less than thirty tons, capable of taking off or landing vertically (VTOL) or on short runways (V/STOL), and able to carry ten tons of cargo in any sort of weather. Aircraft of this sort, added Gavin, could fill the "aircraft gap" produced by the Air Force.[106] General Hamilton Howze, then (1956) the Director of Army Aviation, added that all aircraft flying close to the ground were within the ground environ-

100. U.S. Congress, House, Committee on Appropriations, *Department of Defense Appropriations for 1956*, Hearings before Subcommittee, 84th Congress, 1st Session (Washington: Government Printing Office, 1955), pp. 574–76.

101. Kissinger, *Nuclear Weapons and Foreign Policy*, p. 180.

102. *Air Power Hearings*, p. 1275.

103. Gavin, *War and Peace*, p. 218.

104. *HCA DOD Appropriations for 1958*, Part 2, p. 503.

105. These were the terms of a 1956 directive issued by Secretary of Defense Wilson; this continued a 1952 agreement between Air Force Secretary Finletter and Army Secretary Pace. The initial issue had involved Army and Air Force responsibilities in Korea. Robert Frank Futrell, *The United States Air Force in Korea, 1950–1953* (New York: Duell, Sloan and Pearce, 1961), pp. 533–34; U.S. Congress, House, *United States Defense Policies Since World War II*, 85th Congress, 1st Session, House Document 100 (Washington: Government Printing Office, 1957), p. 44.

106. *HCA DOD Appropriations for 1958*, Part 2, p. 503.

ment of the foot soldier and thus were properly a part of the Army.[107] Finally, said General Matthew Ridgway, all new transports should be able to land in areas without any airfields at all:

> It is difficult to interest the airmen . . . in these specialized needs. We have persistently and consistently importuned the Air Force to provide us with an assault transport to meet the specifications we have laid down. . . . We must have a fleet of global transports capable of ferrying an Army halfway across the world and setting it down in rough airfields, in open country where there are no airfields at all, or perhaps even on water. . . . We need an aircraft that can carry heavy loads, land on very rough fields, or no fields at all, and take off after very short runs. . . .
>
> There is an understandable opposition in the Air Force to the development of these types . . . somebody must man these planes, and the Army, of course, has considered seeking to relieve the Air Force of its unwanted burden. . . .[108]

At various times, then, Army spokesmen favored the predesignation of certain MATS aircraft, reassignment of transports to a new joint command, direct assignment of transports to the Army, and more instead of less emphasis upon short-haul aircraft having many of the characteristics of World War II tactical airlift. All of these stated needs were not totally compatible with each other, and the Army's leaders seemed unable to decide between old-fashioned massive airborne assaults and some new type of tactical nuclear war. In short, Army spokesmen simply were not certain about the type of airlift they sought. They were certain only that they had to control it and that nobody else should use it, peace or war.

This is not to imply that the problem was a one-sided one. Air Force champions of SAC found it difficult to imagine anything but general war, and they were not ready to agree that MATS always was available for Army deployments. The analysis of Army requests has been carried this far to demonstrate the range of the argument and the preoccupation with the old military issue of command and control.

The issue of control was part of Taylor's proposal that a single Chief of Staff replace the supposedly cumbersome JCS committee

107. *Air Power Hearings*, p. 795.

108. Matthew B. Ridgway, *Soldier: The Memoirs of Matthew B. Ridgway* (New York: Harper & Brothers, 1956), pp. 313–14.

system.[109] Yet the problem was less one of having failed to reach decisions than of having reached decisions that were not satisfactory to Taylor. From his point of view the JCS Chairman of the Eisenhower years (Admiral Radford) had too much power already, and giving him still more was less desirable than replacing him.[110] The Army's airlift demands could have been met only by giving it a veto power over any proposed use of strategic airlift, and this was just about what the airborne generals had sought all along. When these Army complaints were coupled with an inept Air Force briefing procedure which left everybody but the Hoover Commission unaware that there was such an organization as MATS, a political issue inevitably resulted. The Army's protests reached an important congressional audience.

SETTING THE STAGE FOR POLITICAL MERGER

The Army and its supporters may have been correct when they insisted that the Eisenhower administration had neglected military airlift, especially in view of the cancellation of the only transport development program of the 1950's.[111] Yet the Army was not concerned about this cancellation because it involved aircraft destined only for MATS, and the Army's airlift calculations (and those of defense scholars) did not include this existing airlift capacity. Had the Army's airlift proposals been completely clear, moreover, any administration probably would have thought them politically unacceptable on grounds of cost. It remains something of a mystery how strategic airlift capacity ever managed to expand as much as it did.[112]

109. Taylor, *op. cit.*, pp. 175–77.

110. Brigadier General T. R. Phillips reported the Army felt itself constantly outvoted by an Air Force-Navy coalition. *St. Louis Post-Dispatch*, November 8, 1959; Taylor himself noted that he had been overridden in the "best traditions of Tammany Hall." Taylor, *op. cit.*, p. 108.

111. This was the Douglas C-132, a cargo transport larger than any in service before or since that time. It was the victim of one economy drive, along with 2 Army divisions, 38 antiaircraft battalions, and 2 atomic support commands. Huntington, *op. cit.*, pp. 94–95.

112. An evaluation is necessary to resolve this apparent ambiguity. To be sure, there was no "rational" plan in the 1950's to move substantial elements of the Army from the U.S. to overseas theaters by means of airlift. But airlift still was in its infancy; the Army really had no acquaintance at all with strategic deployment by air and, on the whole, the Eisenhower administration permitted strategic airlift to expand at a reasonable rate. It was equally clear, moreover, that the airlift could be used to move the Army. One might even

Taken together this chapter and the three preceding ones are intended to demonstrate that World War II and the postwar military and economic policy changes led to a host of new problems and several dissatisfied interest groups. Their attacks upon Eisenhower airlift policies were so disparate in background and theory that it is still a source of wonder that the attackers ever agreed on a single approach.

The international airlines felt neglected by an administration determined to grant unequal concessions to foreign competitors. The supplementals said they had answered the government's calls for assistance in emergencies and should be rewarded by some sort of permanent assistance. The Hoover Commission wanted more centralization in airlift management, less duplication in airlift packages, less expenditure of government funds on its own transports, and more reliance on commercial airlift. The Army, finally, sought more decentralization of airlift management, more duplication of airlift packages, a vast increase in both Army and Air Force expenditures on transport aircraft, and less reliance upon commercial airlift. Despite these substantial differences, these groups managed a partial unification on the issue of military airlift policy, the only area in which each of them had at least some interest.

---

conclude that strategic airlift expanded at a politically maximum rate; much of the criticism (of a business-oriented administration) was aimed at *too much* expansion.

PART THREE

Congressional Policy-Making and the Strange Political Alliance

[7]

Airlift Policy and the Appropriations Committees

DURING THE SECOND TERM of the Eisenhower administration Congress became the important policy-making agency for military airlift. An unusual network of political and economic alliances influenced congressional action. Liberal Democrats found themselves allied with the major airlines, the supplemental carriers, the Hoover Commission, and dissatisfied Army generals. The members of the alliance were not partners in the normal sense of the term; they did not consciously plan strategy and, except for the central position of the congressmen, apparently had little contact with each other. At various times other interest groups joined this loose alliance, among them the CAB, the FAA, and leading academicians seeking an activist role in policy formulation.

The emergence within Congress of the airlift controversy and the gradual cementing of the alliance between 1956 and 1959 are discussed in this chapter. During this period of congressional primacy, the instrument of congressional action were the Appropriations Committees of the two Houses. Other congressmen interested in the subject took their cases to the Appropriations Committees.

When some congressmen attempted, in 1959 and 1960, to advance a more comprehensive airlift program designed to solve both the military problems of the Army and the economic problems of the industry, they found to their embarrassment they had overestimated the alliance. The proposal that wrecked the alliance was a new version of the air merchant marine concept introduced this time by a government agency, the FAA, which opposed the military budget of its own administration. This crucial year is the subject of Chapter 8.

Congressional conservatives rescued military airlift policy in 1960 and, in the process, performed functions generally done within the executive branch. To build a new airlift program, they had to make technical evaluations of several types of transport aircraft that were candidates for the program, then decide how much to spend on each type. The congressional instrument this time was a subcommittee of the House Armed Services Committee, acting as something of a combined JCS and National Security Council. Chapter 9 deals with this final phase of congressional leadership.

All of this activity, especially the construction of the 1960 airlift program, demonstrates that Congress retains a more important, and often indispensable, role in defense policy-making than some observers concede it. Samuel Huntington, for example, concludes that strategy-making is almost wholly beyond congressional competence and that legislators can do little more than lobby with the executive branch which retains the only effective control:

> The military programs have to be weighed against each other, against conflicting interpretations of the security threats and military requirements. . . . No Congressional committee is competent to do this. . . . This can only be done by bodies before which or in which all the conflicting interests can be brought together. . . . The issue of whether the federal government should build greater limited war capabilities is contested by the Army, Navy, Air Force, Marines, Defense Department, State Department, Budget Bureau, Treasury, and a few other officials . . . the interested groups are almost entirely executive agencies. Moreover, they do not function, as often is the case in domestic policy, as representatives of other interests outside the Government.[1]

Some of the important airlift issues, however, have stayed outside of the executive branch since 1934. The constantly increasing separation between government and industry lent a peculiar atmosphere to the airlift controversy of the Eisenhower years, and parts of the dispute caught the executive agencies somewhat off guard. If one approaches defense policy solely from the point of view of the executive branch or within the confines of purely military strategy, the total airlift problem eludes definition, because no executive agency has any responsibility (that it will admit). Congress took the lead because it provided the only political arena accessible

1. Samuel P. Huntington, *The Common Defense* (New York and London: Columbia University Press, 1961), pp. 131–32.

to all the important interest groups and was thus the only branch of government that could reach a decision.

It is possible to oversimplify a complex situation by overemphasizing the actions of specific individuals, but five congressional Democrats played the principal roles in this part of the airlift problem. These were Representatives Daniel Flood of Pennsylvania, Chet Holifield of California, and L. Mendel Rivers of South Carolina, and Senators Symington and Monroney. All except Rivers were leaders of the strange political alliance dominating so much of the period. One other individual was almost equally important to the dispute, General Elwood Quesada of the new Federal Aviation Agency. A Quesada-Monroney partnership produced the air merchant marine proposal that collapsed the alliance.

Congressman Rivers, chairman of a special subcommittee of the House Armed Services Committee appointed in 1960, then stepped into the policy vacuum left by the collapse of the new air merchant marine bill. Rivers and his colleagues were, for the most part, relatively conservative congressmen traditionally friendly both to the military and to business.

1956: CONGRESSMAN FLOOD AND THE HOOVER COMMISSION

Congressman Flood took the first initiative in 1956, demonstrating almost equal interest in Army requests for predesignated airlift and Hoover Commission complaints about MATS. At Flood's suggestion Chairman George Mahon of the Defense Appropriations Subcommittee arranged a special hearing on airlift questions.

Flood used the hearings to attack an Air Force purchase of transports similar in design to those being flown in commercial service. He criticized particularly the Douglas C-118 (DC-6) and the Lockheed C-121 (Super-Constellation). Unlike commercial models the MATS transports had reinforced floors and wide entrance doors to accommodate cargo, but Flood listed several complaints: (1) The aircraft were not predesignated for Army use; (2) some of them had galleys for serving hot meals; (3) enlisted female personnel served as stewardesses; and (4) the aircraft carried passengers properly assignable to the commercial lines.[2] His

2. U.S. Congress, House, Committee on Appropriations, *Department of the Air Force Appropriations for 1957*, Hearings before Subcommittee, 84th Congress, 2nd Session (Washington: Government Printing Office, 1956), pp. 1464–65 (hereinafter *HCA DAF Appropriations for 1957*).

contacts with the Army, said Flood, convinced him that the only useful aircraft were those capable of airlifting both heavy cargo and troops.[3]

Flood also arranged for an appearance by the chairman of the Hoover Commission's Subcommittee on Transportation, P. M. Shoemaker (a railroad executive). The hearing record indicates that his appearance was arranged almost casually, and that neither the airline industry nor the Hoover Commission had planned to carry their arguments to Congress. Shoemaker cited the growth of MATS and the apparently wasteful examples unearthed by the Commission, but Chairman Mahon and others on the committee expressed surprise that they had not been approached directly by the industry.[4] Shoemaker avoided making statements on military questions, but he repeated the Hoover Commission arguments that military traffic should be turned over to commercial carriers.[5] The Air Force quickly put in an appearance, but it apparently had not taken seriously the Hoover Commission accusations.[6] Yet the military witnesses did not defend the existing organization with much spirit and tacitly admitted that military airlift ought not to be used for nonessential items. The subcommittee concluded from all this that the Hoover Commission's arguments were valid and that remedial action was in order.

The Appropriations Committee's 1956 report on defense appropriations was the first formal expression of congressional interest. The committee encouraged the Air Force to manage its airlift operations so as to "assist" the airlines in their attempts to maintain a sound financial position. The committee also adopted as its own the oft-quoted statement from the 1954 ACC report; the govern-

3. *Ibid.* It will be recalled that the Army's only experience with deploying troops by airlift had been confined to the C-124's on "Operation Gyroscope" (see p. 135). In the traditional manner of tactical assault operations, the Army had planned a separate load for each aircraft, with both cargo and troops being carried on each C-124. The view became prevalent within the Army that cargo and troops always should be airlifted together, regardless of the distances involved, thus ruling out the notion that troops could be moved on passenger aircraft and cargo on cargo planes. Interviews.

4. *Ibid.*, p. 1453. Shoemaker's appearance seems to have been only casually arranged. Flood announced he had happened to see Shoemaker in Washington and had invited him over.

5. *Ibid.*, p. 1465.

6. It is obvious from the testimony that the Air Force had not studied the Hoover Commission reports in detail and was ill-prepared to respond. *Ibid.*, pp. 1460–61.

ment was asked once again to use "unutilized capacity."[7] Finally, said the committee, it expected the Air Force to submit a detailed statement each year covering all aspects of MATS operation.

The chairman of the Senate's Defense Appropriations Subcommittee, Dennis Chavez of New Mexico, joined with Mahon in sending a letter to Secretary of Defense Charles Wilson; it endorsed the House Committee's statements and called particular attention to the now-famous ACC paragraph.[8] Despite Flood's interest in Army affairs, then, Congress expressed itself only on the question of military competition against the airlines.[9]

1957: SENATOR SYMINGTON AND THE ATA

The focus shifted from the House to the Senate in 1957, where the ATA succeeded in getting the Appropriations Committee to accept suggested language and write it into the committee report on defense spending. The subject came up again in the House, and Flood produced evidence he had gathered himself and some provided by the CAB.

Flood introduced a personal survey of the scheduled airlines; they reported to him the amount of unutilized capacity for the previous year.[10] Flood was now convinced that MATS was a "billion dollar boondoggle" and should stop using "plushed-up airplanes" to compete with the airlines.[11]

The CAB took two actions in support of increased government expenditures for commercial airlift; one was designed to aid the scheduled carriers, the other the supplementals. CAB Chairman

7. U.S. Congress, House, Committee on Appropriations, *Department of Defense Appropriations for 1957*, 84th Congress, 2nd Session, House Report 2104 (Washington: Government Printing Office, 1956), pp. 45–47.

8. Letter reproduced in U.S. Congress, House, Committee on Appropriations, *Department of Defense Appropriations for 1958*, Hearings before Subcommittee, 85th Congress, 1st Session (Washington: Government Printing Office, 1957), Part 2, p. 1703 (hereinafter *HCA DOD Appropriations for 1958*).

9. A word should be said here about the status of Congressional reports. Their legal status is ambiguous at best, especially because many actions are taken on the floor long after reports are in the record. Important committees (particularly the Appropriations Committees) regularly use reports to outline in detail what they expect the executive branch to do. Reports on defense appropriations are as important as any, because of the explicit constitutional responsibility of Congress in military affairs.

10. *Ibid.*, pp. 1706–9, 1713–16.

11. *Ibid.*, pp. 1772–73.

James Durfee sent Flood a letter pointing out that a great deal of capacity was indeed available for military use.[12] The board also advised the supplementals to petition Congress for an increase in military contract operations.[13] In point of time, this CAB intervention corresponded to peak congressional interest in the supplementals and to the Eisenhower administration's liberal concessions to foreign carriers. Yet the CAB's promotion of the scheduled airlines made little sense and was contrary to what those companies wanted.[14]

Despite Flood's perseverance and testimony from the ATA and the supplementals, the House Appropriations Committee chose to say nothing of the subject in its report. The decision followed a lengthy briefing on airlift problems by JCS Chairman Admiral Radford; he explained at some length the difficulties of trying to meet all the Army's requests and defended existing policies.[15]

The ATA managed, however, to achieve a major victory in the Senate, with the help of a special team put together for appearances before Congress. In addition to President Stuart Tipton, the team included Ramsay Potts and Robert Turner. The former, a Washington attorney retained especially for the occasion, once had represented the supplemental airlines before the Small Business Committee, had worked with the Hoover Commission on behalf of the ATA, and had been an executive officer to Symington during his tenure as Secretary of the Air Force.[16] Turner was an airline executive and a former head of DATA. Given the team's composition, its presentations were quite effective, especially at the Senate hearings where Symington did the questioning as the ex officio representative of the Armed Services Committee.[17]

12. *Ibid.*, p. 1756.
13. *Ibid.*, p. 2115.
14. Durfee cited the over-all airline load factor of 62 per cent. The load factor for some routes was higher, for others lower. There was no reason to assume that the empty space coincided with particular military requirements and, in any case, 62 per cent is not an abnormally low load factor. The industry, moreover, had little interest in trying to fill empty space; the ATA asked for charter flights under military contract. Testimony of Stuart Tipton in U.S. Congress, Senate, Committee on Appropriations, *Department of Defense Appropriations for 1958*, Hearings before Subcommittee, 85th Congress, 1st Session (Washington: Government Printing Office, 1957), p. 1429 (hereinafter *SCA DOD Appropriations for 1958*).
15. *HCA DOD Appropriations for 1958*, Part 2, pp. 2049–97.
16. *Ibid.*, p. 2268.
17. *SCA DOD Appropriations for 1958*, pp. 1423–32.

Airline executives later credited Potts with having devised the ATA formula written into the Senate Committee's report on defense appropriations:

> . . . maximum utilization by the Department of Defense of United States civil air carriers is essential both in the promotion of our free enterprise economy and in the provision of the necessary ready reserve civil airlift for national defense; . . . Government operation of its own transport facilities should be limited to that essential to military security.
>
> . . . In summary we do not feel that sufficient effort has been made by the Department of Defense in the international and overseas field to 'adjust its use of air transportation so as to use existing unutilized capacity of United States air carriers'. . . .
>
> It is the wish of this committee, therefore, that within the 1958 appropriations for operations and maintenance and for military personnel the Defense Department reprogram expenditures for operating MATS and other Government-owned transport activities sufficiently to permit the funds so reprogrammed to be applied toward procuring the services of United States civil air carriers to meet as nearly as possible 40 per cent of the passenger requirements and 20 per cent of the cargo requirements of the Military Air Transport Service.[18]

House conferees on the defense money bill, among them Congressman Flood, accepted this language, and the conference report endorsed it.[19] It is difficult to overestimate the importance of this turning point in the airlift controversy.

Language written into a congressional report transformed the controversy from one between conflicting interest groups into a dispute between the Defense Department and Congress. The ATA never could have secured such broad congressional support on its own. Flood's main interest remained the Army and its airlift prob-

18. The ATA proposal is in *ibid.*, p. 1422; also *HCA DOD Appropriations for 1958*, Part 2, p. 2301. For the views of airline executives, see Hiram C. Caroom, "Management Responsibilities and National Interests: An Analysis of the Practicability of the U.S. Flag International Passenger Airlines' Transporting More Military Traffic in Peace and War," (unpublished doctoral dissertation, Harvard University, Cambridge, Mass., 1959), p. 22; the quotation is from U.S. Congress, Senate, Committee on Appropriations, *Department of Defense Appropriations for 1958*, 85th Congress, 1st Session, Senate Report 543 (Washington: Government Printing Office, 1957), pp. 13–14.

19. U.S. Congress, House, Committee on Appropriations, *Department of Defense Appropriations for 1958*, 85th Congress, 1st Session, House Report 841 (Washington: Government Printing Office, 1957), Amendment #37.

lems, as General Gavin had outlined them.[20] Symington, who equated airlift with paratroopers, said more than once that commercial airliners could not be militarily effective because they could not drop paratroopers.[21] The two legislators, however, had some reputation as students of military affairs, and this overcame doubts advanced by other legislators not quite so ready to adopt the ATA's suggestion.[22] Two significant defects in the ATA proposal went largely overlooked.

ATA spoke only for the scheduled airlines but the supplementals wanted something for themselves, quite beyond the percentages asked by the ATA. Yet the ATA language was general enough to keep the supplementals from speaking out against it. A second defect was that the proposal obscured major differences within the ATA itself. At least one cargo airline was an ATA member, but it was using overseas military traffic to make up for a lagging domestic cargo market. The ATA was far from united behind the proposition that the government should concentrate on "unutilized capacity."

Despite the defects, Congress was bound to consider any failure by the Defense Department to meet the 40–20 provision, as it was called, as a deliberate flouting of congressional intent. This proved the case almost immediately; indications that the Defense Department was not meeting the quota prompted Congressman Holifield to enter the fray.

1958: CONGRESSMAN HOLIFIELD AND THE CRAF

Several small airlines located in Holifield's home district in California called his attention to Defense Department disregard of the

20. See Table VI, note d, for a recapitulation of Flood's contacts with the Army. He credited Gavin with having prompted his interest in airlift; for Gavin's views on his associations with Congress, see James M. Gavin, *War and Peace in the Space Age* (New York: Harper & Brothers, 1958), p. 172. While Gavin insists that he kept his distance from Congress, it was not unusual for Flood to make long speeches on the floor that seemed the result of research within the Army, e.g., *Congressional Record*, July 2, 1959, pp. 11502–5.

21. *SCA DOD Appropriations for 1958*, pp. 897–902; U.S. Congress, Senate, Committee on Armed Services, *Study of Air Power*, Hearings before Subcommittee, 84th Congress, 2nd Session (Washington: Government Printing Office, 1956), p. 839 (hereinafter *Air Power Hearings*).

22. Chairman Chavez, for example, observed that it seemed to him that ATA was advocating that MATS get out of business. *SCA DOD Appropriations for 1958*, p. 1432.

40–20 provision, and he began his own investigation of military airlift. As chairman of the Military Operations Subcommittee of the House Committee on Government Operations, Holifield established a new and direct relationship between military purchases of commercial airlift and airline membership in the Civil Reserve Air Fleet (CRAF).

Holifield also was disturbed at an Air Force program that had aroused industry opposition. That service had announced late in 1957 an experiment with a program known as "bailment"; ten military transports (five C-54's and five C-118's) were to be leased to commercial operators. The Air Force invited all airlines to bid on the project, and it advised them of certain conditions: (1) companies would operate normal schedules for military traffic; (2) they would have to guarantee the availability of both planes and crews for training exercises and war; and (3) they would have to agree to operate within combat zones.[23] At the time, however, commercial companies had many aircraft more modern than the C-54's; MATS itself was then retiring them from active service. Labor representatives described the proposal as an attempt to circumvent prevailing wage rates, in that companies would have to offer a low bid, then try to persuade employees to accept a non-union agreement. As one labor witness put it: ". . . If MATS isn't able to afford the wages and benefits the commercial carriers have negotiated through the process of collective bargaining, it has no business proposing such a plan."[24]

Confronted with united opposition from management and labor the Defense Department quickly withdrew its proposal with a promise to consider it further, and the whole program has not been heard from since. It had a lasting impact, however; when he testified before the House Appropriations Committee, Holifield cited the bailment proposal as evidence that something was wrong

23. U.S. Congress, House, Committee on Government Operations, *Military Air Transportation*, Hearings before Subcommittee, 85th Congress, 2nd Session (Washington: Government Printing Office, 1958), p. 21 (hereinafter *Military Air Transportation, 1958*). The Defense Department issued the initial press release on the bailment program, but identified it as having been initiated by MATS. Both agencies later denied they had favored the idea. The idea probably originated within the office of the Secretary of the Air Force. *Ibid.*, pp. 21, 393, 552.

24. Testimony of James Horst, *ibid.*, p. 288. Holifield agreed and added that the bailment program could lead to "scab" labor conditions. *Ibid.*, p. 289.

with airlift management.[25] In the meantime he had expanded his inquiry into other phases of MATS operations.

Holifield became convinced during his hearings that CRAF was the best available solution to strategic airlift problems, and his subcommittee's report reflected that conviction. Members of the Commerce Department and DATA explained at length the statistical procedures they used to allocate wartime airlift requirements among the CRAF members. These witnesses convinced the subcommittee that the Defense Department should enlarge CRAF and depend upon it.[26] The subcommittee stated its conclusions in simple terms; if CRAF were to fulfill its wartime obligation, military authorities must assist the companies who were members.[27]

The subcommittee did not specifically recommend that the Defense Department reduce the size of MATS, but it did say that MATS should concentrate on "outsize and special-cargo traffic and technical missions" and should turn over everything else to the airlines.[28] Further, argued the subcommittee, it was the "sheerest kind of rationalization" to call MATS a "system" that must be exercised; this could be done with commercial aircraft if it was necessary at all.[29] The subcommittee shied away from asking that MATS lower its utilization rate but added that a "rigorously impartial study" was needed.[30] These conclusions met the airline industry only halfway; the cargo airlines wanted to buy the largest of military cargo transports so they could handle outsize cargo and even

25. U.S. Congress, House, Committee on Appropriations, *Department of Defense Appropriations for 1959*, Hearings before Subcommittee, 85th Congress, 2nd Session, on Advanced Research Projects Agency, etc. (Washington: Government Printing Office, 1958), p. 1101 (hereinafter *HCA DOD Appropriations for 1959*).

26. See testimony of David W. Bluestone, in *Military Air Transportation, 1958*, pp. 216–62.

27. U.S. Congress, House, Committee on Government Operations, *Military Air Transportation, Twenty-Eighth Report*, 85th Congress, 2nd Session, House Report 2011 (Washington: Government Printing Office, 1958), p. 59.

28. *Ibid.*, p. 6. The word "outsize" commonly refers to cargo that cannot be carried in passenger-type aircraft. A Boeing 707 is considered unsuitable for outsize cargo, while a military C-124 or C-133 is suitable. The situation is further complicated by the fact that some military transports have wider doors and other appurtenances not found on their commercial counterparts.

29. *Ibid.*, pp. 88–90. This involves many legal and technical difficulties. Commercial aircraft often use civil airfields to and from overseas destinations, for that is where facilities and mechanics are located. Nor is it feasible to permit military mechanics to work on commercial aircraft, because of dissimilarities in equipment and licensing and insurance problems.

30. *Ibid.*, p. 91.

missiles, and the ATA wanted immediate reductions in MATS.[31] Similarly, the subcommittee stopped short of full agreement with the CAB; its witness had asked that all military airlift be prohibited except when commercial companies could not do the job.[32]

Unlike Flood and Symington, Holifield had relatively little interest in airlift for the Army, giving it only brief mention in his report.[33] While he did not distinguish between various segments of the airline industry, Holifield was generally more sympathetic to the small carriers than to the old-line companies.[34] When he appeared before the Appropriations Committee, however, Holifield spoke only of the complaints that had led to his investigation. He added that the Defense Department should "stop taking directions from MATS."[35] The Air Force had not explained away its failure to comply with the 40–20 provision, he argued, and Congress should remedy the deficiency by earmarking specific dollar amounts for commercial airlift.[36]

Holifield's interest in CRAF embodied several assumptions that have been influential since 1958. It was assumed, for one thing, that the Air Force could insure a larger CRAF by providing more peacetime contracts for the airlines. New aircraft already fully employed on government business, however, must operate at the high utilization rate consonant with normal airline practice, and

31. Testimony of Slick Airways and Seaboard World Airlines in *Military Air Transportation, 1958*, pp. 105, 275. Testimony of ATA at *ibid.*, pp. 13–14.

32. *Ibid.*, p. 329.

33. H. Report 2011, 85th Congress, 2nd Session, p. 81.

34. There were some curious twists in the treatment afforded these carriers. Slick Airways, a cargo airline based in California, had been assigned by the CAB to east–west cargo routes. The company proposed that military contracts for domestic cargo airlift be awarded those companies assigned the routes nearest the military bases involved; such a decision would have given Slick much of the domestic business. Slick's competitors pointed out that Slick wanted to change the procedure and thus secure contracts it had lost to competitors. The subcommittee canceled an appearance by these competitors, but then inserted a favorable comment on the Slick proposal in the subcommittee's report. *Ibid.*, p. 105: *Military Air Transportation, 1958*, pp. 91, 113–18, 203–4.

35. *HCA DOD Appropriations for 1959*, on Advanced Research Projects Agency, etc., pp. 1100–1. Counsel Herbert Roback asserted during the Holifield hearings that MATS had the authority to determine the size and method of its operations. *Military Air Transportation, 1958*, p. 533.

36. At the time commercial airlift was handling about half the traffic contemplated by the 40–20 provision, although it never was clear how the percentages should be measured. Holifield said he did not know where the percentages had come from but that the will of Congress must be observed. *Ibid.*, pp. 420, 504.

this does not allow for emergency expansion. The real source of expansion was bound to remain with those airliners used in commercial activity and the degree to which they could be shifted to military operations.

This assumption rested on yet another one seldom articulated but seemingly prevalent inside and outside the Defense Department. The subcommittee, the industry, and Commerce and Defense Department planners all seemed to accept the notion that what had to be moved in wartime was something *instead of*, but *not in addition to*, cargo already being moved. Yet the Air Force was using airlift as an assertedly indispensable supply line, and the industrial fund supposedly was forcing other services to follow suit. This assumption, however, permitted everyone to look upon any expansion in the size of CRAF as an expansion in wartime airlift capacity.

Yet a third assumption involved what can only be described as an article of faith. It seemed less expensive to guarantee the industry enough business to buy new aircraft for CRAF than to purchase more military transports. It is an old question, of course, whether commercial operations are by their very nature more efficient and economical than their military counterparts. The argument never has been advanced very strongly by the airlines themselves; they have asked that MATS be reduced in size. This would reduce the budget for direct spending on military aircraft, but paying for new commercial transports through the medium of guaranteed military traffic might not automatically be any less expensive.[37]

1958: SENATOR MONRONEY AND THE SUPPLEMENTALS

Senator Monroney, Chairman of the Commerce Committee's Aviation Subcommittee, started yet another investigation in 1958, determined to preserve the supplementals as a valuable national asset. He ultimately became the closest thing to a leader of the alliance of congressmen, airline interests, and Army supporters. When

37. The ATA admitted that military and commercial costs were not comparable. U.S. Congress, Senate, Committee on Commerce, *Study of Military Air Transport Service*, Hearings before Subcommittee, 85th Congress, 2nd Session (Washington: Government Printing Office, 1958), p. 120 (hereinafter *Study of MATS Hearings*). Flying Tiger Line (a cargo carrier) spoke only of the reduced capital costs to the budget if airlines replaced MATS. *Ibid.*, p. 55. Airline executives conceded that costs had not been studied. Caroom, *op. cit.*, pp. 238–39.

he ended his hearings, he gave about the same testimony to the Senate Appropriations Committee that Holifield had given in the House.

Monroney looked upon the supplementals as a military reserve serving without extra pay, and expressed alarm at the unemployment rate among their flight crews. One company had reduced its pilot complement from 400 to 63, another had discharged more than 100 pilots.[38] Monroney told Assistant Secretary of the Air Force Dudley Sharp that it was a military task to keep these airlines, all of them CRAF members, solvent enough to keep pilots from having to turn to "selling shoes."[39] Moreover, said Monroney, these companies provided real bargains; in the system of competitive bidding then in use on military contracts, the military paid only $125 for passenger travel to Europe.[40] He suggested that MATS keep an up-to-date record of these companies and their pilots, because the Defense Department consistently had said it needed them.[41] Monroney listened to all segments of the industry, however, and they spread upon the record their many differences in approach.

The cargo airlines and the supplementals faced the same bitter struggle for survival in a market that had little use for their services. The two groups tended to sympathize with each other but to become bitter individual enemies when one bested another in obtaining a particular military contract. They tended also to use extreme language in describing their quarrels with the government, e.g., one company demanded that Congress curb the "octopus-like growth of MATS."[42] President Stuart Tipton of the ATA gave as much support as possible to the notion that the government should turn over its military traffic to the scheduled carriers already assigned the international routes, at the expense of cargo airlines also belonging to the trade association:

38. *Study of MATS Hearings*, pp. 77, 101.
39. *Ibid.*, p. 23.
40. The normal tourist rate was $300. *Ibid.*, p. 27.
41. *Ibid.*, p. 23.
42. Quotation is from representative of Slick Airways, a former head of the Civil Aeronautics Administration and former member of the CAB. *Ibid.*, p. 179. A spokesman for the supplemental carriers sympathized with Slick Airways, describing it as a company put together with "blood, sweat, tears . . . and $15 million in private funds," only to be "shot down" by the Defense Department. *Ibid.*, p. 179. Two other cargo airlines, on the other hand, continued their dispute on how the government should award domestic contracts. *Ibid.*, pp. 76, 87.

> This committee is certainly familiar with the gracious extravagance with which other foreign airlines have been granted routes not only to gateway points in the United States, but throughout the country.
>
> We can hope that there will be a more realistic appraisal of equitable bargaining procedures in the future. . . .
>
> The inroad being made by foreign airlines is great and is growing as they receive new routes from the United States Government. Thirty-two countries already send their airlines to the United States in competition with United States Flag airlines. . . .
>
> The penetration into the traffic of the United States airlines has been severe. . . .
>
> Therefore, it becomes increasingly important, in the face of the imbalance caused by these landing rights to foreign countries, that every diligence be exercised in curtailing the operations of a totally subsidized military airline.[43]

The ATA agreed that airlines should assign their most modern aircraft to CRAF because any emergency would require the best transports available.[44] Tipton added, however, that the ATA did not want to use its CRAF aircraft on normal military contracts; the government should, instead, use the secondhand piston transports that ATA members were finding difficult to sell during their conversion to jets.[45] Cargo and supplemental carriers, on the other hand, wanted airplanes assigned to CRAF kept busy on government contracts, a proposal that was mathematically impossible to meet.[46] Industry witnesses agreed with each other on only one major point; MATS should be reduced in size, perhaps by as much as 75 per cent.[47]

43. *Ibid.*, pp. 140–43.
44. *Ibid.*, pp. 122–24.
45. *Ibid.*, p. 127.
46. Slick and Flying Tiger proposed, respectively, that 75 and 100 per cent of the CRAF fleet be kept busy during peacetime with military contracts. Since CRAF then included 362 transports, enough business would be required to keep some 270 aircraft fully employed at normal airline utilization rates of at least eight hours. This would have produced about half again the capacity of all of MATS. *Ibid.*, pp. 28, 57, 90–91.
47. Flying Tiger suggested a 75 per cent reduction, *Ibid.*, p. 60. The Chamber of Commerce asked that one MATS airplane be retired for each new aircraft assigned to CRAF. *Ibid.*, p. 109. The ATA was not specific at the time, but it said in 1959 that MATS flying should be 75 per cent less. U.S. Congress, House, Committee on Appropriations, *Department of Defense Appropriations for 1960*, Hearings before Subcommittee, 86th Congress, 1st Session (Washington: Government Printing Office, 1959), Part 6, pp. 615–16.

Monroney insisted in his appearance before the Senate Appropriations Committee that Congress should require the Defense Department and MATS to spend more money for commercial airlift. This was the only way, he argued, to maintain an adequate mobilization base.[48]

1958: THE APPROPRIATIONS COMMITTEES AND EARMARKED FUNDS

Congress took the next step in 1958, making a specific portion of the Defense Department's annual appropriations available only for the purchase of commercial airlift. Both Appropriations Committees were angered by Defense Department failure to comply with the 40–20 provision, and the Senate once again took the initiative. While the House ultimately agreed to Senate action, Flood was at first unsuccessful in securing agreement from his colleagues.

Other members of the House Appropriations Committee disagreed with Flood's position and kept out of the committee's report any mention of the airlift problem. Flood apparently lost his support by stating his case in extremely strong terms: "I am convinced beyond peradventure of intelligent doubt that from the inception of this . . . there has been a deliberate, carefully defended resistance at MATS . . . once this conference committee produced the language that it did last year, then I say, to torture that language was deliberate and planned for the purpose of avoiding it, not executing it. . . ."[49] One colleague noted that the airlines already had received considerable government assistance, especially in the fast tax write-offs granted since the Korean War.[50]

At least four Senators were instrumental in committing the Appropriations Committee to force government expenditures on commercial airlift. Chavez, cautious in his acceptance of the ATA proposal the previous year, had become incensed at the Defense Department's failure to heed the will of Congress. He lashed out at those in the Pentagon who, he said, dismissed congressional reports as only the statements of "a bunch of politicians."[51] He

48. U.S. Congress, Senate, Committee on Appropriations, *Department of Defense Appropriations for 1959*, Hearings before Subcommittee, 85th Congress, 2nd Session (Washington: Government Printing Office, 1958), pp. 1194–95 (hereinafter *SCA DOD Appropriations for 1959*).

49. *HCA DOD Appropriations for 1959*, Advanced Research Projects Agency, etc., p. 829.

50. *Ibid.*, pp. 830–31. The Air Force also entered a strong protest against the implication it should reduce MATS. *Ibid.*, p. 759.

51. *SCA DOD Appropriations for 1959*, p. 978.

bluntly warned the Secretary of the Air Force that "unless something is done, we will write it into the law, sir."[52]

Symington joined Chavez and described his own position in terms of two groups he believed active within the Air Force. One contained those who had "devoted their entire lives to the service," and he believed such individuals when they asked for increases in appropriations. Another group, he added, was involved in the "pretty shoddy business" of MATS and its use of female stewardesses; for this group Symington had only contempt.[53]

Senator John Sparkman, still interested in the supplementals as small businesses, joined Monroney in asking that funds be earmarked. Any such funds, added Sparkman, should be spent on small business, and the congressional reports on defense spending should indicate as much.[54]

The Appropriations Committee added two provisions to the defense money bill, and Congress ultimately approved both. One added $140 million to the budget, to buy new military transports suitable for the airlift of both cargo and troops. This reflected Symington's interest in Army requests for airlift, but it was not not made clear whether the funds were for tactical or strategic transports.[55] The committee also specified that $84 million of the defense budget should be spent on commercial airlift, with $21 million of that amount reserved for small business.[56] All of this could be done, said the Committee, "without any additional cost to the Government."[57] This provision emerged from the conference between the two Houses in a slightly modified form: "Of the funds made available by this act for the services of the Military Air Transport Service, $80,000,000 shall be available only for procurement of commercial air transportation service; and the Secretary of Defense shall utilize the services of civil air carriers which qualify as small business to the fullest extent practicable."[58]

52. *Ibid.*, p. 1091.
53. *Ibid.*, p. 978.
54. *Ibid.*, p. 1198.
55. U.S. Congress, Senate, Committee on Appropriations, *Department of Defense Appropriations for 1959*, 85th Congress, 2nd Session, Senate Report 1937 (Washington: Government Printing Office, 1958), p. 17. On Symington's specific role, see remarks of Armed Services Committee Chairman Senator Richard Russell, in *Congressional Record*, March 10, 1959, p. 3357.
56. The committee merely adopted an Air Force estimate that $84 million would be spent anyway.
57. Senate Report 1937, 85th Congress, 2nd Session, pp. 17, 19.
58. Public Law 85–724, Sec. 634.

SIGNIFICANCE OF THE FIRST PHASE

During this initial three-year period of congressional interest in military airlift, complaints from Army spokesmen and from airline interests gradually led Congress, or at least some of its most important committees, to take the most direct form of action open to legislators. Several aspects of this initial phase deserve further comment.

The controversy was kept alive in its early stages by only the thinnest of margins. The 1956 appearance of the railroad executive was only casually arranged, and the House Appropriations Committee virtually abandoned the entire subject only a year later. The Senate Appropriations Committee was initially somewhat indifferent to the ATA's plea, but Symington's position was crucial. The Appropriations Committee accepted the 40–20 provision on vague grounds of national defense, not because Congress suddenly was ready to embrace the ATA.

The focus of attention very gradually shifted from the scheduled airlines to the supplementals. The Hoover Commission had concentrated on the former, but the plight of the supplementals attracted far more interest from influential congressmen. Many of the small companies sent their own representatives to testify before the Holifield and Monroney committees, but the major companies spoke only through the ATA. The trade association, moreover, was concerned largely with the international carriers, e.g., PAA, TWA, and Northwest. Throughout these several years of congressional involvement, the major domestic airlines showed little interest in the discussion. One result was that when attention shifted from the major carriers to the supplementals, congressional action was somewhat unrelated to the airline problems that first had interested the Hoover Commission.

Yet another significant aspect of this initial phase was the CAB decision to intervene on its own. A career employee of the CAB who testified before the Holifield subcommittee described the board as "a quasi-judicial agency" when he wanted to avoid particular questions.[59] This did not prevent the CAB from supporting all segments of the industry and encouraging Flood to pursue his argument.

The Air Force looked upon airlift as something to be used for the support of SAC and thus overemphasized the importance of

59. *Military Air Transportation, 1958*, p. 336.

split-second response to emergencies. It appeared to some congressmen that the MATS fleet should remain within the U.S. at all times, precisely for this reason.[60] Even while the Air Force was defending MATS as important to preparedness, it abandoned the C-132 cargo development project because there would not be enough cargo available to keep such large aircraft fully employed in peacetime.[61] It was left to Admiral Radford to make a comprehensive defense of existing airlift policies, even if he did not touch on questions about CRAF.

None of these important legislators was ready to demand that MATS be reduced or eliminated, but they chose to accept terms suggested by lobbyists who sought just such reduction or elimination. Liberal Democrats thus became for the first time the political allies of the major carriers, the supplementals acting as sort of a bridge between these congressmen and the other members of the political alliance. Whatever the interests of the legislators, they supported an action having only the most tenuous relationship to military preparedness.

60. *Study of MATS Hearings*, p. 55.
61. *Ibid.*, p. 32.

[8]

The Quesada Plan and Airlift Stalemate

THE ALLIANCE REACHED its greatest strength during the second phase of congressional activity, but almost immediately disintegrated under the pressure of attempts to find agreement on a more comprehensive airlift program. This second phase can be dated from approximately midway in the 1959 congressional session to the early part of the 1960 session.

An important addition to the alliance was General Elwood R. Quesada, administrator of the Federal Aviation Agency (FAA), who presented his own program for the future of military airlift. Quesada's plan, supported by members of the academic community and by two retired Army generals, envisaged a new type of "air merchant marine" built with government-guaranteed loans. The new fleet was to provide a commercially operated answer to Army requests for airlift. This aspect of the plan never was discussed in any detail, however, and it was not made clear how a commercial fleet could have met the Army's specifications. When Congressional hearings began in 1960, the scheduled airlines and some government agencies said they had opposed it all along. The collapse of the plan brought on another policy vacuum, this time in military airlift.

This decisive phase is most understandable in terms of the role played by Quesada, chief of an important and newly reorganized executive agency. He chose first to work with opposition senators in developing legislation that could not get administration approval. Then, in order to advance his own program, he opposed military budgets approved within his own administration.

THE PROPOSED NATIONAL AIR CARGO FLEET AND ITS SOURCES

Quesada's appointment as head of FAA followed a long career in both military aviation and aircraft manufacturing. His Air Force career had been in tactical aviation, and he had become after retirement a director of the Lockheed Aircraft Corporation.[1] When mid-air collisions took on national significance in the 1950's, he became the president's Special Assistant for planning aviation facilities; his principal task was to implement the report made by his predecessor, Edward P. Curtis.[2] The major purpose of this activity and the legislation creating FAA was to consolidate the control of domestic airspace within a single agency.[3] Before leaving the White House, however, Quesada turned his attention to broader questions of aviation policy.

The National Air Cargo Fleet had two intellectual wellsprings. One was a study of airline financial problems done for Quesada by Professor Paul Cherington of Harvard University, the other a memorandum sent Vice-President Nixon in 1958 by Professor William Y. Elliott of Harvard. Cherington explored the types of assistance that might properly be given the scheduled airlines, then in the process of converting from piston to jet operation. This was an extremely expensive conversion, and Cherington analyzed the specific difficulties facing the carriers.

It seemed to Cherington that only the four largest airlines were able to raise the money for the new jets.[4] Several airlines, he said, already had pressured manufacturers to sell the jets at cut-rate prices. Manufacturers were in no better shape, because development costs had exceeded cash reserves; some of them expected to reduce their labor force before reaching a break-even point on new orders.[5] Moreover, argued Cherington, the CAB had promoted too much direct competition; four airlines on any single route were too many, especially in a time of recession.[6] Finally, he added,

1. James Shepley and Clay Blair, Jr., *The Hydrogen Bomb* (New York: David McKay Company, Inc., 1954), p. 176.

2. *New York Times*, June 15, 1957; U.S., Executive Office of the President, President's Special Assistant for Aviation Facilities Planning, *Aviation Facilities Planning* (Washington: Government Printing Office, 1957).

3. U.S. Congress, House, Committee on Commerce, *Federal Aviation Act of 1958*, 85th Congress, 2nd Session, House Report 2360 (Washington: Government Printing Office, 1958), p. 1; John M. Lindsey, "The Legislative Development of Civil Aviation: 1938–1958," *Journal of Air Law and Commerce*, 28 (Winter, 1961–62), pp. 18–45.

4. Cherington estimated a need for $764 million to finance the new purchases. U.S. Congress, House, *Airline Equipment Reinvestment Program*, 85th Congress, 2nd Session, House Document 430 (Washington: Government Printing Office, 1958), p. 27 (hereinafter *Cherington Report*).

5. *Ibid.*, pp. 50–52.

6. *Ibid.*, pp. 48–49.

U.S. international carriers were losing out to foreign competitors.[7]

Cherington suggested that Congress authorize the government to guarantee purchase loans for the new jets. A program of this sort already was helping the local service airlines, and Cherington wanted to extend it to the major carriers as well.[8] He asked also that the problem of MATS be resolved along Hoover Commission lines. He estimated MATS' growth as 300 per cent since 1951, positive evidence to him that the Air Force was downgrading the ability and willingness of the airlines to respond to emergencies. Under no circumstances should MATS acquire new jets, added Cherington, for this would only take more business from the airlines. He regarded this as his most important recommendation.[9]

Professor Elliott's 1958 memorandum was strikingly similar in language to what Quesada and Monroney later said in support of the proposed cargo fleet.[10] The memorandum circulated among several government agencies thought to have an interest in it. Elliott tried to relate the airlift problem both to military strategy and to economic assistance programs for underdeveloped countries. Elliott argued for a vast new program to develop passenger-cargo transports. They could provide modern transportation between the U.S. and the underdeveloped countries, thus enabling the U.S. to recapture the world's imagination by having its flag carried everywhere. In Elliott's scheme the government either would lease the transports to private operators or would buy them outright and then give them to the carriers. This new fleet, he argued, also would solve the military airlift shortage Kissinger had described the previous year, because the Army then could move entire divisions on short notice. The new fleet would employ the pilots that airlines were releasing and, said Elliott, the government could reduce the size of MATS. He predicted a market large enough to support one hundred such transports, if operating costs were low enough.[11]

7. *Ibid.*, p. 19.
8. *Ibid.*, p. 44.
9. *Ibid.*, pp. 37–40.
10. Memorandum from William Y. Elliott to Vice-President Nixon, February 11, 1958.
11. The Congressman representing the district in which the Douglas Company is located made a similar proposal. He urged an increase in airlift capacity because the company had not received its fair share of defense orders, and he supported a proposal by Allen Ferguson of Northwestern University that the Air Force lease its transports instead of buying them. *Congressional Record*, January 26, 1959, pp. A520–1, February 2, 1959, pp. 730–31.

Elliott's memorandum was unique in its attempt to combine in one document advocacy of more military airlift with an argument for less of it. To accomplish this merger of ideas, he had to conclude that commercial airlift could satisfy the Army. This became an important, if undiscussed aspect of the Quesada plan ultimately forwarded to Congress.

Quesada obtained tentative agreement from a group of officials who attended a Pentagon meeting in March, 1959, a meeting called initially for another purpose.[12] The Air Force and MATS had persuaded Monroney to listen to a briefing about MATS but, because of his interest in Quesada's ideas, the meeting became a discussion of the merits of the National Air Cargo Fleet.[13] After the meeting, Symington and Monroney sent a long letter to Defense Secretary McElroy, supporting a joint civil-military cargo transport development program.[14] Such a program, they said, could produce a new cargo "workhorse" for both the commercial companies and MATS.

THE UNVEILING OF THE PLAN AND ITS IMMEDIATE SOURCES OF SUPPORT

Quesada introduced his plan during the 1959 hearings of the Holifield Subcommittee; he gained immediate support from liberal Democrats, retired Army generals, and organized labor. The Defense Department indicated tacit agreement with the program by changing some aspects of MATS operation.

12. This account is based upon a memorandum for the record circulated within one of the agencies present at the meeting. It concluded that Monroney and Quesada were working together.

13. The program apparently was being formulated at the time. In a lengthy speech on airlift made only the week before, Monroney said nothing about it. *Congressional Record*, March 10, 1959, pp. 3357–65.

Symington attended the meeting at Monroney's invitation. Others present were General Tunner of MATS, Air Force Secretary Douglas, CAB Chairman Durfee, and Delos Rentzel of Slick Airways, listed this time as a professional adviser to Monroney's subcommittee. The memorandum for record indicates that Symington and Monroney still were unfamiliar with even the types of aircraft assigned to MATS, and still convinced that its principal job was the operation of executive-type aircraft for travel by government leaders. Both senators were upset at an administration decision to buy three new Boeing jets for the President's use. U.S. Congress, Senate, Committee on Commerce, *Study of Military Air Transport Service*, Hearings before Subcommittee, 85th Congress, 2nd Session (Washington: Government Printing Office, 1958), p. 29 (hereinafter *Study of MATS Hearings*); U.S. Congress, Senate, Committee on Appropriations, *Department of Defense Appropriations for 1960*, Hearings before Subcommittee, 86th Congress, 1st Session (Washington: Government Printing Office, 1959), pp. 1239–40 (hereinafter *SCA DOD Appropriations for 1960*).

14. Reproduced in *Congressional Record*, July 14, 1959, pp. 13293–94.

Quesada tied his program to what he called an "uncompromised" cargo aircraft. The difference between compromised and uncompromised cargo transports, he said, was like the difference between a bus and a truck, or between a pullman and a freight car.[15] The uncompromised cargo plane would have no task but to carry cargo. There was only one such aircraft in production; this was the Canadian CL-44, then being sold to U.S. companies with loans guaranteed by the Canadian Government.[16]

Two significant points emerged from Quesada's presentation. For the first time during the airlift controversy, Congress had to deal directly with technical arguments. Quesada distinguished between different types of cargo aircraft, and he expressed his own preferences in some detail. A second point was Quesada's initial interest only in cargo aircraft; he had nothing to say, at least in 1959, about the airlift of Army troops.[17]

In Quesada's scheme, purely military operation would be limited to "hard core" missions defined as those in "direct support of the execution of military emergency war plans."[18] The commercially operated National Air Cargo Fleet would do everything else, and it would replace CRAF in the bargain. Furthermore, said Quesada, he should manage CRAF—not the Commerce Department.[19] Quesada did not specify the size of the fleet he had in mind, but Monroney called attention to published estimates calling for three hundred civil and one hundred military transports.[20]

The first important support came in a lengthy speech in the House by Representative Charles Porter of Oregon.[21] He did not specifically mention Quesada, but he supported the program and called it the 1959 version of the "air merchant marine" pioneered by Democrats ten years earlier. Citing Gavin, Kissinger, and the Democratic party's Advisory Council, Porter argued that the new commercial fleet should be large enough to move three or four di-

15. U.S. Congress, House, Committee on Government Operations, *Military Air Transportation*, Hearings before Subcommittee, 86th Congress, 1st Session (Washington: Government Printing Office, 1959), p. 201 (hereinafter *Military Air Transportation, 1959*).

16. *Ibid.*, p. 203.

17. Quesada insisted that any attempt to make a single transport suitable for both cargo and passengers would result in a "very, very inefficient aircraft for cargo purposes." *Ibid.*, p. 206.

18. *Ibid.*, p. 202.

19. *Ibid.*, p. 213.

20. *Congressional Record*, March 9, 1959, p. A1858.

21. *Ibid.*, May 20, 1959, p. 7818.

visions to any part of the world. National airlift capacity must be tripled, he said, and this could best be done by expanding the airline industry. Moreover, he argued, the General Accounting Office should study each airlift requirement in detail and should prohibit MATS from carrying anything the airlines could handle.

General Gavin spoke up from retirement to second Porter's speech. Gavin praised the "excellent presentation" that had demonstrated Porter's "courage and willingness to lay the facts before the American people."[22] Gavin was joined by Brigadier General S. L. A. Marshall, military analyst of the *Detroit News*, who had been exchanging letters with Porter on the subject of airlift.[23]

Walter Reuther, president of the United Auto Workers, sent Defense Secretary McElroy a letter praising Monroney and Symington for their support of increased airlift capacity. Reuther promised labor's support if the administration sent the proposal to Congress. He added that such a program would provide badly needed assistance to the aircraft manufacturers, their unemployed workers, and the communities in which both groups lived.[24]

The Defense Department chose the same Holifield Subcommittee hearings to announce three new actions apparently designed to prepare for gradual reductions in military operation as the prelude to expanded commercial operations.[25] One decision restricted the daily utilization of MATS aircraft to a maximum of five hours, instead of the six-hour rate MATS had been trying to reach. The Air Force announced also that it never would permit the peacetime airlift capacity of MATS to exceed its 1959 level. As new aircraft joined the MATS fleet the Air Force would retire older ones, so as to keep MATS at its 1959 level. Because new transports would be larger and faster, the Air Force admitted that it might have to further reduce the daily utilization rate.[26]

A second decision was to acknowledge the need for drastic action to make sure that all of the earmarked funds would be spent. While MATS was being held within its ceiling, therefore, its transports would fly empty; the cargo left behind would be turned over to the airlines in an all-out attempt to spend the funds.[27]

22. *Ibid.*, May 28, 1959, p. A4562.
23. *Ibid.*, May 20, 1959, pp. 7821–22.
24. *New York Herald Tribune*, June 28, 1959.
25. An internal Air Force memorandum suggested that the decisions were made in conjunction with the anticipated Quesada program.
26. *Military Air Transportation, 1959*, p. 15.
27. *Ibid.*, p. 218.

The Air Force promised, finally, that it would use commercial airlines to move at least 10 per cent of all the cargo to be airlifted during the following year.[28] If competitive bidding brought the prices down, the Air Force added, the percentage would go up, for it promised to spend at least as much on commercial airlift as during the current year.

Quesada's external support was not very cohesive. Most of the praise came from liberal Democrats who took care not to openly compliment the head of an executive agency in a Republican administration. Some supporters emphasized Army needs, others spoke only of cargo, still others looked for economic assistance to faltering industries. Gavin and Marshall did not specify how a National Air Cargo Fleet could answer such Army demands as predesignation. Nevertheless, the promise of a new cargo fleet bearing no resemblance to commercial models then in service led Congress to reject the administration's 1959 request for strategic airlift modernization funds.

THE IMPACT UPON THE 1959 APPROPRIATIONS CYCLE

Congress was not immediately free to act upon the Quesada program, for he outlined it only in conceptual terms. He did not send draft legislation to Congress, and his Congressional allies were not ready to introduce legislation on their own. His outline implied that the program would take four or five years to mature, because a completely new aircraft had to be developed. His proposal also appeared after the administration had presented its budget. The net effect was to throw the budgetary process into considerable confusion. An unusual sequence of events, including a significant Senate vote on airlift policy, began with the House Appropriations Committee's annual hearings on the defense budget.

The Air Force asked for an initial purchase of ten jet transports; these would be modified versions of either the Douglas DC-8 or the Boeing 707, the most prominent of the commercial jets then in service. The military versions, described initially as "C-jets," would have swing-tails[29] and would be equally suitable for several purposes: (1) with their great range, they could bypass island bases

28. *Ibid.*, p. 17.

29. This refers to a hinged fuselage which can be folded back to permit quick, straight-in loading of cargo.

destroyed in general war and still meet their post-strike rendezvous with SAC bombers; (2) they could carry normal cargo which could be loaded straight-in through the tail section; (3) they could carry some missiles; and (4) they could airlift Army troops.[30] These jets would be larger in all dimensions and faster then the Canadian CL-44 (also a swing-tail) that Quesada had praised. Whatever the validity of all these specifications (any long-haul transport might have difficulty operating in an environment in which an enemy had knocked out island bases), the transports seemed to answer all the demands made by Quesada, Monroney, and Symington. None of them, however, supported the procurement of C-jets.

Quesada did not openly oppose the C-jets until later in 1959, but his Congressional allies did so and insisted there was no adequate substitute for the proposed air cargo fleet. The first major opposition to the purchase, however, came from Congressman Flood.

The Air Force, in making its presentations to the House Appropriations Committee, did not help its own case. For its yearly report on MATS operation the Air Force chose a lieutenant colonel, accompanied by the Assistant Secretary for Financial Management. This meant to Flood that the Air Force was not too interested in MATS.[31] The briefing concentrated on the financial management of the industrial fund, and the proposed C-jet purchase was largely ignored at the time the Committee had its attention focused on airlift.[32] In the absence of an effective argument, Flood concluded that C-jets were only the first step in an expensive conversion

30. U.S. Congress, House, Committee on Appropriations, *Department of Defense Appropriations for 1960*, Hearings before Subcommittee, 86th Congress, 1st Session (Washington: Government Printing Office, 1959), Part 5, pp. 584–85 (hereinafter *HCA DOD Appropriations for 1960*).

31. *Ibid.*, Part 4, pp. 912–67, especially pp. 913, 938–41.

32. General Tunner of MATS had been in Washington only the day before to appear before another committee studying transportation mobilization. Apparently because MATS did not usually appear before the Appropriations Committee, Tunner was "not asked" to appear. U.S. Congress, House, Committee on Armed Services, *Adequacy of Transportation Systems in Support of the National Defense Effort in the Event of Mobilization*, Hearings before Subcommittee, 86th Congress, 1st Session (Washington: Government Printing Office, 1959), p. 189. One result of Flood's reaction was that MATS has appeared before the Appropriations Committee in each of the subsequent years as an exception to the Air Force practice of presenting only its major combat commanders. MATS does not normally appear before the Senate Appropriations Committee.

program that would cost hundreds of millions. Whatever the C-jets could do, Flood argued, the airlines could do better, and he was determined to eliminate the funds.[33]

Committee members criticized the Air Force for having failed to spend the $140 million appropriated the previous year for new transports.[34] They were not swayed by an Air Force argument that it had conducted further research into the feasibility of swing-tails instead of immediately buying the transports then available. The Committee thus was generally opposed to the C-jet purchase and agreed with Flood. The House report concluded that the Air Force could do without the planes and that CRAF could handle any emergency.[35]

Airline interests may have had some influence on this decision. They accused the Air Force of planning to use new jets on scheduled passenger service, thus damaging the airlines still more. If the Air Force needed the jets to support SAC, argued the ATA, they should be located at the SAC bases and should not be used for other purposes.[36] A witness from the supplementals asserted that the Air Force had demonstrated once again its intention to force small carriers out of business.[37]

The Senate was, if anything, even more hostile to the Air Force request. Because the Quesada plan was introduced after the close of defense appropriations hearings in the House, Air Force Secretary Douglas tried to convince the Senate Appropriations Committee that the C-jets would become a part of the civil-military development program. The Air Force, said Douglas, originally had intended to buy many more than ten C-jets; it now had decided to reduce the total purchase to some fifty C-jets. There would be no massive purchases of transports until the new cargo workhorse appeared on the scene.[38] All of this had little effect; the Senate, following Monroney's lead, turned down the Air Force proposal.

Monroney described at some length what he was seeking to achieve. He wanted, first of all, a substantial increase in forced expenditure on commercial airlift, up to $150 million during the

33. *Aviation Daily*, May 6, 1959.
34. *HCA DOD Appropriations for 1960*, Part 1, pp. 88, 124–25.
35. U.S. Congress, House, Committee on Appropriations, *Department of Defense Appropriations for 1960*, 86th Congress, 1st Session, House Report 408 (Washington: Government Printing Office, 1959), p. 62.
36. *HCA DOD Appropriations for 1960*, Part 6, pp. 585, 598, 653–54.
37. *Ibid.*, p. 620.
38. *SCA DOD Appropriations for 1960*, pp. 1600–1.

following year.[39] He was much annoyed that MATS had spent only $71 million of the $80 million set aside the previous year. Monroney also left no doubt that he did not want the Air Force buying new jets for MATS: ". . . nobody loves jets better than I do. . . . But if this committee permits MATS to buy and begin equipping its fleet with pure jet transports, and this is what they are planning—they are not cargo planes, they are swing-tail transports—you will never be able to develop the reserve air cargo fleet that this nation needs so desperately."[40] He went on to explain the advantages of a new civil-military development program:

> . . . If the military would realize that they can have this emergency airlift for practically no cost whatsoever, by furnishing a seedbed for the creation of an air cargo industry, we would have in 2 or 3 years, and we have talked to enough airframe manufacturers, a cargo airplane that could be designed and built and operated, a fleet operated by the airlines, sold on Government guaranteed loans, perhaps on a 6 or 7-year payout, that would furnish us free of charge and free of cost the capacity needed to airlift two or three divisions anywhere in the world within 18 hours.[41]

Monroney added also that he saw a place for both the supplemental and scheduled carriers; he echoed a position taken earlier by Vice-chairman Chan Gurney of the CAB: ". . . the supplemental carriers have the capacity to handle traffic which is appropriate for commercial carriage to be moved over routes or between areas which no carrier is certificated to serve or over routes where there is a requirement in excess of the capacity of the carriers certificated to serve it."[42]

The Senate ultimately accepted its Appropriations Committee's recommendation that forced procurement be increased to $100 million, but only after a spirited floor debate which revealed the composition of the opposing forces in the airlift controversy.[43] Sena-

39. *Ibid.*, p. 1242.

40. *Ibid.*, p. 1241.

41. *Ibid.*

42. *Ibid.*, p. 1237; for Gurney's position see *Military Air Transportation, 1959*, p. 135. Gurney usually represented the more conservative and cautious approach, e.g., he voted against the 1955 decision declaring the supplemental carriers eligible for scheduled operation. The supplementals praised Monroney's leadership but protested ATA efforts to build a "Chinese Wall" around their established routes. *Ibid.*, pp. 1300–1.

43. The subcommittee recommended the $150 million requested by Monroney. The full committee's report indicated an amount of $150 million but,

tor Strom Thurmond of South Carolina tried to reduce the amount to $80 million on grounds that forced expenditures would weaken MATS and, thereby, national defense. He was joined by Senators Howard Cannon of Nevada and Barry Goldwater of Arizona, both Air Force reservists, who warned against weakening the nation's ability to maintain a strong supply line during emergencies. Goldwater added that supply lines would prove more important in smaller wars than in general war.[44]

Monroney, Sparkman, and Symington supported the increase in forced expenditures, each of them for a different reason. Monroney used Flood's argument; if the money was not spent and the new development program approved, Congress ultimately would have to spend hundreds of millions on MATS.[45] Sparkman favored the expenditure because he felt certain that small business would benefit from it.[46] Symington reversed his earlier argument that the country needed a new cargo fleet. He said this time that if the U.S. maintained control of the air, transports of any age would be satisfactory; all new jets purchased should go to SAC.[47] The vote on Thurmond's proposal was a tie and this, of course, defeated it.[48]

Most of those who might be termed liberal Democrats supported the increase in forced procurement, e.g., Joseph Clark of Pennsylvania, Thomas Hennings of Missouri, Hubert Humphrey of Minnesota, Estes Kefauver of Tennessee, John Kennedy of Massachusetts, Eugene McCarthy of Minnesota, Richard Neuberger of Oregon. Southern Democrats and other conservatives generally agreed with Thurmond, e.g., Everett Dirksen of Illinois, and Bourke Hickenlooper of Iowa. Sparkman, of course, was an exception to the general pattern and, in addition, some liberal Democrats opposed Monroney on grounds that he was not friendly enough to small airlines.[49] Yet another liberal Democrat condemned the "non-

---

on the floor, Chavez reported that the agreed figure was $100 million. U.S. Congress, Senate, Committee on Appropriations, *Department of Defense Appropriations for 1960*, 86th Congress, 1st Session, Senate Report 476 (Washington: Government Printing Office, 1959), p. 24; *Congressional Record*, July 15, 1959, p. 13170.

44. *Congressional Record*, July 14, 1959, pp. 13314–15.

45. *Ibid.*, p. 13315.

46. *Ibid.*, p. 13302.

47. *Ibid.*, p. 13307.

48. *Ibid.*, p. 13316.

49. Senator Clair Engle thought Monroney too restrictive. Engle was interested in yet another classification of carriers, completely outside CAB jurisdic-

sense" that had forced MATS to curtail its flying and turn over the cargo to commercial carriers; he felt obliged to end his record of supporting Monroney in all matters of aviation policy.[50] It remains generally the case, however, that the vote demonstrated the nature of the alliance between liberals and airline interests, and its opposition by conservatives usually friendly to business.[51]

Despite all of these congressional rebuffs, the Defense Department made one more attempt to obtain at least a part of the funds it had asked. Senator Cannon carried to the Appropriations Committee an Air Force request for a supplemental appropriation of $50 million.[52] Secretary Douglas indicated by letter that part of this amount would finance continued development of jet engines, but this fact was not made clear during Senate hearings on the request. Cannon outlined a broader Air Force plan to give MATS three types of strategic transports: (1) C-jets or their equivalent; (2) fifty C-133's for outsize cargo and missiles; and (3) several hundred cargo transports suitable for both civil and military use.

Monroney opposed the request, unaware that $30 million of the total was for engine development.[53] He discovered only later that the engines were intended for use on cargo transports already ordered by some airlines. He quickly reversed himself and tried to reinstate the $30 million while the supplemental money bill was in conference between the two Houses. Senate conferees agreed with Monroney but, for the House, Representative Albert Thomas insisted the item be passed over in favor of a new study of the airlift question during the next congressional session.[54] The conference report gave the first open indication that an important segment of Congress might step into the controversy:

---

tion, known as "contract" carriers. Engle wanted these included, even though Monroney argued that ninety-one companies currently met the criteria for MATS contract operation. Engle was joined by Senators Wayne Morse and Paul Douglas. *Ibid.*, p. 13297.

50. This was Senator John Carroll of Colorado. *Ibid.*, pp. 13304–5.

51. It might be added that the debate included no references to government spending as an issue. The earmarked funds were reduced in conference to $85 million. Public Law 86–166, Sec. 631.

52. U.S. Congress, Senate, Committee on Appropriations, *Supplemental Appropriations Bill for 1960,* Hearings before Subcommittee, 86th Congress, 1st Session (Washington: Government Printing Office, 1959), pp. 803–6.

53. *Ibid.*, pp. 807–20.

54. Senators on the conference committee reported they had made every attempt to get the House to agree to the appropriations. *Congressional Record*, August 19, 1959, pp. 15504–6.

It appears that the Air Transport Association of America is opposed to anything affecting the modernization or strengthening of the Military Air Transport Service. It also appears the Air Force is not ready to purchase new planes, but wants funds instead for further development of the turboprop T-61 engine which it has supported for several years and says requires about $30 million in 1960. To confuse the matter more, the Administrator of the Federal Aviation Agency, who is supporting a program of Government-insured loans for commercial carriers, states that these requested funds should be denied; that private industry should develop its own cargo plane, and intimates that the Military Air Transport Service should be on its way out.

It is suggested that any step which weakens or tends to weaken the Military Air Transport Service would be a serious matter. MATS has an important national defense function and must be preserved. It must be modernized, and its costs of operation reduced to a minimum through efficient and economical operation. To be caught in a national emergency without a working unit of MATS would be like not having guns or planes or ammunition under the same circumstances.

It is disturbing that the Air Transport Association actively enters into the MATS picture every year, and on every other phase of aviation that even remotely affects its interests. Civilian aviation has grown by virtue of the generous subsidy it has received from the Government. It is now a giant grown fat by Government subsidies and high rate charges.

The new Federal Aviation Act continues to promote its progress. The act is extremely broad, and a careful reading will disclose that authority exists therein for the Administrator of the Federal Aviation Agency to do practically everything to support and promote air commerce. This is indicated by Section 312 (b) which even authorizes him to undertake research, development, and service testing of aircraft, aircraft engines, propellers, engines, and other appliances whether these be civilian or military. No funds have been provided for such purposes yet, and the inclination is to withhold any such funds from the Federal Aviation Agency. If the FAA were to enter into this field, it would appear to be an unnecessary duplication and expense in view of the fact that the Air Force, Navy, and National Aeronautics and Space Administration, the successor agency to the National Advisory Committee for Aeronautics, have been given that specific job through the years.[55]

55. U.S. Congress, House, Committee on Appropriations, *Supplemental Appropriation Bill 1960*, 86th Congress, 1st Session, House Report 943 (Washington: Government Printing Office, 1959), pp. 7–9.

This quite critical report and the denial of engine development funds prompted two immediate reactions. Monroney, defending the ATA in a Senate speech, described it as his partner in a joint attempt to build an air merchant marine. He wanted his speech to demonstrate Senate disagreement with the conference report, but Cannon and Thurmond immediately disagreed.[56] A second reaction came from the industry itself, and demonstrated the significance of the deleted funds. Two airlines, Pan American and Slick, canceled their orders for twenty-three new cargo transports then being developed by Lockheed, because the engines would not be available.[57]

The 1959 appropriations cycle thus produced a stalemate that stopped strategic airlift modernization and had a substantial effect upon commercial carriers. This new impasse resembled somewhat the 1934–38 policy vacuum, but there were important differences. Unlike the earlier vacuum, Congress created this one itself; it rejected administration plans to begin modernizing MATS but proposed no alternative of its own. The administration's performance was no better. Quesada's proposal, as of 1959, was not part of the administration program, and it quite obviously contradicted the over-all Defense Department program. The C-jet purchase was part of the official budget, but one gathers the impression that the administration had no real position; each agency was free to present its own, and Congress was left to referee the fight.

Given the problems of lead times for developing and procuring aircraft, the 1959 impasse is an important item of airlift history. It became inevitable that by the mid-1960's there would be a notable gap in strategic airlift capacity. There is no intention here to apportion blame or fix responsibility. It seems clear enough, however, that leading congressional Democrats were just as willing to accept a vacuum in 1959 as the administration had been in 1934.

INTEREST GROUP ACTIVITY: THE ATA, THE INTERNATIONAL AIRLINES, AND THE SUPPLEMENTALS

It was clear to all of the competing interest groups that some action probably would have to be taken in 1960, for continuous denial of appropriations would not suffice. The combination of a

56. *Congressional Record*, August 19, 1959, p. 15504.
57. *Aviation Daily*, July 22, August 19, October 22, 1959; *St. Louis Globe-Democrat*, July 31, 1959; *New York Times*, October 21, 1959.

policy vacuum and near certainty of new action prompted these groups to intensify their efforts to gain support. The climate of the airlift controversy became increasingly bitter, a development best demonstrated by studying the activities of the most important groups at work.

The major companies apparently concluded that political conflict had not proved too satisfactory and that new approaches were in order. Government expenditures for overseas commercial airlift had increased from $14 million in 1955 to $71 million in 1958, an increase traceable to congressional pressure and to greater demands for airlift during the 1958 crises in Lebanon and the Taiwan Straits.[58] The distribution of the increase, however, was not to the ATA's liking, especially since it had been the author of the 40–20 provision. Congress had shifted the emphasis to small businesses, and they had become the chief beneficiaries.[59]

Attempts by the major companies to redress what they thought an undesirable balance proceeded along two lines. One was a continued effort by ATA to speak as the single voice of all its members, an effort difficult to sustain because of the divergent interests of ATA's cargo airlines. The second was an attempt by all the international carriers to negotiate directly with the Defense Department for an exclusive contract to airlift all the military traffic turned over to the industry.

The ATA's principal contribution was a rather typical lobbying campaign to highlight the allegedly unwarranted expansion of MATS and to convince Congress to halt any more of the same. ATA gave wide distribution to a statistical compilation of the traffic handled by MATS and the scheduled carriers. MATS, said ATA, flew 62 per cent of the cargo airlifted in the Atlantic area, foreign airlines carried 21 per cent, but U.S. international companies were left with only 17 per cent.[60] Passenger distribution was

58. These amounts are only those spent by MATS for overseas charters. The total amount spent by the Defense Department for all types of commercial airlift was much greater, e.g., $254 million in 1958. *Study of MATS Hearings*, p. 7.

59. The Air Force and the Small Business Administration had agreed to 25 per cent as an equitable amount to be set aside for small companies, defined in this instance as airlines having fewer than five hundred employees. The Air Force then decided that airlift was the type of business that could more easily be given to small companies; thereafter, 35 per cent became the Air Force target. *HCA DOD Appropriations for 1960*, Part 5, p. 526; *Military Air Transportation, 1959*, pp. 57–58.

60. *Aviation Daily*, August 31, 1959.

25 per cent for MATS, 44 per cent for foreign airlines, but only 31 per cent for U.S. companies. The ATA, however, included within the MATS figures all of the traffic handled by domestic cargo airlines and the supplementals.

A trade journal usually sympathetic to the ATA pointed to the calculations as proof of the extent to which the "state airline" had penetrated the commercial market.[61] The *New York Times* used the same figures in an editorial urging a substantial reduction in MATS.[62] Chairman Celler of the House Judiciary Committee, a long-standing opponent of the major airlines, then agreed with the *Times* that MATS was too large and should stop competing with the airlines.[63] ATA's calculations thus had the desired effect of producing additional support for reducing MATS and helping the airlines, even though Celler and the *Times* hardly would have advocated the exclusion of small business.

Operating outside the ATA, the four scheduled U.S. international carriers approached the Defense Department in 1959, and again in 1960.[64] They sought designation as the prime contractors for all the commercial airlift used by the military, and they agreed to negotiate a special tariff rate below the prevailing commercial level but considerably higher than the rates then being paid through open competitive bidding. The four airlines agreed to buy modern aircraft, to provide additional airlift during emergencies, and to subcontract business to other airlines when that proved necessary. They tried to demonstrate that their proposal would save millions each year, by comparing it against normal commercial rates. This ignored the low rates actually being paid by the government and was, therefore, something of a distorted calculation.

As had occurred with the ATA calculations, the four-carrier proposal gathered some editorial support, this time from those unaware of both the distortions and the status of the supplemental carriers. The *Christian Science Monitor* called the proposition a "unique and far-reaching proposal for reducing tariffs."[65] Another

61. *Ibid.*, September 1, 1959.
62. *New York Times*, September 19, 1959.
63. *Aviation Daily*, September 28, 1959.
64. The companies were PAA, TWA, Seaboard World, and Northwest. The proposal is reproduced in U.S. Congress, House, Committee on Armed Services, *Hearings before the Special Subcommittee on National Military Airlift*, Hearings, 86th Congress, 2nd Session (Washington: Government Printing Office, 1960), pp. 4618–28 (hereinafter *Rivers Committee Hearings*).
65. *Christian Science Monitor*, November 12, 1959.

newspaper concluded that the government could save some money, and yet another said the suggestion was "too sensible to pass up."[66] Chairman Celler added his approval to the proposition.[67] Had Celler and other supporters been aware of the proposal's immediate cause, they might have been less eager to support it.

The international companies approached the Defense Department shortly after losing their share of the military market to the supplementals who took full advantage of the government's contracting procedures. Beginning in 1958, MATS awarded contracts each October for the following year's traffic. PAA and TWA captured 40 per cent of the business in October of 1958, but the procedure then in use called for new bids each year; the major carriers lost out in 1959.[68] A supplemental carrier made exceedingly careful plans to capture the military market in 1959, making conditional arrangements to lease some used DC-7 transports should the company be successful.[69] Fortified with these arrangements, the small company submitted the lowest bid and suddenly became the largest of the military contractors. The four-carrier proposal appeared shortly thereafter, at about the time TWA's president was complaining that the small company's bid was well below the costs of its operation.[70] The four-carrier proposal was a device planned to recapture the market for the international carriers who, significantly enough, did not mention Quesada's plan in connection with their own.

For their part, the supplementals supported the Quesada plan because of the natural appeal of guaranteed loans to carriers having neither established routes nor authorized subsidies. When it seemed that Air Force Secretary Douglas might be attempting to block administration approval of the plan, a trade journal close to the supplementals bitterly attacked him. It described him as a "quiet, reserved, ordinarily conservative" individual who had been

66. *Washington Daily News,* November 13, 1959; *Denver Post,* November 22, 1959.

67. *Washington Daily News,* December 11, 1959. As chairman of the committee that had investigated the airlines, had been unfriendly to the major carriers, and had chastised the CAB for being too stringent in regulating the supplementals, Celler had been one of the strongest supporters of the carriers who stood to lose the most from the four-carrier proposal.

68. Hiram C. Caroom, "Management Responsibilities and National Interests: An Analysis of the Practicability of the U.S. Flat International Passenger Airlines' Transporting More Military Traffic in Peace and War," (unpublished doctoral dissertation, Harvard University, 1959), p. 23.

69. *Aviation Daily,* August 26, 1959.

70. *Aviation Week,* November 16, 1959.

"taken in" by MATS, and converted to "state socialism . . . Government encroachment on private business . . . progressing toward Soviet ideology."[71]

INTEREST GROUP ACTIVITY: THE AIRCRAFT MANUFACTURERS

Of all the interest groups attempting to influence the outcome of the airlift struggle, the aircraft manufacturers were perhaps the most experienced in lobbying as such. All such manufacturers seek to keep both congressmen and key officials of the executive branch constantly up-to-date on the particular advantages of specific items of equipment they can produce.

Important congressmen became the targets of numerous briefings about the jet and turboprop transports then becoming available. Monroney had a long series of meetings in 1959 with both airline officials and manufacturers' representatives.[72] Holifield was impressed by briefings on the Canadian CL-44 (manufactured by a subsidiary of General Dynamics) and the Lockheed GL-207.[73] The briefings emphasized the advantages of turboprop engines and the ability of transports so equipped to use shorter runways than those needed for pure jet transports. This made the Air Force appear to be incorrect in its decision to buy C-jets, and the Holifield Subcommittee so reported.[74] There is no evidence that the Subcommittee explored the subject with the Air Force; it had advised the House Appropriations Committee that the C-jets would have a new type of engine suitable for use on shorter runways.[75]

Some manufacturers set up a complex network of financial relationships with the supplementals, arranging for the latter to operate the piston aircraft discarded by the major companies. The supplemental airline (Overseas National) that captured the military market in 1959 made its conditional leasing arrangements with a subsidiary of General Dynamics.[76] By adopting the relatively new practice of taking old transports as trade-ins, the man-

71. Editorial in *Airlift Magazine*, September, 1959. Report of Douglas' activity is in *New York Herald Tribune*, July 27, 1959.

72. *Aviation Daily*, May 8, 1959.

73. *Military Air Transportation, 1959*, p. 64.

74. U.S. Congress, House, Committee on Government Operations, *Military Air Transportation*, Tenth Report, 86th Congress, 1st Session, House Report 1112 (Washington: Government Printing Office, 1959), p. 13.

75. *HCA DOD Appropriations for 1960*, Part 5, p. 653.

76. The conditional arrangements involved only aircraft. It was up to Overseas National to hire pilots and mechanics, and to set up its organization after getting the contract. *Aviation Daily*, August 26, 1959.

ufacturers placed themselves in a strategic position between the supplementals and the scheduled carriers. As the parent companies for the supplementals, the manufacturers profited from the old transports; this forced the major carriers to advocate using modern aircraft on military contracts, a position they initially had tried to avoid.

It might be noted that these arrangements quietly reinstated a relationship between operators and manufacturers that the Roosevelt administration had condemned in 1934. These more recent connections, some of them still in effect, have gone largely unnoticed; they are, for one thing, quite obscure, and they are hidden behind the facade of small business operation.[77]

INTEREST GROUP ACTIVITY: THE AIR FORCE, MATS, AND THE MILITARY LOBBIES

Lobbying activities by military groups did not form a monolithic pattern. The Air Force staff was divided, for example, with most of it not wholly in accord with Secretary Douglas' support of MATS. Outside groups which usually act as unofficial military spokesmen viewed the controversy as a misguided attempt by business groups to weaken a necessary military function. MATS also undertook a defense of its own, generally designed to increase the use of airlift for military traffic, thus providing enough business for both military and civil transports. Of these activities, those within Air Force headquarters were perhaps the most complex.

Air Force decisions to put a ceiling on MATS flying and to buy only a few C-jets indicated that the Air Force was ready to have industry replace at least a part of the strategic airlift force. The

77. One vice-president of Lockheed described the financial arrangements as "very complicated," with manufacturers taking trade-ins and also helping customers finance their new aircraft. U.S. Congress, Senate, Committee on Commerce, *Government Guarantee of Loans to Certain Air Carriers*, Hearings before Subcommittee, 86th Congress, 2nd Session (Washington: Government Printing Office, 1960), p. 255 (hereinafter *Guarantee Loan Hearings, 1960*). The president of a small cargo airline said later that, among others, General Dynamics, Douglas, Lockheed, and Whitworth Clouster had made such arrangements with operators. He added that "whether this verges on a new form of indirect control, unforeseen in the making of aviation laws . . . are matters I shall not pursue. . . . There is a complex tangle of interests at work here." U.S. Congress, Senate, Committee on Commerce, *Aircraft Guaranteed Loan Program*, Hearings before Subcommittee, 87th Congress, 2nd Session (Washington: Government Printing Office, 1962), p. 81 (hereinafter *Guarantee Loan Hearings, 1962*).

remaining question was the extent to which the military ultimately would support Quesada's program, because he wanted to give the commercial fleet first priority on modern transports. Many in the Air Force looked to airlift only as an auxiliary force for SAC, and it seemed clear enough to them that less and less airlift would be needed as SAC converted to intercontinental ballistic missiles.[78] Seldom has a prediction been wider of the mark. The Air Force, like the Navy, wanted also to keep a few small packages of specialized airlift for its own use and, also like the Navy, was less than an enthusiastic MATS supporter.[79] The military side of the Air Force staff was thus somewhat disposed to accept the Quesada plan.

Secretary Douglas, on the other hand, hesitated to join forces with Quesada. In a White House meeting held to discuss the program, Douglas opposed the implication that MATS must stop modernizing its own fleet.[80] Reports had it that Douglas' opposition "shocked lower elements of the Air Force staff who were in agreement with Quesada."[81] Douglas thus found himself under attack from within the Air Force staff, from those close to the supplementals who thought he had accepted a socialist philosophy and, finally, from Monroney who, in a speech attacking Douglas' hesitancy about the Quesada program, said this:

> . . . let me say a word of personal thanks and encouragement to all those in the Air Force, the Department of Defense, the civilian agencies of the government, and the aviation industry, who know these things to be true, and have fought to hurry the changes which must come.
>
> You have been threatened, insulted, frustrated, and vilified because you have defied the most precious altar—the military

78. Caroom, *op. cit.*, p. 303.

79. One typical Navy argument was that the fleet always had specialized needs that a generalized force like MATS could not handle. *Navy Times*, October 21, 1959. A typical Air Force argument was that the specialized Logistics Support Squadrons of the Air Materiel Command were the only ones capable of airlifting nuclear weapons to overseas Air Force units. *Rivers Committee Hearings*, pp. 4383–86. It was left to congressional committees to point out that MATS carried nuclear weapons for the Army, because it had no airlift of its own, and that MATS ought to be able to do the same job for the Air Force itself. H. Report 1112, 86th Congress, 1st Session, pp. 25–27.

80. Attendees were Quesada, Douglas, CAB Chairman Durfee, Commerce Secretary Mueller, and Presidential Assistant Gerald Morgan. *Aviation Daily*, August 19, 1959.

81. *New York Herald Tribune*, July 27, 1959.

> status quo. But you labor in the nation's interest, and, because of your efforts, some day the word "military" in the name "Military Air Transport Service" will no longer describe an airline's customers, but will again mean "of and pertaining to soldiers, arms of war."[82]

The more traditional military lobbies and MATS took a more simplified approach. Military trade papers announced that MATS was "battling for its life," and asked if Congress wanted to "kill" the command.[83] The president of the Reserve Officers' Association condemned the "vicious" campaign against MATS and suggested adding transport wings to the Air Force reserve as something of a countermove.[84] The journal of the Air Force Association published long editorials ridiculing industry charges of "state socialism."[85] Speaking for MATS, General Tunner advocated the shifting of more military traffic from sealift to airlift, a step calculated to provide enough business for both MATS and the airlines.[86] MATS employed outside advertising consultants to improve its image and, in the words of a Tunner aide, tried to "sell airlift like a Fuller Brush man."[87] Tunner made a round of speeches on the subject, and made his own reply to unfriendly editorials in the *New York Times*.[88] While he carefully put himself on the side of a joint civil-military development program, the MATS campaign gave every indication of a deeply felt struggle for organizational survival and identity.

THE COLLAPSE OF THE QUESADA PLAN

The Quesada plan for a National Air Cargo Fleet came to a test early in 1960, and the alliance could not agree within itself on a

82. *Congressional Record*, August 25, 1959, pp. 15510–11; *Aviation Daily*, August 26, 1959.

83. *Army Navy Register and Defense Times*, December 12, 1959; *Army-Navy-Air Force Journal*, July 18, 1959.

84. *Washington Post*, November 1, 1959.

85. *Air Force Magazine*, October, 1959.

86. MATS News Release, May 24, 1959. Even industry witnesses commented on the Army's hesitancy to use airlift for its normal transportation needs. *HCA DOD Appropriations for 1960*, Part 6, p. 587.

87. *Wall Street Journal*, October 15, 1959.

88. Reports of speeches in *Aviation Daily*, September 8, 1959, *Congressional Record*, September 8, 1959, pp. 17869–70; *New York Times*, October 12, 1959. One trade journal described MATS activity as "frightening" in its implications, something "like the TV scandals." *Aviation Daily*, November 23, 1959.

new airlift program. It became surprisingly clear that those supporting the program were not aware of the complex interests at work. The CAB demonstrated an inability to understand the implications of its suggestions. Monroney was left with a feeling of bitterness toward the airline interests he had expected to support him. Before the hearings began, it had seemed to Monroney, Quesada, and the CAB that the program was certain to be passed.

Between congressional sessions Monroney made several important speeches on the subject, predicting quick passage after the start of the new session. He called the new merchant marine a "lightning-fast left" to accompany the "Super Sunday Punch of the big bombs"; the fleet would be able to go anywhere on six hours notice, thus solving the problems that had forced three generals to leave the Army.[89] Others noted that Quesada was the President's principal aviation adviser and thus had the administration's support as well as that of the Democrats and organized labor.[90]

Yet, well before Congress convened there was reason to believe the Quesada program would face heavy going. The CAB found it quite difficult to solicit support from within the airline industry. Cargo and supplemental carriers seemed ready to support Quesada, but the major companies were not. PAA and TWA wanted to develop their own modernization plans.[91] As the CAB polled other companies, they said about the same thing; if economic studies indicated a profitable market was available for cargo transports, they could find private financing. If any company used private financing, so went the argument, it would be unfair to give any of its competitors the favored treatment of guaranteed loans.[92]

Monroney and Symington, still convinced passage would be relatively easy, introduced the Quesada program shortly after Congress met in 1960.[93] Eight days of hearings in February proved

89. Excerpts from speech at Wings Club Banquet, New York, October 19, 1959, distributed by Monroney's office.

90. Editorial in *Airlift Magazine*, June, 1959. Some reports put the program's cost at $1 billion, with government guarantees covering $750 million. *Wall Street Journal*, July 23, 1959; *St. Louis Post-Dispatch*, July 5, 1959. Using $5 million as a reasonable figure for one airplane, this sum could not have produced more than two hundred aircraft.

91. *Forbes Magazine*, June 15, 1959.

92. *Aviation Daily*, August 31, September 1, 1959; *New York Times*, September 6, 1959; *Business Week*, September 12, 1959. The CAB had encouraged the four international carriers to make their separate approach to the Department of Defense. *Wall Street Journal*, August 31, 1959.

93. *Wall Street Journal*, January 13, 1960.

embarrassing to Monroney, for he was left without substantial support for his program. He discovered that Quesada's only supporter within the administration was the Defense Department.[94] The Commerce and Treasury Departments argued that the logical course was to develop the market and to shift more traffic from military to civil carriers.[95] The CAB tried without success to show that most of the airlines supported the program, but Monroney's colleagues demonstrated that the CAB's testimony was misleading at best.[96] Practically all of the trunklines opposed the program, and even the cargo airlines were not unanimous in support.

The major airlines, for example, held to their position that guaranteed loans would corrupt the regulatory system. The CAB, said a representative of American Airlines, would feel obliged to favor companies flying aircraft bought with government loans, if only to protect the government's vested interest.[97] The airline representative continued:

> We do not believe guaranteed loans will bring about the desired national objectives or that they are necessary. The problem is not one of financing. It is lack of a low operating cost cargo system, including an airplane which the airlines will want to buy.
>
> Once a good enough airplane is available, credit will be available to companies with a good enough business record to buy it because there will be convincing evidence that it will operate profitably and lending institutions will thus have the normal business assurance. . . .
>
> In effect, we feel we have something of the cart before the horse. We are considering methods of providing for capital funds for the purchase of a cargo plane, but we have in this legislation no methods which will assure that such a plane will even be for sale.[98]

The same witness also argued against developing a cargo plane without taking into account the entire airlift system that must accompany such a transport. This was a viewpoint seldom heard

94. The Defense Department was represented by its traffic managers in the Office of the Assistant Secretary for Installations and Logistics. They asked only for assurance that the aircraft be allocated to CRAF and be available in emergencies. *Guarantee Loan Hearings, 1960*, pp. 116–17.

95. *Ibid.*, pp. 409, 431.

96. *Ibid.*, pp. 28–31. Monroney apparently intended to complete the hearings within a week; other senators disagreed, asking that enough time be made available to permit opponents to appear.

97. *Ibid.*, p. 511.

98. *Ibid.*, pp. 513–14.

during the controversy; the ATA spoke in broad economic and political terms, and the supplementals had no experience with large airlift systems. The viewpoint is worth noting here:

> The system is not available yet. It involves mechanized terminals, it involves mechanized loading facilities, and it is probably going to mean that those facilities should fit all airplanes. It involves processing and distribution costs. It involves automation. . . .
>
> As an example of what the mechanized loading involves, here is a figure that may startle you. . . . We figure now that it takes a half-hour to load 10,000 pounds into a DC-6A. The present method is doing it by hand—a half-hour for 10,000 pounds. It will take three hours to put that into a cargo jet by the present methods. You cut the flying time in half, and you increase your loading time by six times. It doesn't make sense. You've got to have both a system and an airplane.[99]

Monroney lost additional ground when he tried to prove that the program would encourage vast increases in the air cargo market. Professor Stanley Brewer of the University of Washington apparently was expected to predict such increases; he proved instead to be a cautious witness, and he advised against overestimating the future of air freight. The market would include, he said, only about a dozen major cities within the U.S. and, because oil, coal, and other ores would move by surface for years to come, the market would grow at a modest rate. Brewer added a significant note of warning; businessmen were becoming hesitant to use air freight because they could not rely upon it. If the cargo airlines continued removing aircraft from scheduled operation to fulfill military contracts, the market never would grow.[100]

Monroney was confronted, then, with disunity among the cargo and supplemental airlines, opposition from the ATA and its members, confusion within the CAB, opposition from important government agencies, and a lack of interest in the House (nobody introduced a companion bill). He had no choice but to abandon the entire project, a decision that left him quite bitter. He announced he was "sick and tired of the airlines' inability to look beyond their nose"; he could wait no longer, he said, and would support immediate modernization of the MATS fleet.[101] Almost

99. *Ibid.*, p. 510.
100. *Ibid.*, pp. 391–96.
101. Allan Cromley in *The Daily Oklahoman*, Oklahoma City, February 28, 1960.

overnight, Monroney became a MATS supporter, and he played an important part in appropriating airlift funds in the 1960 session. He did not abandon permanently his fondness for a guaranteed loan program; it will be seen shortly that he tried in 1962 to secure passage of a modified version of the Quesada plan.

The sudden collapse of the National Air Cargo Fleet meant that some new approach was needed if the airlift stalemate was to be resolved. Whatever the previous opposition, MATS probably would have to modernize itself. Monroney was ready to vote for that, but somebody besides the Commerce Committee would have to supply the initiative. The second phase of congressional activity thus ended with the vacuum unfilled.

SIGNIFICANCE OF THE SECOND PHASE

During this single year of intense activity, the airlift controversy became an almost incredible web of interlocking interests that seemed to defy attempts to resolve the struggle. Some aspects of this important phase deserve more extended analysis.

One is the difference between the intellectual approaches of Cherington and Elliott and the loan program ultimately introduced in Congress. Cherington focused his attention upon the major trunklines because, he said, other carriers made up only 5 per cent of the market and were too specialized for his purposes.[102] When he and Elliott recommended guaranteed loans in conjunction with military airlift problems, however, they entered a political arena dominated by the supplemental carriers. Had the economic analysis included the relevant political considerations, it might have shown that guaranteed loan legislation could pass only if endorsed by the supplementals. But this, in turn, could only force the major carriers to oppose the program out of fear that the program would open the door to an unregulated airline system. While Cherington hoped to alleviate the problems of 95 per cent of the industry, therefore, the proposal ultimately became one favorable only to the other 5 per cent. The Elliott memorandum was defective, in retrospect, because of its failure to account for the international ramifications of its argument.

Elliott, Cherington, Quesada, and Monroney apparently hoped that the new fleet would occupy the international air cargo market, although Elliott mentioned both cargo and passengers. To the ex-

102. *Cherington Report*, p. 10.

tent that the program's supporters spoke in numbers, Elliott advanced a figure of one hundred transports while Monroney implied a need for three hundred. Yet there could not be enough business to and from underdeveloped countries to keep this many airplanes busy, and the U.S. then would have to use the new fleet in the normal international market. In the absence of any explanation as to why other countries would grant enough new route concessions, one can conclude only that the program's supporters envisaged some sort of international price-cutting war, to be followed by abandonment of both bilateral negotiation and the IATA rate structure.

In this instance a more thorough analysis probably would have shown that the cause of airline difficulties was not a U. S. failure to produce suitable transports. Plenty of transports had been available all along; foreign airlines had bought them with Export-Import Bank loans.

The Quesada program's proponents also left important military questions unanswered. The new airplanes would be primarily for cargo, said Quesada, so as to reduce the operating costs. As Professor Brewer pointed out, businessmen would not rely upon such cargo service if they felt the airplanes would be withdrawn for military emergencies. If, as Elliott, Monroney, and Symington suggested, the new transports were the answer to the Army's problems, the situation was even less clear. The new fleet would only replace MATS, and the Army and its supporters already had renounced any fleet not kept constantly available for Army use. Transports fully employed in cargo service, not predesignated for Army use, not confined to the U.S., and not equipped to drop paratroopers, hardly could satisfy the Army's demands. If the transports ever were to airlift passengers of any sort, moreover, this would call for extra facilities and extra costs within the aircraft, e.g., seats, washrooms, pressurized cabins, and kitchen equipment.

There appears to have been no direct relationship, finally, between the announced objectives of the Quesada plan and the legislation introduced to achieve them. Everybody spoke of developing a new transport of vastly improved design, but Monroney introduced a bill dealing only with guaranteed loans. It required only that the loans involve "newly developed" aircraft "designed exclusively for the carriage of property or mail."[103] Because the draft statute did not provide for a research and development pro-

103. S. 2774, 86th Congress, 2nd Session, Sec. 2(d).

gram, one is left to conjecture how to interpret such a legislative provision in the practical world of day-to-day administration. Strict interpretation might have delayed implementation for five years or so, for it takes that long for a "newly developed" aircraft to appear.

It remains difficult, even in retrospect, to account for the affiliations between Quesada, Symington, Monroney, and Gavin, tenuous as those affiliations were, and to understand all of the arguments each advanced. Quesada apparently had broad ambitions for his new agency, to include managing CRAF and research and development programs connected with it. While it is not unusual for an agency head to ask Congress for more funds than allocated in the president's budget, it is somewhat unusual for an agency chief to oppose the requests already approved for another agency, or at least to do so in such a direct manner.[104] Quesada, Monroney, and Symington criticized the lack of airlift and saw dangers to national security, but they were willing to cut off all appropriations for improvement and, perhaps, to wait five years for a new airplane.[105] Gavin's intervention lent a note of military authority to the air merchant marine proposal, but it also appears in retrospect that Gavin had begun his new role as an active Democrat. He announced in his book that the Army, Navy, and troop carrier elements of the Air Force were liberals, and usually associated with the Democratic party: the other elements of the Air Force were conservatives who usually associated with Republicans.[106] To support the proposed program, Gavin suddenly had to abandon the theories he had advanced so forcefully during his congressional appearances on behalf of the Army. He supported

104. It can be admitted, of course, that when one military service or another indicates to congressional committees that its budget should be increased, the statement suggests that Congress should reduce the budget of another service.

105. The opposition to C-jets remains difficult to understand, given Quesada's praise for the Canadian CL-44. Even the president of the airline buying the CL-44 said the C-jets would be far better strategic transports; he had bought CL-44's because they fit his domestic route pattern. Testimony of Robert Prescott in *Guarantee Loan Hearings, 1960*, pp. 447–48.

106. James M. Gavin, *War and Peace in the Space Age* (New York: Harper & Brothers, 1958), pp. 248–53; Taylor said almost the same thing but did not use party labels, in Maxwell D. Taylor, *The Uncertain Trumpet* (New York: Harper & Brothers, 1959, 1960), pp. 102–3; the same categories also have been described as the non-SAC "pragmatists" and the SAC "absolutists." Morris Janowitz, *The Professional Soldier* (Glencoe, Illinois: The Free Press, 1960), pp. 273–75.

Quesada and Monroney, moreover, at just about the time Army Secretary Brucker was insisting that what the Army needed most was airlift predesignation.

1962: A POSTSCRIPT ON GUARANTEED LOANS

After announcing on several occasions that he still favored government loan guarantees for cargo transports, Monroney attempted in 1962 to add the cargo airlines to the legislation which authorizes such loans for local service "feeder" airlines.[107] While unsuccessful, the attempt revealed that significant changes had occurred within government agencies as the result of the 1960 election, and demonstrated also that guaranteed loan proponents still were unclear on the civil-military relationships at stake. The immediate occasion for this new attempt was the renewal of legislation initially enacted in 1957 to help the feeder airlines.

In supporting an amendment by Senator George Smathers of Florida to permit guaranteeing loans made to cargo airlines, Monroney had by 1962 secured support within the Commerce Department, the Air Force, and the Defense Department.[108] His opposition this time came from Najeeb Halaby, Quesada's successor as head of FAA, who looked upon cargo and supplemental carriers as members of a single group. Special treatment for only a few members of the group was illogical, said Halaby; four cargo airlines could not alone provide a substantial increase in the national mobilization base.[109] This attempt to extend guaranteed loans to the cargo airlines attracted little public attention, but it very nearly became law. While Halaby was doubtful, the new proposal got the support of an FAA task force appointed in 1961 to make yet another study of national aviation policy.[110]

The chairman of the House Aviation Subcommittee used Halaby's arguments to prevent passage of the amendment.[111] Congressman Williams argued that only airlines already declared by Congress to be eligible for subsidy should be eligible for guaranteed

107. Accounts of Monroney speeches are in *Aviation Daily*, November 15, 1960, January 6, 1961.

108. *Guarantee Loan Hearings, 1962*, pp. 151–54, 187–89.

109. *Ibid.*, p. 175.

110. U.S. Federal Aviation Agency, Task Force on National Aviation Goals, *Project Horizon* (Washington: Government Printing Office, 1961), p. 190 (hereinafter *Project Horizon*).

111. *Congressional Record*, September 10, 1962, p. 17804.

loans. He pointed out that Congress had placed the feeder airlines in this category, and he thus had no argument with the basic legislation. It would be dangerous, said Williams, to give indirect subsidies to carriers not eligible for the direct variety.

To Monroney and CAB Chairman Alan Boyd, also a supporter of the Smathers amendment, subsidies and guaranteed loans were not directly related. Boyd reported there had been no defaults on either principal or interest in $37 million in guaranteed loans made to local service airlines since 1957; the government thus had incurred "no liability."[112] Elsewhere in his testimony, however, Boyd noted that subsidy payments to the feeder companies had increased from $28 million in 1956 to $71 million in 1962.[113] The CAB estimated in addition that local service subsidies would continue to rise until at least 1964.[114] Neither the CAB nor Monroney, then, was ready to admit that the government might be paying off the guaranteed loans by means of the subsidy payments. Only Halaby took an opposite view, explaining that any extension of guaranteed loans to cargo airlines probably would make the subsidy question exceedingly important.[115]

Without an authorization for explicit subsidy, only a vast increase in the amount of military traffic assigned to the cargo carriers could have protected the government's new investment of guaranteed loans. Whether generally understood or not, guaranteed loans to feeder airlines are possible only because there is corollary machinery (direct subsidy) which lets the government pay off its own loans.

112. *Guarantee Loan Hearings, 1962*, pp. 3–5.
113. *Ibid.*, p. 112.
114. *CAB Annual Report, 1962*, p. 25.
115. *Guarantee Loan Hearings, 1962*, p. 164. Senator Clair Engle had pointed out this factor in the 1960 hearings, but had gotten little response from the CAB witness or from Monroney. *Guarantee Loan Hearings, 1960*, pp. 24–27.

The value of the guaranteed loan programs remains in doubt. Monroney insisted that subsidy payments to the feeder airlines had been lowered as a result, but they actually rose constantly. *Guarantee Loan Hearings, 1962*, p. 2. The Project Horizon task force advocated guaranteed loans for cargo airlines while admitting loans had proved ineffective for the feeder airlines. *Project Horizon*, pp. 169–70.

[9]

The 1960 Breakthrough in Military Airlift

CONGRESS BUILT its own military airlift program in 1960, a task performed, it is concluded here, with dispatch, logic, and reasonableness. How Congress did that job, the elements of the new consensus, and the limitations of that congressional accomplishment are subjects of this chapter.

The legislative branch enacted both a program and the funds necessary to implement it well before the start of the 1960 presidential campaign, largely through the efforts of legislators having little direct interest in election year politics. The Eisenhower administration accepted practically the entire program, without publicity and apparently without reference to the campaign, and took most of the necessary administrative actions before leaving office. The new program probably was the best available under the circumstances, and it was an unusual accomplishment for the Congress.

The result was largely the handiwork of the Subcommittee on National Military Airlift of the House Armed Services Committee, which carried out tasks normally thought the prerogative of the executive. Headed by Congressman L. Mendel Rivers of South Carolina, the subcommittee studied the military airlift problem and reviewed in detail the secret war plans of the JCS. Having satisfied itself that the country needed more airlift, the subcommittee set itself up as a quasi-executive agency and designed a program and a budget. The subcommittee listened to all the experts it could find, examined the technical characteristics of various types of transport aircraft, designed a program compatible with the current political climate, and then took it to the Appropriations Committee for approval.

One need not agree with every item of the subcommittee's program to conclude that Rivers and his colleagues carried off a political tour de force incorporating the best characteristics of executive decision-making and congressional investigating. It should be noted also that both the legislative and executive branches gave the subcommittee impressive assistance. Monroney and Flood, for example, no longer were formidable opponents of strategic airlift procurement, but were new and forceful advocates. Quesada removed his objections to certain military transports provided they had certain characteristics he liked. The subcommittee benefited also from the cabinet level leadership and the closer liaison between the Army and the Air Force that were important aspects of Defense Department activity during the final year of the Eisenhower administration. The fact that some individuals and groups reversed positions taken earlier does not detract from the subcommittee's accomplishment in reconciling the conflicting military and economic interests with which it dealt.

The Rivers Subcommittee did not have complete latitude in building a consensus. While the subcommittee was at work, executive agencies were implementing the earlier recommendations of the Appropriations Committees, Holifield, and Monroney. The Commerce Department, Budget Bureau, and those in the Defense Department charged with business management of airlift, all planned for an increased use of commercial airlift. The CAB chose 1960 to set tariff rates for military traffic which wiped out the competitive bidding system, thus returning to the major carriers some of the military traffic they had lost and accomplishing many of the objectives of the four-carrier proposal, if not its substance. These actions tended to neutralize many of the remaining criticisms of military operation that the airline industry might have raised.

THE IMPETUS FOR ANOTHER INVESTIGATION AND THE REASONS FOR ITS SUCCESS

The immediate stimulus for the investigation was the Army's officially open disagreement with the JCS position on the adequacy of airlift capacity. The JCS had held for some years that enough airlift was available to meet most conceivable emergencies.[1]

1. See, for example, quotation from General Thomas D. White, Air Force Chief of Staff, in U.S. Congress, House, *United States Defense Policies in 1958*, Prepared by the Legislative Reference Service, Library of Congress, 86th Congress, 1st Session, House Document 227 (Washington: Government Printing Office, 1959), p. 19.

As General Nathan Twining, Chairman of the JCS, put it in 1959, one should not speculate that six or eight limited wars were likely to occur at the same time and in different parts of the world.[2] In reply to Congressman Flood's charge that the Air Force had neglected airlift for the Army, General Thomas White insisted that the airlift capacity then available (February, 1960) met the criteria of the JCS. The agency demanding more airlift, he argued, must first secure JCS approval for the increase and then find a place for it in the budget. Whatever the importance of airlift, he added, it had less claim upon Air Force funds than did the strategic deterrent, air defense, and tactical air power.[3] Army Secretary Brucker and Chief of Staff Lemnitzer disagreed; the former cited Taylor's 1958 request for enough airlift to move a two-division spearhead force, and asserted that Lemnitzer "was not getting all the cooperation he ought to get from the JCS."[4]

Chairman Carl Vinson appointed the seven-man Airlift Subcommittee the next day, and commissioned it to undertake a broad study of the problem: ". . . study all aspects of the national airlift, including MATS, the CRAF program, other elements of military and civilian aviation pertinent to the subject, the type, number, availability and adequacy of both aircraft, personnel, and required items of support for a national airlift in terms of national defense, and such additional considerations as may be pertinent to the subject matter."[5] Rivers was at the time the fourth ranking Democrat on the parent committee and, as a congressman from a district having an important MATS base, had been interested in airlift for some years. He had conducted a 1958 investigation culminating in a report favorable to MATS, and had defended as only common

2. U.S. Congress, House, Committee on Appropriations, *Department of Defense Appropriations for 1960*, Hearings before Subcommittee, 86th Congress, 1st Session (Washington: Government Printing Office, 1959), Part 1, p. 101.

3. U.S. Congress, House, Committee on Appropriations, *Department of Defense Appropriations for 1961*, Hearings before Subcommittee, 86th Congress, 2nd Session (Washington: Government Printing Office, 1960), Part 2, pp. 380–81 (hereinafter *HCA DOD Appropriations for 1961*).

4. *Ibid.*, pp. 424–26. Contrast this with Brucker's 1956 evaluation; see Table VI, note g.

5. Letter from Chairman Vinson to Congressman Rivers, reproduced in U.S. Congress, House, Committee on Armed Services, *Hearings before Special Subcommittee on National Military Airlift*, Hearings, 86th Congress, 2nd Session (Washington: Government Printing Office, 1960), pp. 4055–56 (hereinafter *Rivers Committee Hearings*). The Brucker-Lemnitzer testimony was in executive session and did not attract public attention for about ten days. *Washington Post*, February 9, 1960.

sense the use of MATS aircraft to carry military traffic.[6] He had expressed doubts about an air merchant marine, and had criticized the Hoover Commission's general approach.[7] Rivers had had little success in these earlier attempts, because the mainstream of congressional opinion was headed in another direction. His new appointment permitted him to roam almost at will in both commercial and military air transport policy.

Much of the subcommittee's relative success is attributable to its methods and its essentially nonpartisan approach. An important factor was the performance of Robert Smart, Chief Counsel for the parent committee, who doubled as subcommittee counsel. As one of the most experienced of congressional staff members, Smart brought to the investigation much prestige of his own. This was made clear when several members of the Appropriations Committee praised him when he appeared with Rivers before that committee.[8] Smart's principal contribution was detailed preparation for the hearings; he spent about five weeks working with military leaders of all the services, and forced them to jointly prepare their presentations. Enough has been said already to demonstrate that nobody had adequately explained airlift to Congress; Smart had to sort out the airplanes, determine what organizations had what transports, then develop a framework suitable for congressional presentation and congressional action.

A second factor was the unity of approach and opinion that subcommittee members maintained throughout the hearings. The Armed Services Committee is to a large extent comprised of congressmen from what often are called "safe" districts. Unconcerned about the partisan problems of 1960, they brought to the subject a relatively dispassionate approach. Rivers carefully maintained subcommittee unity, and members were equally careful to keep up their attendance at the hearings. Prior to Rivers' appearance at the Appropriations Committee, for example, he made certain he had

6. Rivers' home district contains several Navy installations, a Marine airfield, and one of the major operating bases of MATS. The earlier report is U.S. Congress, House, Committee on Armed Services, *Investigation of National Defense, Phase II*, 85th Congress, 2nd Session, Committee Print (Washington: Government Printing Office, 1958), pp. 2–3.

7. Rivers noted the inadequacy of the merchant marine in the early stages of both World Wars, and he criticized the Hoover Commission for having "skeletonized" Navy yards in the name of efficiency. These were parts of a speech made in reply to Congressman Porter's support of an air merchant marine. *Congressional Record*, June 24, 1959, pp. 10751–54.

8. *HCA DOD Appropriations for 1961*, Part 7, pp. 332–38.

his colleagues' unanimous approval.[9] The Monroney and Holifield investigations were, relatively speaking, more in the nature of one-man enterprises.

Another factor was the care with which the subcommittee avoided criticizing past congressional actions. While Rivers admitted congressional responsibility for the recent debacle, he brushed aside the 1959 denial of appropriations as only something from "yesterday."[10] His colleagues agreed that attempts to find scapegoats would only make it more difficult to break the impasse; to construct a new program, they must ignore side issues. It was not unusual for subcommittee members to begin analyzing reasons for the stalemate, then suddenly decide to drop the subject for the sake of consensus.[11] Rivers chimed in by issuing statements in praise of MATS' former critics for their "bigness" in changing their positions after learning the facts.[12] He and Smart invited to the hearings anyone who might be able to lend a hand; these included administrative assistants to Senators Engle and Monroney, staff investigators from the Holifield Subcommittee and the Senate Appropriations Committee, and representatives from the Commerce Department.[13]

Rivers was less restrained in his criticisms of the executive branch, but he did not openly attack any individual or agency. The administration, he said, had permitted MATS to "to go to hell on a one-way ticket," especially those in the Pentagon who gave the command only "lip service."[14] Had the administration put up a better fight the year before, he added, Congress would not have turned down the C-jet purchase.[15] It was, of course, good congressional politics to criticize agencies outside Congress while being gentle with Congress itself. This approach thus made it comfortable for Flood, Holifield, and Monroney to join in support of the purchase of new strategic military transports.

Yet another factor was the subcommittee's emphasis upon airlift as an important function of military aviation, a stature it had not attained up to that time.[16] In most previous hearings, the Defense

9. *Rivers Committee Hearings*, p. 4252.

10. *Ibid.*, p. 4847.

11. *Ibid.*, p. 4325.

12. Interview quoted in *Christian Science Monitor*, May 18, 1960.

13. *Rivers Committee Hearings*, p. 4143.

14. *Ibid.*, pp. 4162, 4632.

15. *Ibid.*, pp. 4323–24.

16. Deputy Secretary of Defense Douglas observed that these were the first hearings covering the entire field in sufficient detail. *Ibid.*, p. 4058.

Department had sent to the hill its civilian experts in traffic management, financial programming, and procurement; some of them were appointive officials, and some of them were civil service employees. Whatever their expertise in specialized fields, they could not convince Congress of the need for new appropriations.

The military departments provided the subcommittee an unusually high degree of support, especially the Army. Chief of Staff Lemnitzer remained in attendance through six weeks of hearings. Whatever the merits of each of his arguments, and some will be questioned here, this was an outstanding performance by the Army's highest officer. It mobilized the Army's congressional supporters behind the Rivers airlift program, and Rivers observed that he never had known such a high official to spend more time before a single congressional committee.[17]

There was, in addition, a closer working relationship between Lemnitzer and White than between previous Army and Air Force chiefs, thus making it easier to secure agreement between them. This seems to have reflected the prevalent Defense Department atmosphere of 1960, usually credited to Defense Secretary Thomas Gates. The year was notable for improved administration within his department, and for smooth and productive working relationships between civilian and military heads of the services.[18] Gates' chief assistant was James Douglas, now Deputy Secretary of Defense, who represented the department in the hearings.

All of these factors, significant to any congressional investigation regardless of subject matter, contributed to the subcommittee's ultimate achievements. If there appears to be too much emphasis placed here on the development of unity within and outside the committee, it is worth noting that the task of the Rivers Subcommittee was to add funds to the military budget that the administration would agree to spend. The mechanics of such a task involve the most careful co-ordination with the internal administrative pro-

17. *Ibid.*, p. 4815.

18. ". . . during the year 1960 the improvement and speedup in the work of the Joint Chiefs of Staff, because of Secretary Gates' decision to sit in personally at their weekly meetings, has astounded oldtimers and has been hailed by knowledgeable Congressional leaders of both parties as the greatest single improvement in our defense system since the passage of the National Security Act of 1947." Article by George Fielding Eliot in *Army Times*, December 17, 1960, reproduced in U.S. Congress, Senate, Committee on Armed Services, *Military Procurement Authorization, Fiscal Year 1962*, Hearings, 87th Congress, 1st Session (Washington: Government Printing Office, 1961), pp. 266–67.

cedures of the executive branch to make sure the funds are truly usable in terms of dollar amounts, available technology, production schedules, and similar considerations. This is a vastly more difficult task than the ostentatious addition of funds which cannot logically be spent and which, because of the political atmosphere in which the funds are added, the administration is bound to oppose.

THE RIVERS SUBCOMMITTEE AND THE WAR PLANS

The subcommittee studied the war plans because the Army had openly disagreed with the JCS, and also because Defense Department policy statements were, at the outset, somewhat vague. Deputy Secretary Douglas, for example, pronounced the existing airlift capacity "marginally capable" of meeting wartime requirements, provided that CRAF could be used immediately.[19] General White added that the combined civil and military fleets could meet the airlift needs of the first thirty days of a general war.[20] The subcommittee decided to make its own study, and it spread upon the public record an account of one phase of JCS war planning.

The subcommittee described the assumptions and definitions the JCS used to determine the relationship between types of emergency situations and the requirement for airlift. The subcommittee discovered, for example, that war planning up to that time was based upon three general assumptions. One visualized a mobilization period of six months followed by sixty days of general war; the subcommittee rejected this as unrealistic and said no more about it.[21]

A second assumption envisaged the same sixty-day general war, but without any time for mobilization; the subcommittee thought this more realistic and it was, in any case, the only assumption the JCS actually had used in preparing war plans. Airlift requirements for the emergency were divided into time periods extending throughout the hypothetical sixty-day period. The JCS thought the first twenty days the most critical, and subdivided this period into segments of five, five, and ten days. Within each time frame, airlift requirements were designated as "passenger" or "cargo," and

19. *Rivers Committee Hearings*, p. 4059.
20. *Ibid.*, p. 4880.
21. U.S. Congress, House, Committee on Armed Services, *Report of Special Subcommittee on National Military Airlift*, 86th Congress, 2nd Session, Committee Print (Washington: Government Printing Office, 1960), p. 4030 (hereinafter *Rivers Report*).

further classified as "hard core" or "civil eligible." "Hard core" items were those needed for immediate support of SAC and requiring movement aboard military aircraft; the list included nuclear warheads, key personnel, and recovery teams. In 1960, however, the term "hard core" was in the process of being redefined to include "tactical employment." The Army took this to include the movement of troops anywhere in the world, whether for general, limited, or some other type of war.[22]

The subcommittee concluded that less than enough airlift was available for general war, even in the absence of plans to move the Army.[23] The plans assumed also that CRAF would be totally effective within forty-eight hours after the start of war, but Rivers thought five days the best that could be expected.[24]

A third JCS assumption used as a case study the resumption of limited war on the Korean peninsula; there was no approved plan for this contingency, but White and Lemnitzer were trying to arrive at one.[25] Lemnitzer and the subcommittee agreed, however, that any plan based upon Korea might be unduly optimistic. The U.S. had troops there, not to mention a substantial infrastructure of facilities and forces in nearby Japan. Even so, there was a shortage of airlift for the Korean plan.[26] The subcommittee's study thus permitted it to report that airlift was in short supply, a necessary first step if Congress was to approve an airlift appropriation. This conclusion, however, points up some of the long-term puzzles of the airlift controversy.

JCS war planning followed, up to 1960 at any rate, the traditional dichotomy between peace and war. The term "hard core" referred only to traffic airlifted during the war itself. When the airlines, Quesada, and Holifield asked that MATS move only "hard core" traffic, the request left still unanswered the old question of what the transports should do in the absence of war.

22. *Ibid.*, pp. 4131–34; *Rivers Committee Hearings*, pp. 4080, 4124; the definition of "hard core" is from U.S. Department of Defense, Assistant Secretary for Supply and Logistics, *The Role of Military Air Transport Service in Peace and War* (Washington: Department of Defense, 1960), p. 2 (hereinafter *MATS in Peace and War*).

23. *Rivers Report*, p. 4032.

24. *Rivers Committee Hearings*, p. 4093.

25. *Ibid.*, p. 4886. The new relationship between Army and Air Force Chiefs was noted by several analysts. Brigadier General T. R. Phillips in *St. Louis Post-Dispatch*, March 23, 1960; Hanson Baldwin in *New York Times*, March 22, 1960.

26. *Rivers Committee Hearings*, p. 4071; *Rivers Report*, p. 4034.

JCS and subcommittee airlift calculations also depended upon other assumptions that may or may not have been adequately studied. Whether one agrees that a shortage did, in fact, exist depends upon the validity of the data base. The idea of using airlift instead of sealift to move whole Army divisions across the oceans was new then and, in a sense, still is. Assuming the likelihood of conventional wars, it remains valid to ask how much the U.S. is ready to pay for split-second strategic delivery of whole divisions and their equipment. It is worth repeating that airlift capacity did not stand still in the 1950's; it expanded by a factor of three or four, depending upon the data source. Yet the road ahead seemed distressingly long.

Nor did the subcommittee and the JCS include any calculations on the need for tactical, or battlefield, airlift. There seemed to be deficiencies there also, said the subcommittee, but it was not certain. It will be seen in the next chapter that the questions of how much airlift there should be within combat zones, to what degree it should replace surface transportation, and what military department should operate it, have become the prevailing controversies of the 1960's. In sum, the subcommittee's study was less broadly conceived than it might have been, but accomplished its intended objective of breaking the airlift stalemate.

To emphasize its findings, the subcommittee pointed to the approaching obsolescence of the current fleet. Taking MATS and CRAF together, said the subcommittee, only eighty modern transports were on hand, and fifty-eight of those were commercial passenger jets.[27] This left the subcommittee with a final and most significant paradox. A mathematical deficiency in airlift capacity for general war became the basis for producing a fleet of new transports for use in conventional war. The general war deficiency was the only measurable one, yet very little airlift might be needed in a general war between thermonuclear powers. As with the other complications, the subcommittee did not explore this question at any length, perhaps because it would have taken the investigation far afield into theoretical problems that could only defeat the current objective. The subcommittee took the only evidence immediately available, whatever its logical shortcomings, and built a program. It is worth demonstrating the true magnitude of this task.

27. *Ibid.*, p. 4032.

THE AIR FORCE, TAC, AND MATS

The hearings came at a time of much confusion within the military airlift complex, especially within the Air Force. This was the sequel to Air Force decisions taken the previous year to prepare for the Quesada program by limiting MATS operation. Because the air merchant marine was to get initial delivery on the new cargo airplane, MATS would have only fifty of them by 1965.[28] The Air Force also was planning to shift some of its airlift units from MATS back to TAC, in an attempt to make airlift more of a truly combat function, a move generally in accord with traditional relationships between airborne divisions and troop carrier commands.[29] TAC was modernizing its wings with the Lockheed C-130B, a four-engine turboprop transport. In emergencies, these planes would airlift packaged forces of tactical air power (Composite Aerial Strike Forces, or CASF's) to overseas theaters, and would thereafter operate as an intratheater airlift command.[30]

The Army liked the C-130B because it could airdrop sixty-four paratroopers as well as items of heavy equipment, could carry ninety-two soldiers on normal flights, and needed runways only twenty-five hundred feet in length.[31] The Marine Corps was planning to acquire thirty-six of the same transports by 1963.[32] The aircraft itself, despite its four engines, was something more than a tactical transport and something less than a strategic aircraft. The Marine Corps used it only for short-haul operations within the combat zone (including the refueling of Marine fighters), and neither the Air Force nor the Marines expected to depend

28. *HCA DOD Appropriations for 1961,* Part 2, pp. 294–95.

29. *Ibid.,* Part 4, p. 530.

30. Air Force Chief of Staff White implied that these CASF's moved themselves overseas. U.S. Congress, Senate, Committee on Appropriations, *Department of Defense Appropriations for 1960,* Hearings before Subcommittee, 86th Congress, 1st Session (Washington: Government Printing Office, 1959), p. 267 (hereinafter *SCA DOD Appropriations for 1960*). TAC representatives implied the same thing. *Rivers Committee Hearings,* pp. 4244–46. The C-130B, however, did not do well across long distances, with the result that TAC depended upon MATS to a considerable extent, thus raising once again the old question of who would get first crack at the airlift during emergencies. During the Taiwan crisis of 1958, for example, 144 of 224 MATS trips from the U.S. to the Far East were in support of TAC. *Ibid.,* pp. 4169–71.

31. *Ibid.,* p. 4223.

32. *Ibid.,* p. 4401.

upon the C-130B for strategic operation. The hybrid nature of the aircraft made it difficult for the subcommittee to build its airlift program, and the complex relationship between TAC and its CASF's, the Army, and MATS, led the Air Force to undertake an internal reorganization which complicated the matter still further.

The Army's airlift traditions included the First Allied Airborne Army of World War II, the 18th Air Force formed at the end of the Korean War, and constant attempts by airborne experts to organize yet another joint tactical command. When the Army asked the Air Force in 1959 to make one of its commands responsible for meeting Army airlift needs, the Air Force chose TAC, presumably because of the old relationship between TAC and the airborne divisions. Called the "single point of contact," the arrangement gave TAC control of the planning and operation of both strategic and tactical airlift exercises.[33]

MATS retained, of course, the job of managing strategic airlift for the Defense Department, but TAC now had a share of that responsibility. As a command whose operations were confined to the U.S., however, TAC had neither organizations nor bases located overseas. As a command charged with training forces for use overseas, TAC surrendered its control of those transports as soon as they entered the overseas theater, e.g., General Lauris Norstad took charge of them as they arrived in Europe. To treat strategic airlift in this fashion, however, would only reinstate arrangements discarded at the outset of World War II.[34]

As Hanson Baldwin pointed out, the TAC-MATS arrangement implied a gradual technological merger between short-haul paratrooper airlift and strategic airlift, the C-130B representing a bit of both.[35] The new association between TAC and the Army implied also that an important future task of strategic airlift would be the delivery from the U.S. directly to a battle zone of thousands of paratroopers. Yet another implication was that the new air merchant marine would airlift nonparatrooper Army forces to overseas staging areas.[36] The Rivers Subcommittee slowed these trends,

33. *New York Times,* March 25, 1960.

34. For these earlier difficulties with overseas commanders, see Chapter 3.

35. *New York Times,* March 22, 1960.

36. The confusion over the use of commercial airlift continued into 1960. Generals Gavin and Marshall, it will be recalled (Chapter 8) had supported Congressman Porter, but Brigadier General Phillips observed that any notion of using commercial airplanes in emergencies was "spurious." *St. Louis Post-Dispatch,* November 9, 1959, March 27, 1960.

all of them partially attributable to the Quesada plan, but it did not succeed in immediately reversing them.

A major factor in emphasizing the unique nature of strategic airlift was "Operation Big Slam," an exercise MATS had planned for some time and which coincided with Rivers' hearings. The exercise tested MATS' ability to surge from five-hour to eight-hour utilization, thereby producing the reserve airlift capacity needed for emergencies. As the result of direct negotiation between MATS and the Continental Army Command, the exercise included the strategic deployment and return to the U.S. of a suitable Army force (21,000 troops and 11,000 tons of equipment).[37] The force moved from the U.S. to Puerto Rico, went through a short training period, then moved back. Budgetary considerations had prevented such exercises up to that time, because a single exercise of this sort cost some $10 million.[38] Given 1960 conditions, however, the training exercise helped spend the money earmarked for commercial airlift.

Significantly, MATS and the Army maintained a direct relationship throughout the exercise, largely because of the practical necessity of having the commands actually involved in the exercise do the planning and operating in tandem. Also significantly, MATS made the most of the exercise as a public relations device for propagandizing strategic airlift. MATS gained widespread support for strategic airlift modernization, principally by making certain that prominent military analysts were on the scene. Mark Watson, Brigadier General T. R. Phillips, and Hanson Baldwin, among others, noted the efficiency of the exercise but, they added, nothing could obscure the age of the airplanes in service.[39] Other publications called MATS a "stepchild," wedged between the Air Force (dedicated to SAC) on one side, and the Army, Congress, and the airlines on the other.[40] For its promotional activity, MATS won the yearly prize for the best public relations program in the military field.[41]

37. Military leaders agreed the force was somewhat unbalanced, in that the tonnage was less than needed by that many troops in an actual situation. *Rivers Committee Hearings*, pp. 4167, 4174.

38. *Ibid.*, p. 4167.

39. Phillips, *St. Louis Post-Dispatch*, March 20, 1960; Watson, *Baltimore Sun*, March 27, 1960; Baldwin, *New York Times*, March 22, 1960.

40. Clay Blair, Jr., "Uncle Sam's Orphan Airline," *Saturday Evening Post* (June 4, 1960); *Time*, March 28, 1960.

41. This was the award of the American Public Relations Association. *Army-Navy-Air Force Journal*, June 3, 1961.

MATS managed during the exercise to reach the eight-hour utilization that had been its yardstick for so long, and to reach that objective on routes considerably longer than those of the World War II "Hump" and the Berlin Airlift. The use of Puerto Rico made the exercise appear less strategic than tactical, but many of the troops came from bases in the western U.S., e.g., Fort Lewis in Washington; the distance thus approximated that from the U.S. to Europe.

The operation was in the tradition of the Berlin Airlift, i.e., aircraft flew at intervals carefully controlled by an integrated schedule. Once the airlift pipeline was full, the same transports were used over and over to accomplish the total maneuver, in contrast to the Army's previous experience with massive formations of transports. Prior to the exercise, military leaders of all services had not tended to look upon the Berlin experience as the type of operation useful within a wholly military context. "Big Slam" thus demonstrated that the combination of high utilization and conveyor belt operation produced somewhat more airlift capacity than anyone had thought actually available. In a sense, then, Big Slam was as useful for education within the military as for public relations. Gavin, for example, had asserted that strategic airlift exercises were impossible before 1965.[42]

All of this changed the airlift climate enough to permit MATS to advance its own modernization program, outlined as follows by General Tunner:

Swing-tail jets for support of SAC	45
Other Swing-tails for interim modernization	49
Outsize C-133's	50
Civil-military cargo workhorse	188
Total new aircraft	332[43]

Tunner estimated the cost as $250 million for each of the succeeding eight years.

Tunner argued for "interim modernization" to fill the gap caused by the halt in airlift procurement and the expected delay

42. James M. Gavin, *War and Peace in the Space Age* (New York: Harper & Brothers, 1958), p. 284.

43. *Rivers Committee Hearings*, pp. 4183–88; *HCA DOD Appropriations for 1961*, Part 4, pp. 423–25.

in building the new workhorse aircraft.[44] He continued, however, to reflect the Air Force preoccupation with general war; all new transports, he argued, should be able to bypass island bases that might be destroyed in nuclear war.[45] Tunner added that commercial airlift would become increasingly less important, and perhaps superfluous, upon enactment of his program.[46] This assertion brought an immediate administration reaction.

The Air Force and the Defense Department opposed giving MATS so many new transports; both appeared still committed to

44. It can be argued that constant modernization is not extremely important in the transport field. Military transports have a tradition of becoming more efficient and dependable after about a decade of operation, and the need to attain high utilization makes dependability the most important characteristic. The twin-engine C-46 was a "killer" and a "plumber's nightmare" when new in World War II, but it was the most dependable transport of the Korean War. Wesley Frank Craven and James Lea Cate (eds.), *Services Around the World* (*The Army Air Forces in World War II* [Chicago: The University of Chicago Press, 1958]), VII, 24–26; Robert Frank Futrell, *The United States Air Force in Korea, 1950–1953* (New York: Duell, Sloan and Pearce, 1961), pp. 530–31. The twin-engine C-119 was "temperamental" when new in Korea, but it became the most efficient of transports while assigned to the reserves in 1960. *Ibid.*, pp. 528–29; *Rivers Committee Hearings*, pp. 4241–42, 4251. The four-engine C-124 never was very useful in Korea because of mechanical difficulties, but it became the strategic workhorse transport and so dependable that it has yet to be retired from active service. Futrell, *op. cit.*, pp. 530–31. The classic case is, of course, the Douglas DC-3, still quite dependable after three decades.

45. *Rivers Committee Hearings*, p. 4163. Other things being equal, increased range usually is an asset, although the Air Force argument implied that the U.S. could depend on sanctuaries in the form of bases within the combat zone but could not depend upon island bases. The emphasis on increased speed and range has had a profound effect over the years. During the Berlin Airlift, General Kuter warned against overemphasis on speed in transports. Speech reproduced in Eugene M. Emme, *The Impact of Air Power* (Princeton: D. Van Nostrand Company, Inc., 1959), pp. 382–83. Tunner spoke of the need for transports having an interior width of 164 inches. *Aviation Week*, January 17, 1949, p. 6. During Korea, Tunner mentioned "flying freight cars," and added that 150 miles per hour was fast enough. Major General William H. Tunner, "Technology or Manpower," *Air University Quarterly Review*, 5 (Fall, 1952), 17. Given the emphasis of the 1950's on SAC and nuclear war, the Air Force discarded these views in the attempt to make transports keep up with intercontinental bombers, and the widest transport produced after the Berlin Airlift was the C-124, with its interior width of 156 inches; later models are narrower yet. *Interavia*, 17 (January, 1962), chart pp. 55–56. One airline has noted that the best cargo plane for defense needs would be a flying boat capable of lifting 1,000 tons. U.S. Congress, Senate, Committee on Commerce, *Government Guarantee of Loans to Certain Air Carriers*, Hearings before Subcommittee, 86th Congress, 2nd Session (Washington: Government Printing Office, 1960), p. 66.

46. *Rivers Committee Hearings*, pp. 4831–32.

some version of the Quesada plan. Assistant Secretary of the Air Force Philip Taylor told the Holifield Subcommittee he was certain that the need for CRAF would continue as before, because MATS would not get the fleet it wanted.[47] Taylor, charged with managing the industrial fund and commercial airlift aspects of MATS, asserted that Tunner was incorrect in claiming that an increase in military airlift operations would bring on a decrease in costs. Commercial airlift, Taylor added, always would be cheaper and more efficient than its military counterpart. This exchange demonstrated that the Rivers Subcommittee could not count on unqualified Air Force and Defense Department support for a massive airlift program. The best source of political support in 1960 was the Army, by this time a significant force in congressional politics.

THE ARMY AND ITS ATTEMPTS TO REACH A DECISION

The Army was the important customer whose demands must be met before Congress as a whole would act on a new program.[48] The subcommittee, however, found it difficult to find out exactly what sort of airlift program the Army would support, and only extensive direct negotiation between the subcommittee and the Army settled the matter. Lemnitzer still insisted on predesignation even though the Army was not following that practice in managing its own transports.[49] When asked what specific types of transports he wanted, Lemnitzer asked for more C-130B's.[50]

Chief counsel Smart convinced Lemnitzer in an extended discussion that the C-130B, given its limitations, would not answer Army needs for long-haul airlift. With Lemnitzer holding to the view that the C-130B's abilities to perform many battlefield tasks and use short runways were virtually indispensable, it was clear that the

47. U.S. Congress, House, Committee on Government Operations, *Military Air Transportation*, Hearings before Subcommittee, 86th Congress, 2nd Session (Washington: Government Printing Office, 1960), p. 73 (hereinafter *Military Air Transportation, 1960*).

48. The Rivers Subcommittee concluded that the Strategic Army Corps was the only major combat organization that did not have its own strategic airlift. In view of the limitations on the C-130B, this was not a wholly correct assertion. *Rivers Report*, p. 4036.

49. *Rivers Committee Hearings*, p. 4075. The Army had by this time begun to acquire the twin-engine *Caribou* for airlift within the battle area. They were managed by the highest level of field Army command, so as to achieve the greatest flexibility for all Army subcommands. *SCA DOD Appropriations for 1960*, pp. 126–27.

50. *Rivers Committee Hearings*, p. 4086 *et passim.*

subcommittee had to come up with an aircraft that combined strategic and tactical characteristics. The Army would not support a purely strategic transport, and the subcommittee was unwilling to abandon strategic airlift.

QUESADA AND THE NEW DEFINITION OF UNCOMPROMISED AIRCRAFT

Quesada reversed in 1960 his original definition of uncompromised cargo transports, disagreed with MATS' claims, and defended with spirit his far-reaching proposal. In describing uncompromised aircraft, he now ruled out the Canadian CL-44 he had mentioned in 1959. It was not a suitable transport, he argued, because its high fuselage made it useless for operation into remote areas where sophisticated loading facilities were not available. The same criticisms, he added, applied to all swing-tail jets then being proposed.[51]

The best candidate for procurement before 1965, argued Quesada, was the Lockheed C-130B, suitably modified for long-haul operation.[52] It could be produced almost immediately, he said, because the production line was in being. These repeated references to Lockheed led Rivers to inquire why no mention was made of other production lines in existence, e.g., the Douglas C-133 had a low fuselage, was larger than the C-130B, and could be modified to carry passengers.[53]

The Defense Department should reduce MATS flying by a factor of two-thirds, Quesada asserted, and MATS should not fly at all except for exercises of the type just concluded.[54] A five-hour utilization was just too high, he said; pilots could stay well trained with much less flying. Whatever the merits of the utilization argument, Quesada had modified his earlier arguments; it now was clear he would not object to the purchase of the Lockheed C-130.

THE SUBCOMMITTEE AND THE INDUSTRY

Rivers and his colleagues used three tactics to discourage the airlines from mounting an opposition to the subcommittee's airlift

51. *Ibid.*, pp. 4410, 4416. The CAB, on the other hand, still considered the CL-44 an "uncompromised" cargo plane. *CAB Annual Report*, 1962, p. 13.

52. *Rivers Committee Hearings*, pp. 4411, 4415.

53. *Ibid.*, p. 4426. Quesada then withdrew endorsement of the C-130B.

54. *Ibid.*, p. 4408. It is clear from the testimony that Quesada did not realize the maneuver had depended upon high utilization similar to that achieved in the Berlin Airlift.

program. One was to single out the major international carriers for specific praise, then absolve them from complicity in the attacks on MATS. A second was to criticize ATA for having championed with too much spirit the drive against MATS. The third was to treat all industry witnesses, including those from the supplemental carriers, in the most gracious manner imaginable while searching for issues on which those witnesses could agree.

The treatment of the major international companies marked something of a turn in relationships between those companies and Congress. The presidents of PAA and TWA, Juan Trippe and Charles Thomas, appeared as their own spokesmen, and the subcommittee even postponed specific hearings until Trippe could appear. It was a rare occasion for Trippe and his company to meet with a friendly reception from a congressional committee. Rivers praised the pioneering contributions of the major airlines and said he was certain they had not been prime movers in the airlift controversy.[55] Trippe and Thomas, a former Navy Secretary, spoke principally in support of the four-carrier proposal; they had just submitted it once again to the Defense Department.[56]

Subcommittee generosity did not extend to ATA; when Stuart Tipton tried to compare MATS with TAC and SAC, Rivers cut him short. The witness knew nothing of existing war plans and the missions of all these commands, Rivers asserted, and he should not try to address military questions.[57] When Tipton announced that ATA favored modernizing MATS, if on a modest scale, Rivers ticked off the long history of ATA opposition to new airplanes for MATS. Tipton next suggested that MATS might buy some C-130B's; he then agreed quickly with Rivers that a business lobby should not advocate particular types of military aircraft.[58]

The ATA understandably had nothing to say about the Quesada program, but the supplementals were not yet ready to surrender their hopes. Their representative deplored the actions of the major companies in "scuttling" the plan, and reiterated that the best approach still was one that would force the government to buy aircraft, lease them to the carriers, then turn over any excess

55. *Ibid.*, pp. 4724–26.

56. Trippe also offered an explanation for the apparent reluctance of industry to buy jet cargo transports. PAA's passenger jets had cost $400 million, too much to permit simultaneous financing of cargo jets. The available engines also were unsuitable for cargo hauling. *Ibid.*, p. 4734.

57. *Ibid.*, p. 4531. *Aviation Daily*, April 4, 1960.

58. *Rivers Committee Hearings*, pp. 4532–35.

aircraft to MATS.[59] This was the most extreme version yet advanced, for it would have left MATS with only aircraft that the airlines would not accept.

Most industry witnesses agreed with the subcommittee that the competitive bidding system then in use had caused military tariff rates to fall to levels so uneconomic as to threaten an imminent decline in air safety (this was only a year before the Richmond crash).[60] The number of companies submitting bids for military traffic had increased from thirty-eight to ninety within only a three-year period. It seemed to the subcommittee that forced procurement, competitive bidding, and complicated leasing arrangements for second-hand aircraft had not helped either the industry or national defense; Rivers called the whole process "ridiculous."[61] The subcommittee decided to go a step beyond the Holifield Subcommittee; it recommended that the government award contracts only to CRAF members and that it negotiate tariff rates with industry if this proved necessary to restore a more balanced approach.[62]

THE AIRLIFT PROGRAM AND ITS PROGRESS THROUGH THE APPROPRIATIONS CYCLE

The airlift hearings occurred while the Appropriations Committees had under consideration the next year's defense budget, thus forcing Rivers and his colleagues to meet stringent time limitations. Rivers asked Air Force generals in attendance to draw up a program for off-the-shelf transport procurement, i.e., aircraft that would be only slight modifications of models then in production. He asked for precise information based upon careful analysis of what manufacturers could reasonably produce and what was militarily acceptable: "I don't want to ask for something I can't get or ask for something you can't use. If the industry can produce X-number of airplanes, that is how much money I want to ask for, this year."[63]

It was clear to the subcommittee that it could accept neither the MATS nor the Army proposal as it stood. The only possible compromise that had any promise of quick acceptance by all concerned was the modified C-130B, newly designated as the C-130E.

59. *Ibid.*, pp. 4542–43.
60. See Chapter 5.
61. *Ibid.*, pp. 4198, 4533, 4545, 4609, 4688.
62. *Rivers Report*, p. 4046.
63. *Rivers Committee Hearings*, p. 4352.

It proved acceptable to both the Army and Air Force, as a tactical transport with a range at the lower levels of the strategic spectrum. Except for a slight increase in speed and an improvement in paradrop capability, however, it represented no significant advance over the C-124 introduced during the Korean War.[64] The C-130's internal dimensions were smaller than the C-124's, thus limiting the new plane to the airlift of only medium-sized equipment. It can be argued that the C-130E was simultaneously the best political compromise available and something less than the best military choice, for it was not a complete answer to either strategic or tactical requirements.

Rivers appeared before his Appropriations Committee during a break in his own hearings, and he met with instant success. He admitted it was unusual for Congress to recommend procurement of a specific aircraft, and he felt obliged to point out he had no connection with any aircraft manufacturers. He asked for an earmarked account of $337 million, set aside for fifty C-130E's and fifty modified jets.[65] To speed deliveries and maintain the integrity of a long-term development program, Rivers suggested discarding the swing-tail feature in favor of minor alterations to make the jets useful for support of SAC.

Appearing as the current congressional airlift expert, Rivers got immediate support from Flood, who condemned seven years of "conspiracy . . . gross and appalling incompetency and evasion" that had thwarted Congress.[66] While supporting Rivers, the committee made some alterations to his program while retaining the principle of earmarked funds. The committee designated $370 million for airlift, an amount seemingly $33 million above Rivers' request, but the total included $50 million for aircraft development (the residue of the Quesada plan) and $70 million for continued C-130B procurement. Rivers had assumed these expenditures to be outside his program. In effect, the committee reduced the Rivers proposal by some $100 million and asked the Defense Department to give first priority to the C-130E.[67]

The Appropriations Committee was less willing than Rivers to

64. The C-130E could carry 17.5 tons for 2,000 miles. *Rivers Committee Hearings*, p. 4814.

65. *HCA DOD Appropriations for 1961*, Part 7, pp. 332–33.

66. Other committee members disagreed. *Ibid.*, p. 338.

67. U.S. Congress, House, Committee on Appropriations, *Department of Defense Appropriations for 1961*, 86th Congress, 2nd Session, House Report 1561 (Washington: Government Printing Office, 1960), pp. 16–18.

forego criticizing the administration. It chastised military agencies for neglecting airlift over a long period, while praising the Army, the MATS-Army exercise, and the Rivers Subcommittee for having solved the problem. The committee also added specific instructions; the new jets were "solely" for the airlift of SAC and could not engage in scheduled passenger service.[68] The committee added $80 million for commercial airlift, something of a tacit admission that forced procurement had reached a ceiling. The $80 million was available only for CRAF members, and the committee warned the Air Force against depending upon irresponsible marginal operators.[69] Given this new provision, the Air Force had no further difficulty in encouraging individual airlines to enter into contracts formalizing their participation in CRAF.

The Senate Appropriations Committee treated MATS in an extremely cordial fashion. Monroney promised General Tunner all the support "humanly possible."[70] Two of the senators who had led previous opposition to MATS demonstrated the change in climate:

> SENATOR CHAVEZ: . . . If they need airplanes . . . why do they not tell us about it?
>
> SENATOR MONRONEY: This comes back to the budget, Senator, and we do not have our military requirements set by the military. We get them through the Budget Bureau.
>
> SENATOR CHAVEZ: I would like to find out from the military whether enough money is being made available for an adequate airlift.
>
> SENATOR MONRONEY: I agree . . . I would at least double the amount requested this year if I had my way.[71]

After Senate consideration and the ironing out of differences with the House bill, the Rivers program was at $140 million for C-130E's and only $60 million for modified jets, plus funds for C-130B's and a development program; the grand total was $311 million.[72] The conference report modified the instructions to the Defense Department and permitted MATS to use the jets for air-

68. *Ibid.*
69. *Ibid.*
70. U.S. Congress, Senate, Committee on Appropriations, *Department of Defense Appropriations for 1961*, Hearings before Subcommittee, 86th Congress, 2nd Session (Washington: Government Printing Office, 1960), p. 818 (hereinafter *SCA DOD Appropriations for 1961*).
71. *Ibid.*, pp. 818–19.
72. Public Law 86–601, Sec. 531.

lifting either SAC or the Army, still a highly restrictive provision.[73] In addition the conference report restored the earmarked account, earlier withdrawn by the Senate at the Defense Department's behest.[74]

While this appropriations cycle was in progress, Rivers finished building his own consensus. He held his final hearings in the atmosphere set at the beginning; the military leaders involved were participants in a round-table discussion. Lemnitzer announced his support of the subcommittee's program, and White indicated tacit agreement.[75] Aside from the recommendations to buy new military transports, the Rivers Subcommittee listed these conclusions:

> The Air Force should review its separate airlift packages in SAC and elsewhere, and consider their consolidation within MATS.
>
> MATS utilization should not be less than half the rate required in wartime.
>
> The President should have power to activate CRAF in situations short of war. CRAF members should execute employee agreements with their employee unions providing for the uninterrupted airlift of defense traffic during employee disagreements [this had been done by PAA]. CRAF members should be required to purchase new cargo aircraft.
>
> Commercial airlift procurement should be limited to CRAF members. If competitive bidding does not prove workable with this new limitation, rates should be negotiated. Long-term contracts should be awarded.
>
> The program to equip the Air National Guard with strategic transports should be continued and expanded.
>
> There should be more joint training exercises like Big Slam.
>
> There should be more emphasis on plans for limited war.
>
> MATS is a weapons system and should be recognized as such. A more appropriate name for the organization would be "Military Airlift Command."[76]

73. U.S. Congress, House, Committee on Appropriations, *Department of Defense Appropriations for 1961*, 86th Congress, 2nd Session, House Report 2040 (Washington: Government Printing Office, 1960), p. 3. This caused Rivers to make a speech announcing that Congress did not intend to prohibit the airlift of TAC, the Navy, and the Marines. *Congressional Record*, June 30, 1960, pp. 14105–6.

74. *SCA DOD Appropriations for 1961*, pp. 1580, 1670–73.

75. *Rivers Committee Hearings*, p. 4848 *et passim*, especially pp. 4851, 4880.

76. *Rivers Report*, pp. 4051–55.

The Rivers Subcommittee did not meet head-on the question of whether MATS should spend a minimum amount each year for commercial airlift. The subcommittee seems to have assumed that Congress would earmark such funds each year and that airlift exercises would help spend the money.[77] In any case the subcommittee encouraged most of the principal interest groups involved. The important losers were the supplementals; the agreement between the Airlift Subcommittee and the Appropriations Committee that competitive bidding was not beneficial led the Civil Aeronautics Board (CAB) to set its own tariff rates for military contracts.

THE CAB INTERVENTION AND MINIMUM RATES

The CAB intervened at a time when the political climate was such as to permit an action detrimental to the small companies, although the board had spent two years considering the rate question. The CAB had begun discussions with the Defense Department shortly after the 1958 Holifield Subcommittee hearings. The first complaint came from small companies which charged that industry giants (PAA and TWA) were using unfairly low rates to force their small competitors out of business. The CAB announced that: . . . the substantial disparity between the certificated carriers and their competitors for military contracts may give the former a substantial economic advantage.[78] The CAB concluded, however, that government traffic was international in nature and that the board had no authority to intervene. If rates were set, the CAB added, every company would use the same rate, and the Defense Department would have no means of choosing between bidders.

After further discussions with the Defense Department, the CAB announced in March, 1959 that it proposed to set minimum rates for government traffic. Military contracts were then exempt from economic regulation, but the board said it would begin to withhold exemptions from carriers which submitted bids below the new

77. *Ibid.*, p. 4053, Recommendation #6(a).

78. CAB Order #E13040, September 30, 1958, reproduced in U.S. Congress, House, Committee on Government Operations, *Military Air Transportation*, Hearings before Subcommittee, 86th Congress, 1st Session (Washington: Government Printing Office, 1959), pp. 121–24 (hereinafter *Military Air Transportation, 1959*).

rate floor.[79] The board chose to describe its rate as only a proposal, however, and decided that Defense Department approval was needed. When that department declined to accept responsibility for a rate the CAB said it could not establish on its own, the board withdrew its proposal and announced once again that it could not undertake unilateral action.[80]

When the subject came up again in 1960, the old positions were reversed. With specialized leasing arrangements worked out with manufacturers and their subsidiaries, and with the help of a liberal interpretation of the small business laws, the supplemental and cargo companies virtually had swept the market.[81] The Appropriations Committee now was criticizing small companies, and the major carriers were pressing their four-carrier proposal. The CAB suddenly advised Defense Secretary Gates in June that it no longer would grant exemptions to its economic regulations. A practice begun during the Korean War, said the CAB, no longer was valid; the scheduled airlines had enough capacity to help the military and nobody need be exempt. In its initial form, the CAB announcement was virtually an endorsement of the four-carrier proposal.[82] After six more weeks of negotiation with the Defense Department, the CAB changed position once again.

The board announced in August that it was establishing a new rate floor, a decision which brought immediate and bitter protests.[83] It was clear enough that the new rates would be high enough to permit the major companies to recover their operating costs, a factor that encouraged them to once again seek out military business. Sparkman and Monroney immediately asked the CAB to continue

79. The rate was to be 3.4 cents per passenger mile, compared with a tourist rate of 5.38 cents. The Government then was paying about 2.5 cents. Proposed Part 399, CAB Economic Regulations, March 10, 1959, reproduced in *ibid.*, pp. 154–60.

80. *Ibid.*, pp. 232–33; CAB Press Release 59–21, June 11, 1959.

81. Small businesses had captured about 90 per cent of the market. The Small Business Administration prohibited major companies from bidding on 35 per cent of the business, but the small companies were free to bid on everything. *Rivers Committee Hearings*, pp. 4828–29.

82. The international lines were to file tariffs which they could change upon thirty days notice or which the CAB could determine to be unreasonable. This would have made prime contractors of the international carriers who were the only ones mentioned by the CAB. Letter from CAB Chairman Whitney Gilliland to Defense Secretary Gates, June 24, 1960 (Mimeographed, CAB). Congressman Rivers protested the proposed action. *Aviation Daily*, July 1 and 5, 1960.

83. The rate this time was 2.9 cents per passenger mile. *New York Times*, August 9, 1960.

the existing system so the small companies could survive.[84] Senator Estes Kefauver sent the CAB a long series of questions implying that the new rates would encourage the monopoly conditions feared by the Celler Committee, and he warned that a new investigation might be in order.[85] The military agencies carried an appeal to the White House, where the president reportedly declined to intervene.[86]

The Defense Department went about developing a means of choosing between companies submitting identical bids. The result was a complex formula which included factors such as CRAF membership, the newness of the aircraft being used, and the additional capacity each company promised for emergency use. Applied rigorously, these factors almost inevitably would favor the large companies.[87] The Defense Department adopted what was, in reality, a compromise designed to divide the business among as many as possible of the eligible claimants.[88] This left no objective standard by which government officials could make choices.[89]

The central logic of the CAB decision seems inescapable; if the government is to rely upon commercial carriers in both peace and war, it must pay rates high enough to permit reasonable profits and the purchase of new transports. To be sure, the CAB decision trapped the Defense Department in some apparent contradictions. It insisted that rates should be only high enough to permit low-profit operation of second-hand aircraft, but asserted also that carriers should buy modern aircraft for military use.[90] The Defense Department thus repeated the role played years earlier by the Post Office, when that agency protested that its responsibility for mail delivery did not include the subsidization of passenger travel. The traditional separation between government and industry once again prevented the former from accepting what clearly seems to have

84. *Aviation Daily*, August 17 and 19, 1960.

85. *Ibid.*, September 21, 1960.

86. *Aviation Daily*, August 19, 1960. The Air Force wanted to retain complete control over rates. Testimony of Assistant Secretary Taylor, *Military Air Transportation, 1960,* pp. 72–73. General Tunner opposed any action that might restrict competition. *Rivers Committee Hearings,* pp. 4823, 4839–40.

87. The large companies were the ones with facilities, personnel, and equipment based around the world.

88. *Aviation Daily*, August 26, 30, October 31, 1960.

89. For the dangers inherent in this arrangement, see Chapter 10.

90. The Air Force had pressured the carriers having the most modern aircraft to join CRAF and use the aircraft on military contracts. *Rivers Committee Hearings,* pp. 4655–56, 4679.

been its responsibility.[91] When the CAB imposed rates in a situation where the administration would not act, some typical airlift problems were left unresolved.

The Defense Department appeal to the White House demonstrated that the interaction of overseas airlift, military policy, and the commercial industry was as confused as ever. The appeal implied that the president could reverse the CAB decision if he chose to do so, presumably because it affected international aspects of airlift policy. This implied, in turn, that the decision ought to have gone to the White House for approval in the first place—as in all international cases. Conversely, if the CAB could act on its own, the president could not reverse the decision.

The CAB demonstrated its own confusion by admitting the following year that it had acted in a "gray area" of its authority.[92] The CAB decision also might have raised more questions about the supplementals. Because the rates were high enough to be profitable for major companies with substantial overheads, the profits of the supplementals became quite high; this led to the questionable traffic in these carrier companies mentioned earlier.[93] Indeed, it is difficult to avoid the impression that the catalyst of the CAB decision was the immensely complex connection between supplemental carriers, aircraft manufacturers, and second-hand aircraft. This was a factor never explicitly considered, but it had forced the major carriers out of the military market and had brought on the price-cutting war for military traffic.

Whatever the unresolved questions, somebody had to make the decision to restore a minimum level of stability. In 1960 as in 1930, competitive bidding was somewhat out of place in a situation involving suppliers regulated by public agency, even if other suppliers were unregulated. The decision was part of a general trend toward long-term stabilized contracts. This was the only realistic approach to the use of commercial airlift, even if it required tacit guarantees to individual companies that business would continue. Congressional and CAB actions seem to have followed from an assumption, or tacit agreement, that commercial airlift was a per-

91. The Rivers Subcommittee suggested that the Air Force negotiate its own rates directly with the industry, and not subject to CAB intervention. *Rivers Report*, p. 4046.

92. *Military Air Transportation, 1961*, p. 91.

93. See Chapter 5.

manent auxiliary to military operation, and administration policy statements followed the same pattern.

THE NATIONAL SECURITY COUNCIL AND THE DEFENSE DEPARTMENT REPORT OF 1960

The administration released early in 1960 a comprehensive report on military airlift policy; the subject had been under detailed scrutiny in the executive branch since 1958, when the president directed the Secretary of Defense to make such a study.[94] When the Quesada program led to controversy within the administration in 1959, the Defense Department study received National Security Council attention. The council reviewed the study at the end of 1959, the president permitting the Defense Department to release it.[95]

The nucleus of the report was nine "Presidentially approved courses of action."[96] Their literal adoption would have restricted MATS to flying its transports only in wartime, for the report argued that hard core (wartime) requirements were so important that strategic transports should do no other flying except when commercial airlift was not available.[97] The report added that an increase in commercial capacity could be translated into a reduction in MATS capacity, but external evidence indicates that the conclusion did not follow from any examination of the need to deploy Army forces.[98] The report implied clearly that the administration had agreed to a cutback in strategic airlift capacity as a response to its critics.[99]

The result was immediate speculation that the Defense Department would turn over more of its passengers and cargo to private

94. Testimony of Mr. R. M. Kenney, a career employee in the office of the Assistant Secretary of Defense for Supply and Logistics, apparently a principal author of the study. *Military Air Transportation, 1959*, p. 75.

95. *Aviation Daily*, December 31, 1959.

96. *MATS in Peace and War*, pp. 5–6.

97. *Ibid.*, p. 8.

98. The report quite clearly was based upon the only officially approved war plans, those concerning general war.

99. Congressman Hardy of the Rivers Subcommittee found it difficult to believe the report had not been written by someone from the industry. *Rivers Committee Hearings*, p. 4359. Holifield said the report was full of "gobbledygook, doubletalk and ambiguity." *Military Air Transportation, 1960*, p. 122.

operators.[100] Even the *Army-Navy-Air Force Journal* announced that the president favored a reduction in MATS.[101] Quesada and Monroney added their approval, Quesada asserting that a 25 per cent reduction was in sight.[102] A trade journal sympathetic to the supplementals hailed the apparent promise that government traffic was to be "returned from socialistic bureaucracy to private enterprise."[103] The ATA has since described the report as the major turning point in its relationship with the government.[104] To keep the sequence straight, it should be noted that the government issued the study during Monroney's hearings on the Quesada program, just a few days before the collapse of the program. Of the reports on airlift policy appearing in 1960 this was the most important, despite its dependence upon an assumption that Quesada and Monroney would succeed. Two other reports, however, deserve mention here.

One was a Commerce Department product, and culminated a two-year study of transportation problems. It generally supported industry claims, pointing out that surplus transports then in the hands of the domestic airlines could meet all conceivable emergency needs.[105] The Budget Bureau and the Defense Department jointly prepared a second report; it is still classified, but an accompanying letter from Defense Secretary Gates and Budget Director Stans outlined its conclusions. Principal among them was the statement that MATS could be reduced as soon as the civil industry acquired some modern cargo aircraft suitable for military use.

All of these reports demonstrated a general administration willingness to accept what seemed inevitable. The studies leading to the reports had begun during the period of intense congressional activity, but the government agencies involved were, to put it simply, operating at a pace that did not keep up with shifts in congressional opinion. The individual government departments thus produced reports wholly compatible with the congressional climate between 1956 and 1959, but almost totally out of step with the

100. *St. Louis Post-Dispatch*, February 10, 1960. *Journal of Commerce*, February 10, 1960.

101. *Army-Navy-Air Force Journal*, February 13, 1960.

102. *St. Louis Post-Dispatch*, February 11, 1960.

103. *Airlift Magazine*, March, 1960.

104. Interview.

105. U.S. Department of Commerce, *Federal Transportation Policy and Program* (Washington: Government Printing Office, 1960), p. 10. The Commerce Department apparently assumed that domestic transportation needs would diminish during war. *Ibid.*, p. 24.

situation in 1960. When Congress decided to reinstate the purchase of military transports, the administration proved itself flexible enough to accept the decision.

THE EISENHOWER ADMINISTRATION'S ACCEPTANCE OF THE PROGRAM

The Air Force had included within its own budget $50 million to begin development of the civil-military cargo transport. The administration, however, retained its hope that industry and government would agree on a particular aircraft. The Army, the Air Force, and FAA all took a hand in drawing up specifications, much of the work being accomplished before Rivers began his hearings.[106] The FAA established liaison with various segments of the airline industry in the attempt to find agreement. As various agencies added new specifications, the proposed aircraft became more of an all-purpose aircraft than originally planned. The new transport was to fly at 440 miles per hour, but also at speeds low enough to permit the dropping of paratroopers. It was to use runways of intermediate length, i.e., 5,000 feet. The fuselage was to be close enough to the ground to permit drive-on loading and cargo handling in areas not having loading equipment.

Despite these requirements, the hope was to produce an aircraft with a low enough operating cost to make it attractive to commercial operators. Military and civilian agencies decided, in effect, that a single transport had to meet every strategic, tactical, and commercial requirement, and at a lower cost in each category than any other aircraft could produce. The specifications, moreover, ruled out the swing-tail jets then under consideration, and made it inevitable that the aircraft would be very much an enlarged version of the Lockheed C-130. The project did not proceed far enough to warrant the award of a contract before the new administration took office.

The Eisenhower administration took action to begin buying the transports specified by Congress. The Air Force very nearly awarded contracts for modified jets, but it decided to return the proposed designs to manufacturers with a request for simpler modifications of existing models.[107] This occurred after the election, indicating that normal administrative action soon would result

106. *Rivers Committee Hearings*, pp. 4309–10.
107. MATS spokesman apparently expected immediate approval. *New York Times*, November 13, 17, 18, 1960.

in acceptance of a modified jet. The administration also began buying the C-130E's urged by Congress.[108] Taken together, these actions were the final step in resolution of the airlift stalemate and the resumption of the airlift modernization program halted so abruptly in 1959.

THE BROADER IMPLICATIONS OF THE CONSENSUS

This third phase of congressional policy-making completed a cycle begun by Congressman Flood's support of the Hoover Commission. The Rivers Subcommittee's performance in filling the congressional policy vacuum in such a way as to pull the administration along with the solution remains a performance that has broader implications for both strategy and politics.

The subcommittee's willingness to dissect JCS war plans and to explore the Army's position in detail made it possible to discover that the Army did not know precisely what it wanted. Its affinity for tactical transports capable of dropping paratroopers made clear for the first time that there was little or no connection between Army requests for airlift, the types of aircraft it wanted, and airline complaints against MATS. The subcommittee actually explained to Army representatives the kinds of airlift they sought, and thus found a consensus that had eluded both the Army and the Air Force. The two services had failed either because their Washington staffs knew little about strategic airlift, because of a lack of communication between those staffs, or both.

Whether the military services should be held responsible for this type of deficiency is an open question. After all, the organizational pattern of the military departments now requires that dialogues between them be subordinated to a rather disjointed discourse between each service and the overhead structure of the Department of Defense. The services, since the creation of the overhead department in 1947, are under only minimal pressure to reach direct agreement; the organizational process encourages them instead to make relatively extreme and separate demands of the overhead agency.

Rivers and his colleagues were able to supply the type of expertise that comes from long experience with the agencies involved. Rivers, for example, had been a member of the Armed Services

108. Recapitulation of Eisenhower administration activity in *Aviation Daily*, January 26, 1961.

Committee for about two decades and had thus dealt with such questions for a much longer period than had any of the military or civilian officials who appeared before him, with the possible exception of Deputy Secretary of Defense Douglas. Rivers and his colleagues had the experience to reject simple solutions to complex problems, and the result was the type of investigation that Congress is unusually well equipped to perform.

The subcommittee gave every interest group the opportunity to be heard and, even when subcommittee members disagreed with witnesses, they did so within a framework of mutual respect for conflicting points of view.[109] The subcommittee shunned publicity except when it could make a measurable contribution to consensus and, perhaps, because no political prize was immediately at stake. These congressional veterans brought to their investigation a maturity of judgment, stability, even wisdom, all virtually indispensable to resolution of the problem. Their performance was in contrast to that of their apparently more decisive colleagues, and it contradicted some familiar stereotypes of congressional behavior.

The image of Southern Democrats and other conservatives as only obstinate opponents of progress seems, in this instance at least, wide of the mark. Using a rather patient approach, the Rivers Subcommittee was able to see more than one side of the problem and to avoid acting in too great haste. It is possible, of course, to dismiss Rivers as a legislator motivated only by his interest in Charleston Air Force Base in South Carolina. Such a narrow view, however, could not have made possible the broad view taken by the subcommittee as a whole. The new airlift program favored MATS only to the degree that the subcommittee assumed that some sort of strategic airlift force would be a continuing necessity. The subcommittee did not, of course, solve all the outstanding problems.

There was no recommendation about what to do if the combination of more strategic airlift and continuing guarantees to the industry produced too much airlift in nonwar situations. The decision to accept the Army view that all transports should be able to drop paratroopers did not answer questions about validity of the strategy and control of the airplanes. The broader problems of airlift policy, especially the supplemental carriers, the CAB's status

109. See especially the exchange with Stuart Tipton, *Rivers Committee Hearings*, pp. 4535–36.

with respect to military and commercial airlift, and the continuing difficulties of the scheduled airlines hardly were touched at all, despite the subcommittee's willingness to listen to everybody. Nor could the subcommittee see in 1960 that by combining high-utilization strategic operations and low-utilization paratroop missions in the same types of aircraft (C-130E and the new civil-military transport), new organizational conflicts had become inevitable. The fact remains, however, that Congress resolved the immediate military airlift problems, appropriated the funds, and got the program well underway before the change in administrations. The campaign itself and the Kennedy administration's initial actions in the defense policy area were, in contrast, the outcome of an assumption that nothing had as yet been done.

PART FOUR

Military and Commercial Policies, Today and Tomorrow

[10]

Military Airlift in the 1960's and 1970's

MILITARY AIRLIFT POLICY played some part in the 1960 Democratic attack on Eisenhower administration defense policies, and the Kennedy administration gave early attention to the airlift question. The Democrats were committed on the one hand to providing the Army with more airlift and with a new joint command and, on the other hand, to Hoover Commission principles of increased centralization within the Defense Department and greater attention to business management. The organizational decisions, the intensification of conflict within the Defense Department, and the prospects for the future of military airlift are examined in this chapter.

The Democrats' early concentration on military airlift only served to obscure further the steadily declining position of U.S. international carriers, a problem of perhaps greater significance to air transport policy as a whole. U.S. carriers have, in the 1960's, continued to expand their operation, but their relative position with respect to the other international airlines of the world has not improved and, in some cases, has declined. Airline regulation and export promotion have remained separate policies, separately administered. Chapter 11 analyzes the sharp contrast between U.S. success in selling transports abroad and the effects of that success upon U.S. carriers. The same chapter also includes an analysis of the theoretical approaches which seem so important to the whole problem.

THE CAMPAIGN AND THE FIRST ACTIONS OF THE NEW ADMINISTRATION

The administration's first actions reflected the 1960 campaign; airlift was not a major issue, but it was one of several examples pointed to as demonstrative of defense policy deficiencies. Kennedy asserted during the campaign that the U.S. did not have the airlift to move more than one division and that more jet transports were needed. Sixteen divisions held within the U.S. had little value, he argued, if the U.S. could not move them around the world within one or two days.[1]

Airlift was at the top of the new president's list of military actions needed immediately, or so he listed it in his first message on the State of the Union. He told Congress that: "Obtaining additional air transport mobility—and obtaining it now—will better assure the ability of our conventional forces to respond, with determination and speed, to any problem at any spot on the globe, at a moment's notice."[2]

Secretary of Defense McNamara took the same approach during his first press conference, announcing immediate steps to improve airlift capacity. The administration, he announced, was buying thirty Boeing jets, and he added that a step-up in the Lockheed production line would deliver fifty C-130E's by March of 1963 instead of September. His estimate was an immediate net increase of fifty-three transports above the previous administration's final program.[3] Congressional questioners brought out later that the C-130E's actually substituted for C-130B's originally scheduled for production; the net increase was only twenty-nine transports, and most of that was in Boeing jets.[4]

Several congressmen, including Senators Goldwater and Smathers, were among the first to congratulate the administration for its action, but Representative Flood was not so sure the action was the correct one. None of the transports but the Boeing jets would be

1. U.S. Congress, Senate, Committee on Commerce, *Freedom of Communications, Remarks, Press Conferences, and Statements of Senator John F. Kennedy, Aug. 1 through Nov. 7, 1960*, 87th Congress, 1st Session, Senate Report 944 (Washington: Government Printing Office, 1961), Part 1, pp. 650–51, 1012 (hereinafter *Kennedy Speeches*).

2. *Department of State Bulletin*, 44 (February 13, 1961), 211.

3. *Aviation Daily*, February 3, 1961.

4. U.S. Congress, House, Committee on Appropriations, *Department of Defense Appropriations for 1962*, Hearings before Subcommittee, 87th Congress, 1st Session (Washington: Government Printing Office, 1961), Part 3, pp. 18–19 (hereinafter *HCA DOD Appropriations for 1962*). McNamara's budget statements at the time were worded so as to imply that he was asking for 129 transports that had not been programmed before the advent of the new administration. U.S. Congress, Senate, Committee on Armed Services, *Military Procurement Authorization Fiscal Year 1962*, Hearings, 87th Congress, 1st Session (Washington: Government Printing Office, 1961), p. 17. Professor Kaufmann later used the same figure in the same way, indicating the whole program was new—but without documentation. William W. Kaufmann, *The McNamara Strategy* (New York: Harper & Row, 1964), p. 58.

available soon enough, he said, and they were not suitable for cargo. The U.S., he added, should purchase some of the swing-tail jets being produced in Canada.[5] In answer to some of Flood's other comments, Air Force spokesmen said the Boeing jets would not engage in scheduled passenger operations. They would be the most austere aircraft imaginable; ordinary tankers would be modified only by adding heavy floors, troop seats, ventilating equipment, and washrooms. In what since has become a standard practice, the new transports would not even have windows.[6] This initial purchase of jet transports later was expanded to a total of forty-five, before the administration decided their limitations as cargo carriers made it inadvisable to buy any more.[7]

The White House announced that Lockheed Aircraft Corporation had won the contract to develop the new civil-military cargo transport.[8] Designated as the C-141, the new transport will become the standard military cargo airplane by the late 1960's; it will carry 34.5 tons for a distance of 4,000 nautical miles, airlift and drop 118 paratroopers.[9] It will be able to bypass some island bases, to land in rough fields, and to take on cargo by straight-in loading (under the tail) from very low platforms. There are indications, however, that this all-purpose transport may not be wholly satisfactory to either industrial or military users.

The ATA maintained liaison with the FAA until late 1961, then announced the C-141 project had become too military in orientation to warrant further industry participation in the planning

5. Goldwater praised the Kennedy administration: Smathers praised Symington and Monroney. *Washington Post*, February 5, 1961; *Christian Science Monitor*, February 4, 1961; *Congressional Record*, February 24, 1961, pp. 2542–43. Flood's position was, of course, a sharp reversal of his earlier opposition to larger swing-tail jets. *HCA DOD Appropriations for 1962*, Part 2, p. 1132, Part 3, pp. 76–77, 484–85. On Rep. Flood, see pp. 174–75.

6. The development of windowless transports, solely to make certain they do not resemble commercial models, may produce odd results for Army troops. Commercial operators agreed that even supersonic transports would need windows, principally to avoid claustrophobia among passengers. *New York Times*, April 21, 1961.

7. The C-135's are typical of the confusion that has surrounded the airlift question. The Kennedy administration bought them principally to help the Army, but that Service had little interest in them. Rivers had asked for them only to support the post-strike mission of MATS to rendezvous with SAC bombers.

8. *New York Times, Washington Post*, March 14, 1961.

9. *Interavia*, 17 (January, 1962), 88; U.S. Congress, Senate, Committee on Appropriations, *Department of Defense Appropriations for 1963*, Hearings before Subcommittee, 87th Congress, 2nd Session (Washington: Government Printing Office, 1962), p. 36 (hereinafter *SCA DOD Appropriations for 1963*).

phase.[10] A number of airlines concluded that the cargo compartment was too large for the present commercial market or, alternatively, that costs had risen too high for that size airplane.[11] From industry's point of view, the C-141 probably is not much of an advance over less expensive jets already in service. The remaining supplemental carriers, on the other hand, may well be ready to operate the C-141 (or anything else) with the help of guaranteed military contracts.[12]

Military difficulties with the C-141 may become even more complex. The transport is a quantum increase in mathematically available airlift capacity, but it will be much smaller in important dimensions than the C-132 development project canceled in 1956 or, for that matter, the old C-124. It will be an elongated C-130, in that the C-141 cargo compartment has most of the C-130's dimensions save length.[13] The C-141 may not be an important advance over the jets first proposed in 1959, and its ability to drop paratroopers may lead only to new, even bitter, controversy.

The paratroopers still seek a joint command having its own airlift or the guaranteed predesignation of transports that can do that job. The C-141 will not show up well at all in the economic analyses now so important to the Defense Department unless the transport can operate in a strategic environment at a high utilization rate. Secretary McNamara decided in 1962 to study the feasibility of operating the aircraft at a nine-hour rate during emergencies.

10. There were several reports of difficulties in the civil-military liaison. *New York Times*, October 5, 1960; *Business Week*, March 25, 1961. On ATA withdrawal, see *New York Times*, November 20, 1961.

11. *Interavia, op. cit.*, p. 89.

12. The C-141 seemed likely to cost $5 million to $7 million per airplane. *Washington Post*, March 14, 1961. One sidelight indicates the confusion surrounding the attempt to build a transport for both the civil and military market. The Air Force, with Congressional approval, included within the contract a provision that commercial operators buying the aircraft would have to pay their share of the development costs. *SCA DOD Appropriations for 1963*, p. 1383. Carriers could be expected to buy the aircraft, however, only if: (1) loans are guaranteed by the government; or (2) the government has guaranteed to provide the traffic to keep the aircraft busy; or (3) the company has a certificate entitling it to subsidy. The government ultimately must make up any losses.

13. C-141 cargo space is 7,156 cubic feet; the C-132 had 15,662. The C-132 would have airlifted 50 tons for 3,500 miles. *Interavia, op. cit.*, p. 89; U.S. Congress, House, Committee on Appropriations, *Department of the Air Force Appropriations for 1958*, Hearings before Subcommittee, 85th Congress, 1st Session (Washington: Government Printing Office, 1957), p. 4 (Statement on C-132).

As he pointed out, such an increase over the traditional eight-hour rate could produce a 12 per cent expansion in airlift capacity without the purchase of an additional transport.[14] The Air Force has emphasized this trend by installing automated cargo handling systems in important air cargo terminals.[15] But the Defense Department may yet have to choose between the high utilization long-haul approach to strategic airlift and the low-utilization one-function concept of the paratroopers.

It is difficult to see, in other words, how the C-141—no matter its versatility—can be made to serve all the operational and commercial requirements ostensibly built into it. If the Defense Department tries to realize a return on its investment by concentrating on strategic operations, the airborne divisions once again will be unhappy with the result. Sustained strategic airlift remains possible only between airfields equipped with instrument facilities, cargo handling equipment, and integrated ground teams that can make quick repairs. If, on the other hand, the Defense Department concentrates on the airborne assault role, sustained strategic operations probably are out of the question. The airdropping of paratroopers and operation into rough fields implies a lack of corollary facilities, losses to enemy action, and high dollar cost per flying hour. The C-141 thus exemplifies, in terms of hardware, the old military controversy that has endured for so many years.

Subsequent sections of this chapter detail the Democratic administration's organizational decisions which seem to contradict each other because they stand, at least in the context of airlift, on both sides of the controversy. The Army, moreover, has added still more to the confusion by developing new theories of mobility and proposing new airlift programs which imply that the C-141 will not operate into combat zones, despite the capabilities built into the aircraft at Army request.

To put them in perspective, the administration's 1961 procurement actions were only the final phase of congressional resolution of the problem; the outcome of a Republican victory would have

14. U.S. Congress, Senate, Committee on Armed Services, *Military Procurement Authorization, Fiscal Year 1963,* Hearings, 87th Congress, 2nd Session (Washington: Government Printing Office, 1962), pp. 151–52 (hereinafter *Military Procurement, 1963*).

15. This is known as the "463L system." U.S. Congress, House, Committee on Appropriations, *Department of Defense Appropriations for 1963,* Hearings before Subcommittee, 87th Congress, 2nd Session (Washington: Government Printing Office, 1962), Part 4, pp. 457–60 (hereinafter *HCA DOD Appropriations for 1963*).

been just about the same. The administration doubtless intended that its immediate announcements reflect its concern with limited war capabilities and redeem campaign pledges to that effect.[16] The administration's organizational actions, by contrast, were wholly its own.

SOURCES OF REORGANIZATION: THE AIRBORNE GENERALS AND THE DESCENDANTS OF THE HOOVER COMMISSION

Two relatively distinct groups shaped the intellectual content of Kennedy administration defense policies, and the theories of the two groups tended to contradict each other. The administration reorganized the Defense Department along lines recommended by both the airborne generals and the Hoover Commission; both approaches were present during the campaign and the interregnum.

Taylor and Gavin became important military advisers to the Kennedy campaign organization. Both belonged to an advisory group, headed by Paul S. Nitze, which also included General S. L. A. Marshall of the *Detroit News*.[17] During the campaign Kennedy referred often to Taylor, Gavin, and Ridgway, implying his total agreement with them.[18] Several members of the group later joined the new administration as defense policy-makers (Taylor and Nitze) or in other capacities (Gavin's ambassadorship to France).

Also influential was a defense policy study group headed by Senator Symington, one of several task forces the president-elect used after the election. The Symington group's approach was much like that of the Hoover Commission. Roswell Gilpatric, a member of the Symington task force and also of the 1957 Rockefeller panel, later became Deputy Secretary of Defense for the administration. The Hoover Commission, it will be recalled, advocated removal from direct military control of all supply and transportation activities within the Defense Department, and the transfer of those functions to businesslike organizations managed by civilian specialists.[19] The Rockefeller and Symington task forces used different phraseology but recommended about the same solution.

16. Kennedy also had referred to airlift as the major problem of the Lebanon crisis. *Kennedy Speeches, op. cit.*, p. 981. See also Chapter 6, note 86.
17. *Ibid.*, p. 61.
18. There were at least twenty-two such references. *Ibid.*, Index.
19. See Chapter 6.

Both groups argued that the Defense Department should be organized into two distinct lines of command and, at least in theory, this has been the case since the 1958 defense reorganization. The line of operational command, said the Rockefeller Report, should run from the president and the Secretary of Defense through the Chairman of the JCS to functional unified commanders. The Symington group named the most important of these commanders as those of the proposed Strategic, Tactical, and Defense Commands, while the Rockefeller panel mentioned commands charged with all-out war, limited war, and continental defense.[20] The military services should be outside this operational line of command, said both reports, and should concentrate on administrative and logistic support functions. The line of logistic command would run from the president and the Secretary of Defense to the Service Secretaries. Once accomplished, this reorganization would free the Secretary of Defense from his role as a mere arbitrator among the services. Defense planning, so went the argument, had become only a series of compromise positions reached through prolonged debate, discussion and negotiation between service chiefs holding stubbornly to parochial views. Both reports thus proposed distinct boundaries between combat and support functions, much in the manner of the old Army bureau system. As Samuel Huntington observes, the removal of the services from a role in combat operations might transform them into "English regiments on a grand scale—administrative organizations rather than fighting organizations."[21]

These all-important similarities of the Hoover Commission, Rockefeller, and Symington approaches were less easily noticeable in 1960. Critics of the Symington group noted that he, Gilpatric, and Thomas K. Finletter, yet another member of the group, all had

20. *New Frontiers of the Kennedy Administration,* The Texts of the Task Force Reports Prepared for the President (Washington: Public Affairs Press, 1961), pp. 1–5 (hereinafter *Symington Report*). *International Security, The Military Aspect,* Report of Panel II of the Special Studies Project, Rockefeller Fund (Garden City: Doubleday & Company, Inc., 1958), pp. 31–33 (hereinafter *Rockefeller Report*).

21. Samuel P. Huntington, "Interservice Competition and the Political Roles of the Armed Services," *American Political Science Review,* LV (March, 1961), 52. McNamara adopted as his own, in 1962, the theory that "logistics" and "operations" were separable into two lines of command. U.S. Congress, House, Committee on Armed Services, *Hearings before the Special Subcommittee on Defense Agencies,* 87th Congress, 2nd Session (Washington: Government Printing Office, 1962), p. 6693.

been associated with the Air Force.[22] True enough, Symington used Air Force terminology to describe the tasks of his proposed major commands, but his proposals also were quite close to what Taylor had asked. Taylor wanted increased authority for the Chairman of the JCS, to overcome what he saw as the sluggishness of committee deliberation; he wanted also a new joint command charged with the planning and conduct of limited war.[23] It must have appeared to the new administration that Defense Department needs were well known to all; beginning with the Hoover Commission almost everyone studying the problem had suggested the same reorganizations.

The administration chose to act immediately, probably before anyone had carefully measured the details of these proposals against each other. Military airlift has turned out to be perhaps the most important single military activity caught in the middle between all of these approaches and some others as well. Taylor looked upon airlift as an integral part of a new joint command and thus a combat function. The Hoover Commission looked to airlift as a logistic function best managed by civilian experts. Taylor's was the standard airborne view; the Hoover Commission equated airlift with airline transportation and thought commercial companies could do most of the work. The Rockefeller and Symington reports were vague enough to be taken as support of either of the other approaches.[24] The administration adopted both these approaches, encouraged the Army to develop bold new programs for limited war operations, and reorganized the budgetary structure of the Defense Department in a fairly major way. These four important decisions require separate analysis and discussion in the next four sections of this chapter.

THE NEW JOINT COMMAND: STRICOM

In establishing a new joint command, the administration went beyond the limits of previous experiments of that type.[25] Yet, be-

22. See articles reproduced in *Military Procurement Authorization, 1962*, pp. 262–86.
23. See Chapter 6.
24. It will be recalled that Henry Kissinger, principal author of the Rockefeller Report, had not studied the airlift question in any depth; like Symington, he apparently was unaware of the airlift assigned to MATS.
25. The First Allied Airborne Army of World War II and the 18th Air Force formed at the end of the Korean War.

ginning with the administration's first announcements on the subject, it has demonstrated some uncertainty about the precise nature of the organization it sought to build.

The *New York Times* reported in March, 1961, that major elements of the Army's Strategic Army Corps and the Air Force's Tactical Air Command would be combined within a new limited war command headed by an Army general, an action that finally would give the Army control of close support fighter aircraft.[26] The Defense Department said nothing until September, then announced the appointment of General Paul Adams to head STRIKE Command (STRICOM) and to be responsible for: ". . . the integrated employment of the land and tactical air forces assigned, and the training necessary to weld these forces into an efficient land-air team."[27] McNamara explained in a press conference that the Marines already were an integrated team having its own close support and airlift, hence were not involved in the reorganization.[28]

The mention of the Marine Corps was the first indication of Defense Department uncertainty about STRICOM. If the new command was to parallel the Marines, it would have tactical, but not strategic airlift. Marine airlift, moreover, does not contemplate massive airborne assault, but only a relatively small airlift force (twelve aircraft per division) for short-haul supply as an auxiliary to surface transport. In the jargon of the Rockefeller and Symington reports, specifically adopted by McNamara, the Marine Corps belong within the logistic chain of command. Like other services the Marines recruit and train individuals and divisions, then send them along to become component parts of the overseas commands which control combat activities. The Commandant of the Marine Corps does not expect to personally lead his troops into battle. STRICOM, on the other hand, is a new unified command (combat organization) wholly within the operational chain of command. It has proved extremely difficult to fit the new organization into the global system of political alliances and regionally unified military commands.

The nature of the problem is best illustrated by listing situations in which the U.S. might assemble combat forces: (1) A training exercise or military crisis within the U.S.; (2) A training exercise or crisis within an area not within the jurisdiction of one of the

26. *New York Times,* March 27, 1961.
27. *Department of Defense Press Release* 1006–61, September 19, 1961.
28. *New York Herald Tribune,* September 20, 1961.

existing unified commands; or (3) A training exercise or crisis in an area of the world within the jurisdiction of one of the existing unified commands.

In the first situation, STRICOM might function as the principal combat command for either an exercise or a real emergency. In the case of exercises, e.g., the airdropping in North Carolina of thousands of paratroopers, STRICOM now acts as the operational command in charge of the maneuver. In an actual crisis, however, a more comprehensive command structure might supersede STRICOM; air defense of North America, for example, already is part of a unified command which includes Canadian elements, and STRICOM is too limited in scope to manage such operations.

In the second situation, STRICOM also might function as the principal combat command, but this might not automatically be the case. Some geographical areas might logically require an organizational arrangement that STRICOM cannot provide. In the 1958 Lebanon crisis, for example, U.S. operations were commanded by a Navy admiral. It might be argued, of course, that STRICOM should include Navy and Marine components as well as Army and Air Force.[29] To put all the forces in the U.S. into a single command, however, would be tantamount to creating a single military service and wiping out all professional distinctions between land, sea, and air forces. One can question if this type of monolithic structure would be an appropriate solution.

It seems likely that most strategic training exercises and crises will continue to occur within or very close to areas in which the U.S. already has forces deployed or, at the minimum, has some sort of alliance system in being. Because of the geographical environment, an Army general traditionally commands U.S. forces in Europe, although Air Force General Lauris Norstad was a prominent exception. Similarly, a Navy Admiral usually controls forces in the Pacific. These senior officers control many units from all the services. In both training exercises and crises, these commanders decide what additional forces they need and so advise the JCS and higher authority. To whatever extent discussions determine precisely what forces are sent, the JCS, the Secretary of Defense and the president conduct them. When forces leave the U.S., the area commander takes control of them almost immediately. Arrangements of this sort blanket much of the world, and the Cuban mis-

29. This argument is in Frank Harvey, *Strike Command* (New York: Duell, Sloan and Pearce, 1962), pp. 154–60.

sile crisis of 1962 was a case in point. The Atlantic Command (a Navy Admiral) would have controlled any military operation in its area of jurisdiction. General Taylor, by that time Chairman of the JCS, reported later that "STRICOM was not appropriate to the command structure that had been planned."[30]

The lines drawn above became less distinct in 1963 than they were at the time of STRICOM's birth. The new command took jurisdiction over several areas distinguishable by the absence of U.S. forces. Over Navy objections STRICOM became responsible for the Middle East, Africa south of the Sahara, and the western coast of the Indian Ocean.[31] If this later action was a step toward U.S. withdrawal from overseas deployments, then broader military and political questions are at stake.

Apparently because of its difficulties in finding a significant niche within the existing geographical structure, STRICOM has redesigned its mission. The command calls itself a "packaging and delivery service" which brings together highly trained Army and Air Force units, combines them into packages, then delivers them overseas.[32] This emphasis upon delivery is a sharp turn away from Marine Corps experience, because Marines depend for delivery upon ships or planes from other services. The concept of delivery, combined with old airborne concepts, has encouraged STRICOM to seek the command or control of the strategic airlift used to move its packages across the oceans. The command's permanent structure, for example, consists mainly of a planning staff which does not take control of combat units until the start of an exercise or an actual emergency. The Army general heading STRAC and the Air Force general commanding TAC then become principal assistants to the commander of STRICOM, and they bring along with them the units being sent overseas. Using such a scheme it seems logical enough to those within STRICOM that all the airlift it needs be assigned to TAC and then turned over to the consolidated overhead organization.

The new emphasis on airlift is only a strategic application of traditional airborne philosophy. General Adams has spoken of plans to quickly marshal the existing airlift capacity and exploit it

30. U.S. Congress, House, Committee on Appropriations, *Department of Defense Appropriations for 1964,* Hearings before Subcommittee, 88th Congress, 1st Session (Washington: Government Printing Office, 1963), Part 1, p. 452.

31. Mark S. Watson in *Baltimore Sun,* October 30, 1963; John G. Norris in *Washington Post,* October 29, 1963.

32. *Army Magazine,* April, 1962.

to a degree never before realized.[33] In phrases reminiscent of the Army's predesignation theory, he adds that Army battle groups will be moved "overnight" after only two hours notice.[34] As the Air Force contributor to STRICOM, the TAC commander has announced that he looks forward to having the new C-141's within his command, and TAC also seeks the C-130's assigned to MATS.[35] There is little doubt that the new command hopes to get and keep its own strategic airlift, but it has proved virtually impossible to permit the command to manage its own airlift.[36]

Practice deployments of Army troops from the U.S. to overseas areas have become commonplace since 1960.[37] The first truly long-haul exercise was "Operation Long Pass"; held in 1961, it involved the airlift of an Army battle group (one-fifth of a division) and a CASF from the U.S., and other units from Hawaii and Okinawa, to the Philippines.[38] The exercise occurred within the Pacific Command's area of jurisdiction, and the Navy Admiral's staff determined the structure of the exercise and the schedule of delivery. Exercises of this sort gradually have become more simplified in execution, especially with the storage at overseas depots of stocks of heavy equipment. By 1963 there was enough equipment in Europe for two Army divisions; in General Norstad's phrase, troops can be flown from the U.S. "just like airline passengers."[39]

33. *Army Information Digest*, May, 1962.

34. *Time*, August 17, 1962.

35. *Army-Navy-Air Force Journal*, June 9, 1962.

36. One knowledgeable individual (not in the military) told the author that STRICOM's objective was to acquire some 400 strategic transports for its own use.

37. The Kennedy administration added $65 million to the 1961 budget for such exercises. U.S. Congress, Senate, Committee on Appropriations, *Department of Defense Appropriations for 1962*, Hearings before Subcommittee, 87th Congress, 1st Session (Washington: Government Printing Office, 1961), p. 31 (hereinafter *SCA DOD Appropriations for 1962*). Thirty-eight such exercises were in the 1964 budget, five of them described as "large-scale." *Armed Forces Management*, April, 1963.

38. This was described as the "most significant limited war exercise yet held." John G. Norris in *Washington Post*, February 15, 1961. Lemnitzer said later the exercise had proved that the U.S. could deploy troops very quickly. U.S. Congress, House, Committee on Armed Services, *Military Posture Briefings*, Hearings, 87th Congress, 1st Session (Washington: Government Printing Office, 1961), p. 655.

39. U.S. Congress, House, Committee on Armed Services, *Military Posture Briefings*, Hearings, 88th Congress, 1st Session (Washington: Government Printing Office, 1963), p. 695 (hereinafter *Military Posture, 1963*). General Norstad used the airline analogy in his final interview before leaving NATO. CBS-TV interview, November 11, 1962.

"Operation Big Lift," the first exercise of this type, was held in October, 1963; an entire division of Army troops flew to Germany and joined up with the equipment stored there. The maneuver implied that the U.S., having demonstrated its ability to quickly airlift troops anywhere in the world, might choose to withdraw some forces from overseas locations. The reaction persuaded the administration to remind its allies, especially the Germans, that it did not contemplate any such withdrawal.[40]

In all such exercises it still is necessary to use the control mechanisms first adopted in World War II. Generally speaking, the JCS allocates the airlift and MATS manages it; it is not subject to direct control by the overseas commanders. As was the case in World War II, the theater commander can take control of the transports long enough to make reasonable changes to the schedule—but only if the hypothetical or actual combat situation in his area forces him to do so. The theater commander, on the other hand, recognizes that his authority is limited; the transports must return to the U.S. for use on other missions for all services.[41] There are at least two reasons why this pattern does not fit the STRICOM scheme.

One reason is that almost any emergency will be large enough to force strategic transports to operate at the highest utilization attainable. Secretary McNamara's notion that C-141's should operate at nine-hour utilization is sufficient evidence of that. If the Marines must deploy, the same transports will have to do the job and, in addition, carry supplies to everyone overseas. Even during training exercises the demands for airlift do not cease, especially the continuous supply of forces already in place. A second reason is that commercial airlift, already fully employed on government contracts, can do some of the work in such training exercises if the plans are drawn up in advance.[42] It may not be feasible to turn over management of these aircraft to a combat commander who may himself leave for the combat area.

Whatever the Democratic administration's pride in its new com-

40. Jim Lucas in *Washington News*, October 25, 1963; Jack Raymond in *New York Times*, October 26, 27, 1963; *Baltimore Sun*, October 26, 1963; *New York Herald Tribune*, October 27, 1963; Richard Fryklund, *Washington Star*, October 27, 1963; Brigadier General T. R. Phillips, *St. Louis Post-Dispatch*, October 27, 1963.

41. See Chapter 3.

42. Some officials advised the author that STRICOM had expressed an interest in using commercial airlift for some exercises.

mand, it admitted tacitly that it could not give STRICOM the full-time airlift it seeks. STRICOM's own concept conflicts with the global structure of unified commands and with McNamara's search for higher utilization, but the administration overlooked these problems in pointing to the new command as an outstanding example of its defense policy innovations.

Journalists, their accounts admittedly based upon extended interviews with high Defense Department officials, reported that airlift was indeed a major consideration in forming STRICOM. *Life*, for example, described the command as one that solved the Army's airlift problem.[43] Theodore H. White put it this way: "Eighteen months ago, the Army would have negotiated with the Air Force for airlift; now STRICOM commands its airlift, and airlift has doubled, will shortly triple in capacity."[44]

McNamara asserted on more than one occasion that the administration's programs would by 1967 have tripled the airlift capacity available in 1961. He released no precise calculations to prove such a claim, but he seemed to be assuming that strategic transports will operate at high utilization.[45] The Defense Department thus calculated its airlift capacity in a way that makes it impossible to give STRICOM its own, told journalists it had in fact given STRICOM the transports, and greatly encouraged the new command itself. One major outcome of these activities was the intensification of organizational conflict within the Air Force.

The TAC-MATS struggle had a substantial effect upon MATS, as it sought to prove it is well equipped to be a combat organization in every sense of the word. It trained its flight crews to fly close formation and to airdrop paratroopers and equipment.[46] The training program began with the C-124, was extended into the C-130E, and probably will involve the new C-141. In the attempt to

43. Louis Kraar, "The Two Lives of Robert McNamara," *Life*, 22 (November 30, 1962), 96.

44. Theodore H. White, "Revolution in the Pentagon," *Look*, 27 (April 27, 1963), 481. Kaufmann listed the organization of STRICOM as one of the major changes introduced into defense organization after 1960. Kaufmann, *op. cit.*, p. 191.

45. U.S. Congress, House, Committee on Merchant Marine and Fisheries, *Review of Merchant Marine Policy, 1962*, Hearings before Subcommittee, 87th Congress, 2nd Session (Washington: Government Printing Office, 1962), p. 88 (hereinafter *Merchant Marine, 1962*); also *HCA DOD Appropriations for 1964*, Part 1, p. 517; Kaufmann uses a McNamara statement that airlift capacity will have increased 400 per cent by 1967. Kaufmann, *op. cit.*, p. 307.

46. *HCA DOD Appropriations for 1962*, Part 2, p. 1098; *1963*, Part 3, p. 525.

overcome its airline image, MATS announced that its heavy cargo aircraft had increased their specialized operations to 75 per cent of the total.[47] The command even tried at one point to rename its subordinate organizations so as to erase its traditional geographic structure. The Eastern and Western Transport Air Forces were to be renamed the Seventh and Eleventh Air Forces, in accordance with the standard method for designating combat organizations. "Air transport" wings were to be called "combat airlift" wings. Press releases announced official approval of the new titles, but they were suddenly withdrawn without explanation.[48]

MATS managed to acquire new popularity in previously unfriendly circles.[49] Its attempts to associate itself with combat airlift led also to increasingly bitter conflicts with TAC which, understandably enough, must look upon itself as STRICOM's airlift supplier. Before this conflict can be settled and STRICOM's ultimate position determined, the Defense Department and the JCS will have to face at least five important questions they have managed to avoid thus far.

The first of these is the old question of airlift and the airborne divisions, raised now to the level of the newest and longest range airplanes available. The STRICOM experiment hardly is a new one; it never has taken hold permanently because, at least in part, it is too expensive to let the airborne divisions keep their airlift exclusively to themselves. It remains as unclear as ever, moreover, whether these airborne divisions are the core of U.S. strategy for conventional warfare.

A second question, closely related to the first, is to what degree any such command must constantly conduct strategic training exercises. The airborne divisions are the only ones requiring constant training for both troopers and flight crews, but this calls only for short-haul operation.[50] Most strategic exercises do little more

47. About 80 per cent of MATS' operation formerly had been on a scheduled basis. *Ibid.*, p. 524.

48. *Air Force Times*, June 30, 1962. As of 1964, MATS still was organized along the regional pattern first developed in World War II, with the exception that only two broad regions remain, the Eastern (Atlantic) and Western (Pacific).

49. Congressman Flood announced his pleasure at finding MATS to be a "red-hot special airlift outfit . . . with the hottest and best and the mostest airlift." *HCA DOD Appropriations for 1963*, Part 3, p. 561.

50. Most transports can airlift and airdrop fewer paratroopers than they can carry as normal passengers. Substantial capacity is not used when paratroopers are carried from the U.S. to overseas areas so as to permit airdrop.

than give soldiers practice in riding as passengers in strategic transports or, perhaps, on commercial airliners. If it is not clear to others that this does not take too much practice, it is clear enough to the Commandant of the Marine Corps. He admitted that his troops might have to use strategic airlift upon occasion but, he added, they do not have to go "running all over the country."[51] It is sufficient, he argued, to have individual transports visit his bases once in a while so that Marines can learn how to load their equipment. Indeed, one might even argue that constant exercises contradict the central logic of STRICOM. If troops must be ready to move anywhere on short notice, it can be dangerous to give them too much practice; they might be on the opposite side of the world when crisis erupts.

The third question is the persistently troublesome utilization problem. By encouraging STRICOM while refusing to give it transports of its own, the Defense Department only reinstated old conflicts. The internal Air Force struggle between MATS and TAC is only an updating of the old conflict between the airborne divisions (now represented by TAC) and the remainder of the Army and, perhaps, the other Services (represented by MATS).

The fourth question, quite a complex one, concerns the subtle shift in military and political priorities that accompanied the organization of STRICOM and the TAC-MATS struggle. The prominence of constant readiness, instant deployment, and repeated training exercises gave Army units within the U.S. a stronger claim upon available logistic support than is permitted the units overseas. It became more important to conduct training exercises for STRICOM than to maintain without interruption the flow of supplies to those overseas units, be they Army, Navy, Marine, or Air Force. The emphasis upon STRICOM may have led to a reversal in the most important of strategic priorities. Yet, so long as the U.S. maintains forces in the overseas areas that seem most susceptible to attack, those forces probably will have to absorb the initial shock of aggression. It seems logical enough to insist that the first order of business for strategic airlift remain what it was in the first place; the maintenance of the logistic pipeline to overseas commands that must be kept at the peak of combat readiness. This is an admittedly difficult specification to meet. The government plans training exercises so as to achieve specific political objectives,

51. *Military Procurement, 1962,* p. 308.

and it is not always possible to tell in advance if a given exercise will damage the forces already overseas. If there are real threats in these overseas areas, however, there is little choice but to treat U.S. forces there as the first order of business and to resist the temptation to cut off their supply line.

There is, finally, the broad question of what types of conventional emergencies the U.S. is likely to confront in the 1960's and 1970's. A strong case can be made that conflicts will be oriented toward some combination of land, sea, and air forces, with the U.S. seeking to maintain a military foothold on some land mass, or to re-establish such a foothold after having been forced to withdraw. The relatively recent crises in Korea, the Taiwan Straits, Lebanon, and Cuba have had a strong maritime flavor of a sort which calls for more than a narrow unification of paratroopers and airplanes. Amphibious warfare may yet turn out to be much more important than airborne assaults, and it may be more of a necessity to increase communication between traditional military services than to create commands overcommitted to a single strategy.

There is, it should be noted, a particular political and strategic context which STRICOM seems to fit. If the U.S. should withdraw its military forces from overseas locations and, at the same time, abandon its overseas command structure, STRICOM might be one suitable method for organizing U.S. military forces within the country. It might be possible to create a single unified command from elements of all three services. Its commander could be within the operational chain of command and could be directly responsible to either a single military chief or the Secretary of Defense. This sort of arrangement might work reasonably well in an environment unencumbered by alliances or overseas commands. In the broadest context of political and military strategy, therefore, STRICOM is the concomitant of a policy of isolation and withdrawal.[52] As of 1964, however, U.S. security seems more dependent upon commands such as those in Europe and the Pacific, already

52. "In the future, STRICOM can become just a training command, preparing joint forces for combat duty with the unified commanders overseas. Or it can win a much greater battle role than it has today. If the U.S. should ever decide to pull back its overseas forces (which today number more than 750,000 men) into a Fortress America, then U.S.-based STRICOM, and U.S.-based nuclear ICBM's and Polaris-firing submarines, would form the bulk of America's striking power in the world." Photographic captions written for Theodore White's article; White said nothing of these possibilities. White, *op. cit.* p. 46.

organized and staffed with the finest units and the most respected commanders available.

THE ARMY'S NEW AIRLIFT PROGRAM

In responding to the administration's attention to conventional and limited war forces, the Army produced an entirely new airlift theory which contradicted the assertions that led to the purchase of C-130E's and C-141's. This new theory was a hopeful one in some respects, because it was Army's first attempt to conceptualize airlift as something designed for purposes other than the airdropping of paratroopers. The theory seems overoptimistic for the foreseeable future, because airlift systems themselves do not seem ready to meet such ambitious requests.

The Army proposed that all airlift operating within the forward battle area be assigned to the appropriate ground force commander and that airlift do most, if not all, of the supply job for those forces. This amounted to a suggestion that an airlift system of the sort used during the Berlin blockade operate between rough landing fields within areas having none of the facilities normally associated with high density airlift.

A board of Army officers headed by General Hamilton Howze, a former Director of Army Aviation, developed the new proposal which based its suggestions upon a scenario used to describe hypothetical wars of the future. During a crisis, said the Howze Board, strategic transports (C-141's) would deliver supplies to major air terminals located overseas, just as in World War II and the Korean War. From those terminals, Air Force tactical C-130's would deliver the same supplies to other terminals a bit closer to the forward areas. Because forward areas would be within the jurisdiction of Army Corps or Field Army commanders, the Army would provide its own airlift to distribute the supplies to the combat divisions themselves. To accomplish this important final step, the Army would need its own short-haul transports. One example might be the delivery of cargo offloaded in Turkey by the Air Force to unimproved areas in Iraq.[53] The entire concept was described as one including two levels of "wholesale" delivery by the Air Force and one "retail" Army delivery.[54]

53. *Army Magazine*, November, 1960; *Aviation Daily*, April 19, 1961.

54. Quote from Secretary McNamara in *Armed Forces Management*, April, 1963, p. 54. For brief but comprehensive reviews of the Army's theories, see

The basic combat unit of this new scheme would be the "air assault division" having enough aircraft to provide some of its own airlift and much of its close air support. The Army hoped these aircraft would replace trucks, ground fire support, and armored combat vehicles, each division having some 460 aircraft.[55] The most important parts of the new airlift system, however, would not belong to the air assault division; higher levels of command (Corps) would control the "air transport brigades" providing most of the airlift.[56] The plan was to have as many of these brigades as air assault divisions by 1970, all part of a broader program that would give the Army more than 8,000 aircraft by that time.[57]

In defending the new program the Secretary of the Army spoke of "tactical mobility" and "air mobility concepts" which emphasize the projected increases in fire power.[58] The Army's Director of Aviation insisted that the expansion would include only those planes suitable for "pop up and down" flying and "nap of the earth" support, a function the Air Force had forgotten in its preoccupation with high altitude jets.[59] This concentration upon questions of close air support and communications within the combat zone diverted attention away from the nature of the Army's suggested airlift program.[60]

While McNamara did not commit himself to the Army concept, he agreed to form a single experimental division and to give it the

---

U.S. Congress, House, *A Compilation of Material Relating to United States Defense Policies in 1962*, prepared by the Library of Congress, 88th Congress, 1st Session, H. Document 155 (Washington: Government Printing Office, 1963), pp. 42–43; *Ibid., United States Defense Policies in 1963*, 88th Congress, 2nd Session, H. Document 335 (Washington: Government Printing Office, 1964), pp. 33–36.

55. *Army-Navy-Air Force Journal*, September 29, 1962; *Time*, February 8, 1963.

56. The mainstay of the air transport brigade was the twin-engine *Caribou*, which the Army began buying during the Eisenhower administration. The transport has many of the characteristics of World War II C-47's; it can carry 4 tons of cargo, 32 troops, or 24 troopers, and weighs slightly under 15 tons. *Interavia*, 17 (No. 1, 1962), 77; *Flying*, 70 (May, 1962), 61–62.

57. *Ibid.*, p. 27. The Army objective is five such divisions by 1970. *Military Posture, 1963*, p. 433.

58. *Army-Navy-Air Force Journal*, February 16, 1963. *U.S. Defense Policies in 1963*, p. 34.

59. Interview with Brigadier General Oden, *DATA Magazine*, 7 (November, 1962), 13.

60. The Marine Corps and the Air Force have disagreed with Army attempts to use helicopters as combat aircraft. Questions of this sort received much more attention than airlift. *Military Posture, 1963*, pp. 976–77; *U.S. Defense Policies in 1963*, p. 35.

appropriate number of aircraft.[61] He thought, in addition, that air assault divisions ultimately might replace the traditional airborne divisions.[62] His statements implied that the test of a single such division might not include a test of the air transport brigade, but only a test of that sort could show clearly that the Army envisages a truly massive battlefield airlift system.

The Howze Board's theories are similar in magnitude to the Berlin Airlift of 1948 and 1949, perhaps the most comprehensive such undertaking in history. If airlift does the entire supply job for five air assault divisions, the task will amount to some 5,000 to 7,500 tons each day.[63] At the peak of the Berlin operation, daily deliveries reached about the same level—7,800 tons.[64] In actuality, the total operation for the supply of five air assault divisions would be much greater.

The airfields from which Army transports would operate would themselves be supplied by air, with gasoline and oil constituting a major share of the supplies. While no exact comparisons are possible, gasoline amounted to 60 per cent of everything carried across the India-China route in World War II, unique because those virtually static forces in China depended almost entirely upon air supply.[65] The Air Force share of the airlift probably would be half again as much in tonnage as the Army's, merely to account for this fuel. The Army would thus operate the equivalent of the Berlin Airlift, the Air Force would operate something much larger, and both would function within a shifting combat environment. Even if U.S. air superiority were to protect all these transports from attack, they would have to use airfields without runways, radar, or facilities of any sort. By almost any calculation, then, small Army transports would fly at eight-hour utilization, change airfields quite often, and do without mechanics and spare parts.[66]

61. *Ibid.*, p. 33.

62. *Military Posture, 1963*, p. 433.

63. McNamara reported that a single division required 1,000 to 1,500 tons of supplies each day. *Merchant Marine, 1962*, p. 88.

64. And this figure is based on the highest thirty-day level of operation. *The Berlin Airlift*, Air Ministry and Central Office of Information (London: His Majesty's Stationery Office, 1949), p. 37.

65. William H. Tunner, *Over the Hump* (New York: Duell, Sloan and Pearce, 1964), pp. 43–135.

66. Statement of Brigadier General Robert Williams (Army), *Army-Navy-Air Force Journal*, April 27, 1963. One Defense Department official, in a study of limited war, has implied that this type of utilization is quite attainable. Seymour J. Deitchman, *Limited War and American Defense Policy* (Cambridge: The MIT Press, 1964), p. 163, table 13.1. The question of

The Howze Board apparently had little to say about airborne assaults. Yet the new C-130E's and C-141's are being purchased throughout the decade on the assumptions that all transports must be able to operate into combat zones, use rough fields, and drop paratroopers. The general intention is that C-130E's and C-141's will deliver troops from the U.S. to the forward areas and then drop them there. If all the airlift programs are put together, therefore, the Army requested: (1) 460 aircraft within each air assault division; (2) perhaps as many as 300 Air Force strategic transports to drop paratroopers; and (3) 134 aircraft in each air transport brigade. This amounts to some 800 aircraft which could well be in the air within the battle area assigned to a single division, all of them operating at altitudes of less than say, 4,000 feet. It may be somewhat difficult to control that many aircraft within a primitive area, or one is which sophisticated communications gear has been a battle casualty, but within which the enemy may not be impotent.

The Army's new theories also seem to sacrifice strategic mobility for the sake of a presumed tactical mobility. The new air assault divisions would be completely dependent upon airlift, but they would have to move from the U.S. to overseas areas on strategic transports if they were to be quickly effective. Many of the Army aircraft, especially the transports, are too large to fit into the strategic transports being purchased and must instead fly the oceans on their own. When this occurs, however, adverse winds and weather impose all the risks and delays that make transoceanic flight so uncertain for small airplanes. The troops of the air assault division would thus move in modern aircraft, at jet speeds; their own transports would arrive much later. Under these circumstances the division would be overseas in a matter of a few days, but there would be no way of keeping it supplied. The significant point is that the Army's 1962 mobility concepts, at least with respect to airlift, contradicted the logic of earlier Army proposals. If the Air Force is to provide four-engine strategic transports capable of quickly deploying Army divisions, it makes little sense to tie those same divisions to supply lines that cannot move with them. There is another way of looking at this problem; the Army has left the Air Force with the task of providing airlift for the slightly more

utilization remains critical and it is very expensive to test, but there is no historical evidence available to support acceptance of an assumption that anything of this sort can be done.

than two airborne divisions still in being, but it has volunteered to provide its own transports for all the other divisions—in the 1960's, some twelve to fourteen. What are the causes of these contradictions?

One is the Army's relative inexperience with airlift; this has led to apparent overoptimism about the abilities of aircraft to accomplish so many difficult tasks. The Army's approach to aviation in the 1960's is, in part, like that of the Air Force of the 1930's. Today's Army aviators find it difficult to explain to their fellow ground officers exactly what Army aviation is designed to accomplish.[67] An aircraft that can land in open fields doubtless is attractive, but it may not be able to do as much flying as the average airline transport unless it is supported with a sophisticated logistics complex.

A second reason is the Democratic administration's relationship with the Army since 1961. The Army began its purchase of short-haul transports during the Eisenhower administration, in technical violation of the boundaries between Air Force and Army aviation. The Wilson directive of 1956 and the earlier Pace-Finletter agreement of 1952 may or may not have been reasonable. They were left on the books, however, while everyone admitted they were dead issues, a factor that removed both theoretical and policy limits to the expansion of Army aviation.[68] The only substitute available is the hallowed military principle of *unity of command*, but this provides no practicable standard for policy-making.[69] The

67. Statement of Army Director of Aviation, *DATA Magazine*, 7 (November, 1962), 13.

68. The Army Chief of Staff said that his Service remained under restrictions as to the size and weight of aircraft, but was buying new aircraft that exceeded the restrictions. McNamara said simply that he was not observing the restriction. *HCA DOD Appropriations for 1963*, Part 2, pp. 33–35.

69. *Unity of command* has usually resulted in the "conviction that a commander must control, and therefore the service department to which he belonged must own, all the means necessary for the accomplishment of his combat mission." Lawrence J. Legere, Jr., "Unification of the Armed Forces," (unpublished doctoral dissertation, Harvard University, 1950), quoted in Paul Y. Hammond, *Organizing for Defense* (Princeton: Princeton University Press, 1961), p. 103. The principle has not been too useful for a function, such as airlift, which constantly moves back and forth between two levels of command. Even total unification would not answer the functional question. During World War II, the airborne divisions believed that Eisenhower had subverted their *unity of command*. The principle could be used to defend turning over the new C-141's to the division commanders in whose areas the paratroopers are dropped, or to defend leaving the aircraft with the theater commander, or to defend leaving the aircraft under centralized control from the U.S.

administration took office in 1961, moreover, generally committed to supporting the Army in its attempts to correct what it considered overemphasis on strategic retaliation. This encouraged the Army to submit its own proposals, and both the Howze Board's report and the Air Force's reply were developed in complete secrecy from each other.[70] The administration's approach inadvertently prevented the services from reaching an accommodation with each other and forced all major decisions upward to the Defense Department. The administration thus produced unification where it seems not to be needed (STRICOM), while preventing unification where it is needed most, as in inter-service study groups (not the JCS) which develop new strategy. The argument here, it should be added, does not mean at all that the Air Force view of these problems is the correct one.

A third reason is the Army's difficulty in producing theories which can simultaneously cope with tactical nuclear war and conventional (or guerrilla) war. The Army's new transports fill the aircraft "gap" that was once a minor issue in the airlift controversy.[71] If small units are to be dispersed within a huge battle area, it makes sense to supply them through the use of small transports. The Howze Board, in contrast, went to work at a time when the Army once again was concentrating on conventional warfare on the order of World War II. There may be room within the hypothetical nuclear battlefield for as many aircraft as are involved in these new theories; it seems doubtful that the conventional battlefield will accommodate them.

The Army may be reacting also to its experience in areas such as Viet Nam and Laos, which seem to call for short-haul transports and unimproved fields. In Southeast Asia, however, relatively small units are involved, and they use fixed airfields. There is no need to airlift huge amounts of gasoline into Army airfields, nor is it necessary to change location every day because of quick shifts in a fluid combat environment. In a large-scale battle area

70. This problem intensified after 1961, with the new realignment within the Defense Department. General White, Air Force Chief of Staff while Taylor was Army Chief, said that Taylor, as the new Chairman of the JCS, was a "political appointee," not a military strategist, and that many military leaders agreed with that assessment. General Thomas D. White (ret.), "Strategy and the Defense Intellectuals," *Saturday Evening Post*, 236 (May 4, 1963), p. 10. The point is that the administration rewarded those who had broken ranks during the previous administration. In that context, the Army was not obliged to integrate its views with those of the Navy and Air Force.

71. See Chapter 6.

(World War II or Korea) the ebb and flow of the battle will make logistics extremely difficult, a factor which recalls General Ridgway's observation:

> It is easy for people to dream up intriguing tactical schemes . . . it requires only a modicum of military knowledge, and even professionals of long standing engage in this game from time to time. What throws you in combat is rarely the fact that your tactical scheme was wrong—though, of course, history is replete with examples of faulty tactical planning—but that you failed to think through the hard cold fact of logistics. You failed to ask yourself, "How am I going to get Force A from X to Y and how am I going to supply and sustain it once it gets there?" There is always a great temptation to think only of the objective to be attained, to ignore the basic planning in the hope that in some way the Lord will provide.[72]

AIRLIFT, CENTRALIZED MANAGEMENT IN THE DEFENSE DEPARTMENT, AND THE RELATIONSHIP WITH THE SUPPLEMENTAL AIRLINES

The Army's new airlift program and the organization of STRICOM, if somewhat in conflict with each other, were decisions taken within the operational line of command. The Defense Department's third important decision, however, was totally different in character. The new administration brought to the Defense Department the business management philosophy expressed so forcefully by the Hoover Commission and endorsed by the Rockefeller and Symington reports. Except for the intervention in 1962 of the House Armed Services Committee, the Defense Department might well have turned over to civilian management all airlift and sealift. This would have been only the latest example of a trend in defense organization that preceded the Democratic administration by some years.

The trend is traceable to the National Security Act of 1947, but the Curtis-McCormack amendment to the 1958 Reorganization Act accelerated that trend. It authorized the Secretary of Defense to consolidate supply or service activities in the interests of effectiveness, economy, and efficiency, provided only that the activity in question be common to more than one service. When he estab-

72. Matthew B. Ridgway and H. H. Martin, *Soldier: The Memoirs of Matthew B. Ridgway* (New York: Harper & Brothers, 1956), pp. 277–78.

lished in 1961 the Defense Communications Agency (DCA) and the Defense Supply Agency (DSA), McNamara brought to five the specialized agencies charged with such operations.[73] The increased use of such agencies drew the attention in 1962 of the House Armed Services Committee.

In its study of DSA the Hardy Subcommittee looked into the most extensive use yet made of the Curtis-McCormack Amendment. The subcommittee discovered also that McNamara intended to form more such agencies. It was a relatively simple matter, he argued, to determine mathematically if a proposed agency would meet the economic objectives of the amendment.[74] He supported at the same time the separation between operational matters (unified commands) and logistics affairs (military services).[75]

These new agencies are subject in some cases to diluted control by the JCS and, in others, are not subject to military control at all. The Chief of DSA, for example, is a military officer, but he is responsible only to the Secretary of Defense and, in some not clearly defined manner, to the Assistant Secretary for Installations and Logistics. McNamara understandably has no time to pay close attention to the affairs of the agency, and the Assistant Secretary could devote (in 1962) only three hours of each week to DSA.[76] It appears also that the concept of the Curtis-McCormack Amendment has been stretched too far to suit the Armed Services Committee.

DSA has taken over procurement of electronic components of particular weapons, even though the agency supposedly is barred from dealing in weapon systems.[77] The Defense Department argued in addition, that DSA and similar agencies need not limit themselves to managing items common to more than one service, but could take on individual items or functions if that was the most efficient solution.[78] The Hardy Subcommittee disagreed and

73. U.S. Congress, House, Committee on Armed Services, *Report of Special Subcommittee on Defense Agencies*, 87th Congress, 2nd Session, Committee Print (Washington: Government Printing Office, 1962), pp. 6596–6605 (hereinafter *Hardy Report*).

74. U.S. Congress, House, Committee on Armed Services, *Hearings before the Special Subcommittee on Defense Agencies*, 87th Congress, 2nd Session, (Washington: Government Printing Office, 1962), p. 6693 (hereinafter *Hardy Hearings*).

75. *Ibid.*

76. *Hardy Report*, p. 6619.

77. *Hardy Hearings*, p. 6844.

78. *Hardy Report*, p. 6613.

gathered support from Chairman Vinson of the parent committee.[79] It was during these hearings that the subcommittee came across evidence that a new transportation agency might soon become a companion to the specialized organizations already in operation.

After the Assistant Secretary had insisted that the idea of a new transportation agency was not under study, one of his chief subordinates admitted that just such a study had indeed been undertaken.[80] He described the effort as less than formal, but the Assistant Secretary clearly was caught off guard by the contradictory testimony.[81] The official in charge of the study, a career employee occupying an appointive position, asserted that a transportation agency was only the logical next step. MATS, he said, should become only a technical agency within a larger centralized transportation agency. It should manage all industrial funds, the procurement of all commercial transportation whether land, sea, or air, and the military organizations which perform those functions.[82]

The same official argued elsewhere for a policy-making agency that would merge all forms of transportation into a system of "distribution-as-a-whole" equally valid for peace or war.[83] Airlift, sealift, CRAF, and domestic transportation were fit subjects for the latest computer techniques and other forms of automation. This sophisticated complex, he added, would be as indispensable as missile systems. Several choices were available: (1) hardened sites could make computers invulnerable to attack; (2) computers could be made mobile, as in the case of *Polaris;* or (3) alternate systems could provide insurance against missile attack.[84]

The necessary base for a new transportation agency already was in DSA. The Army's Transportation Corps largely had been dis-

79. *Ibid.*, p. 6616. Congressman Curtis, one of the amendment's authors, supported McNamara in any such reorganizations he cared to make. Curtis specifically included transportation as a possible next step. *Congressional Quarterly*, August 24, 1962, p. 1418.

80. *Hardy Hearings*, pp. 7017–18, 7067–68.

81. Interviews.

82. *Hardy Hearings*, p. 7073. After the Hoover Commission Reports of 1955, there was a steady and substantial rise in the number and rank of civil service positions in the Defense Department. In the top four grades, where salaries gradually rose to more than $20,000 per year, the increase was from 3,009 in 1959 to 3,950 in 1962. Hanson W. Baldwin, "The McNamara Monarchy," *Saturday Evening Post*, 236 (March 9, 1963) 11.

83. Vincent F. Caputo, "Remembrance of the Year Past," *National Defense Transportation Journal*, 18 (March–April, 1962), 24.

84. *Ibid.*, 27. For Caputo's more recent (if unchanged) views, see *Aviation Daily*, September 15, 1965.

mantled and its airlift procurement function assigned to DSA. The Military Traffic Management Agency had for many years managed domestic transportation for the armed forces, including short-term contracts for the charter airlift of Army recruits between domestic bases. Transferred to DSA, the function was renamed the Defense Traffic Management Service (DTMS). It awarded most of its contracts to supplemental carriers, the Richmond crash of 1961 occurring during the performance of one such contract.[85] In the wake of that accident, the chief of DSA (with an assist from Congressman Flood) barely succeeded in preventing assignment of the entire function to MATS.[86] The DTMS remains the nucleus around which a consolidated transportation agency might be constructed.

If it were to follow the DSA example, the transportation agency would be largely outside military control, would have no direct responsibility for the conduct of military strategy, and would largely be free of policy control. One student of defense organization has challenged at length the logic behind such agencies:

> . . . The defense agencies of today . . . have limited and highly specialized responsibility, and they do not need to worry about criticisms for jeopardizing strategy or combat readiness. Highly specialized agencies are not judged in terms of overall, end-product performance. They are subject only to management criteria such as costs. In short, they lose the very "national view" that was used to justify their existence. . . . Under the presently evolving system, no provision exists for lateral coordination or control among the defense agencies, except by the Secretary himself. This is an impossible job for one man, as Marshall discovered in World War II.
>
> When the current defense agencies are viewed collectively —the nine combat commands, the assistant secretaries who give orders, and the defense agencies,—they bear a striking resemblance to the War Department before the Root organization of 1903. . . .
>
> The fatal weakness of the pre-1903 organization—lack of coordination and central control—was demonstrated only under the stress of the Spanish-American War. . . .[87]

85. See Chapter 5.

86. The Air Force and MATS apparently asked that MATS be given the task of handling domestic contracts as well. The airlines feared that this would intrude MATS into domestic operation and thus might pose a long-term threat. *Aviation Daily*, December 4 and 8, 1961; *Army-Navy-Air Force Journal*, November 28, 1961.

87. John C. Ries, *The Management of Defense* (Baltimore: Johns Hopkins Press, 1964), pp. 191–92.

A transportation agency, or bureau, would be part of the logistics chain of command (if not managed by one of the services), but it never has been clear whether airlift is operational, logistic, or both. The theoretical justification for MATS is the necessity to carry out JCS war plans by deploying troops and equipment overseas. Once accepted, the logic of a centralized transportation agency might call for its acquiring the airlift assigned TAC, the Marine Corps, even the Army. Assignment of MATS alone might give this independent transportation agency responsibility for paratrooper operations in C-130E's and C-141's. If these arrangements might seem peculiar, so might the possible relationship between such a new agency and the supplemental airlines still dependent upon government contracts.

Heightened international tensions and an increase in training exercises have caused a steady rise in the commercial airlift purchased since 1960, and the negotiations between the government and the carriers have become exceedingly complex. The Air Force has negotiated individually with these carriers and has favored those agreeing to buy new aircraft.[88] With $25 million at stake in yearly contracts, even PAA purchased additional transports rather than risk losing its standing.[89] The carriers pledge additional capacity for emergency use, but the basis for such calculations is distressingly vague.[90] The new aircraft, moreover, are only slight modifications of normal commercial models, less advanced than the ones Congress vetoed in 1959.[91]

Air Force spokesmen took credit for having encouraged or pressured various airlines to buy seventeen modern transports.[92] Some

88. Testimony of Mr. E. J. Driscoll, Office of Assistant Secretary of the Air Force for Materiel, U.S. Congress, Senate, *Aircraft Guarantee Loan Hearings*, Hearings before Subcommittee, 87th Congress, 2nd Session (Washington: Government Printing Office, 1962), p. 152 (hereinafter *Guarantee Loan Hearings*).

89. Interviews.

90. While the CRAF program began in 1952 with an agreed objective of ten-hour utilization for civil aircraft used during emergencies, little has been said in more recent years about that objective. The carriers submit their own estimates of "expansion capability," and the government has published no standards of its own.

91. Senator Monroney, however, supported the new arrangements. *Guarantee Loan Hearings*, p. 161.

92. U.S. Congress, House, Committee on Government Operations, *Military Air Transportation, 1963*, Hearings before Subcommittee, 88th Congress, 1st Session (Washington: Government Printing Office, 1963), pp. 17–18, 42 (hereinafter *Military Air Transportation, 1963*).

belong to the supplemental carriers who remain the profitable subsidiaries of giant financial complexes. Their profits are, in effect, unregulated; CAB must set a rate which protects the big carriers.[93] During fiscal year 1961, MATS contracts totaled $74.5 million, and the amount soared to $142 million the following year; between July and October, 1962, expenditures reached a peak, indicating that the yearly sum would not decrease.[94] With all of the expenditures has come an understandable improvement in government-airline relations. Various segments of the industry have stopped protesting, and an ATA official has praised the "growing civil-military partnership."[95]

The principal decision-making authority in this area remains with the Assistant Secretary of the Air Force for Installations and Logistics, an official within the logistic line of command. This office might be merged with its counterparts in the Department of Defense in the event a new transportation agency were established. The whole arrangement then would combine authority in an office having no operational responsibility, busy supervisors unable to oversee what is being done, lucrative contracts awarded on vague criteria, uncontrolled profits giving the misleading appearance of regulation, and carriers that look like small businesses but are, in fact, representatives of hidden interests.

The present relationship has at least the advantage of being within a military department, where the office which awards the

93. Rates were adjusted periodically after 1960—usually downward. *Aviation Daily*, September 27, October 12, 1961; *Military Air Transportation, 1963*, p. 81.

94. During the 1961 Berlin crisis, the U.S. stood ready to begin another airlift. Interviews with General LeMay in *Los Angeles Times*, July 25, 1961, General Kelly of MATS in *New York Times*, September 24, 1961, General Tunner in *Time*, August 18, 1961. By 1962, the Air Force and MATS had provided $20 million in military airlift for the UN operation in the Congo, *HCA DOD Appropriations for 1963*, Part 3, pp. 574–75. Both military and commercial airlift were used in Southeast Asia, including one civil carrier which conducted short-haul tactical operations, apparently in conjunction with the Central Intelligence Agency. *Washington Post*, *Chicago Tribune*, January 22, 1961, *Aviation Week*, May 21, 1962. The 1962 Cuban crisis required a quick expenditure of $3 million to replace military airlift withdrawn for the crisis, *New York Times*, November 29, 1962. Other summaries from *Military Air Transportation, 1963*, p. 66, and U.S. Department of Defense, Military Air Transport Service, *Airlift Service Management Report*, pp. 20–21. (Mimeographed: July, 1961–June, 1962.)

95. *Aviation Daily*, April 3, 1961, April 11, 1962. The MATS commander announced that he had "some pretty well satisfied contract airlines," *HCA DOD Appropriations for 1963*, Part 3, p. 543.

contracts must maintain liaison with the military side of the Air Force. Even so, the arrangements seem dangerous, because they might easily be subverted to private purposes. There is, of course, no evidence that anything of this sort has been attempted or considered, but the possibility is there. This factor vaguely emerged during the 1963 hearings of the Holifield Subcommittee.

Congressman Holifield apparently was ready to concede that the supplementals, as the outcome of legislation enacted after the Richmond crash, had little future in the commercial market. Air Force spokesmen, however, tried to convince him otherwise. When Holifield observed that military traffic would gradually shift to the scheduled industry, one official advised him that several supplementals were showing substantial increases in commercial business, with as much as 30 per cent of total volume being in charters.[96] He gave no examples in support of this assertion which, viewed against the entire history of charter airlift, seems questionable on its face. Congress is advised that these companies purchase transports completely suitable for military use, that they hold in reserve huge amounts of airlift capacity available upon demand, and that they have a great deal of profitable nonmilitary operation. Unfortunately, none of these statements is wholly correct, and it remains unclear why Defense Department spokesmen have made them.[97]

The recent concentration upon the allegedly necessary separation of operational ("military") and logistic ("business") affairs is only the latest manifestation of a theoretical approach to defense questions that dates back many years.[98] Without attempting to trace the trend or to analyze the validity of the approach, it is

96. *Military Air Transportation, 1963*, p. 31.

97. Government witnesses advised the Holifield Subcommittee that in 1963, MATS contracts kept 34 aircraft employed while the airlines were holding 102 in reserve for emergency use. At other places in the testimony, however, it was quite clear that no such expansion capability existed at all because, when the Government attempted to "expand" the assigned contracts, the predicted capacity was not available. *Ibid.*, pp. 12–16, 66. The testimony clearly confused the subcommittee, and it defies the imagination to believe that vast numbers of aircraft were kept available at no cost. No company could afford this.

98. During the 1920's, for example, businessmen and administrative theorists provided the major impetus for attempts to unify the services. Then as now the idea was to economize by applying the principles of business management. For a discussion of the problems involved in equating military services to businesses, see Hammond, *op. cit.*, pp. 87–92; William R. Kintner, in association with Joseph I. Coffey and Raymond J. Albright, *Forging a New Sword* (New York: Harper & Brothers, 1958), pp. 190–92.

worth noting that its advocates are most successful in times of peace (or Cold War) when civilian management seems to promise economy.[99] It attracts bipartisan support for that reason. The Defense Department may or may not decide to add transportation to the list of functions now managed by relatively independent defense agencies. The same airplanes, however, do both an "operational" job (paratroopers) and a "logistic" job (supplies). When any administration argues that the functions are separable, it sets the stage for rather bitter custody fights between "operators" and "managers."

AIRLIFT AND THE "PROGRAM PACKAGES"

Secretary McNamara quickly installed the concept of "program packaging," something of a budgetary equivalent of both the Hoover Commission's principles of business organization and the military suggestions advanced by Symington and Taylor.[100] Because the Defense Department has used this management framework to make important budgetary decisions, some contradictions in the framework have intensified disputes within the Defense Establishment, particularly those between the Air Force's TAC and MATS.

Charles J. Hitch, new comptroller of the Defense Department and a former RAND economist, brought the "program packages" to his new assignment. Hitch explained that Defense Department budgets now are constructed around distinctive military functions, not around the traditional military services.[101] Hitch called this the

99. The Hoover Commission Reports remained particularly attractive because they seemed to promise military preparedness at cut rates. During the 1962 hearings of the Hardy Subcommittee, outside groups, still quoting the Hoover Commission, sent many letters to subcommittee members, arguing against any attempt to disband the independent defense agencies (especially the DSA). The letter campaign included estimates that the agencies could save up to $2 billion per year. The validity of such estimates, whether made within or without the government, never can be determined. See *Hardy Hearings*, pp. 6863–64, 6881–83. The *New York Times* finally protested in 1964 against what it described as the Defense Department's tendency to identify reductions in inventory as "savings." *New York Times*, July 9, 1964.

100. Kaufmann, *op. cit.*, pp. 168–203; White, *op. cit.*, 46.

101. Testimony in U.S. Congress, Senate, Committee on Government Operations, *Organizing for National Security*, Hearings before Subcommittee on National Policy Machinery, 87th Congress, 1st Session (Washington: Government Printing Office, 1961), I, 1007 (hereinafter *National Security Hearings*); Charles J. Hitch and Roland D. McKean, *The Economics of National Defense in the Nuclear Age*, (Cambridge: Harvard University Press, 1960).

"mission approach," agreed that the packages resemble Taylor's proposals for "functional budgeting," and acknowledged that the packages could lead to reorganization of the Defense Department. If budgets are to be functional packages, he said, it might make sense to reorganize the services to match the packages.[102] While the titles change from time to time, these have been the important packages:

Strategic Retaliatory Forces
Continental Air and Missile Defense Forces
General Purpose Forces
Airlift and Sealift Forces
Reserve and National Guard Forces
Research and Development
General Support[103]

When he defended the new approach to Congress, Hitch described their two most immediate advantages.[104] One would be the new validity of the programming data attainable through more careful attention to actual operating costs. If B-52 squadrons were at issue, he said, it was just as important to know the costs of training pilots as to know the price of the planes themselves. A second advantage would be the ability to make a rational choice between clearly defined alternative programs. *Polaris* missiles, B-52's, and *Minuteman* missiles would be parts of a single package, Hitch argued, and would compete against each other for the funds used for that purpose. In budgetary terms, therefore, *Polaris* was a part of the same military service as the B-52's. Significantly, Hitch described and defended his packages largely in terms of the homogeneity of strategic weaponry.

The first three of these packages clearly followed the familiar pattern of strategic, defensive, and tactical forces, as the Rockefeller and Symington reports and General Taylor had outlined them. If, as Hitch observed, the packages ultimately might lead to reorganization, the situation would be less clear. The military departments no longer would control the logistic function, nor would the independent defense agencies such as DSA. The Hitch approach thus

102. *National Security Hearings*, pp. 1020–24.
103. *Ibid.*, pp. 1039–46; 1963 terminology listed is from *SCA DOD Appropriations for 1963*, pp. 10–56.
104. *National Security Hearings*, pp. 1010–14.

seemed to look ahead to some eight or ten services instead of the traditional three or four, especially in view of a management approach which played down the necessity for interservice agreement and pushed all decisions up to Defense Department level.[105] By superimposing the packages upon a military structure organized along other lines, the comptroller succeeded in gaining for himself almost complete authority over every programming decision made within the defense establishment.[106] An increase in the number of detailed administrative decisions made at Defense Department level is not, of course, objectionable solely on that basis. The real question is whether the program packages made as much functional sense and arrayed the choices in as rational a manner as the comptroller claimed they did.

The packagers apparently faced no problem in combining bomber and missile forces into a single strategic package, although it might have been argued that *Polaris* and *Minuteman* are not wholly comparable.[107] Similarly, it semed not too difficult to devise a package for defense forces, especially by defining the defensive task in narrowly continental terms. In retrospect, however, the remainder of the packages present something of a muddled picture, and this is nowhere more the case than with airlift.

The assignment of airlift units to one or another of the packages followed no recognizable pattern. MATS, for example, was in the Airlift and Sealift package, along with the TAC troop carrier wings and the Military Sea Transport Service. Marine transports were in the General Purpose package, even though they were the same models as used in TAC and did largely the same job—supply of troops in the battle area. Navy and Army transports also were in the General Purpose package. There were significant airlift units in the Reserve and National Guard package as well; some worked with TAC, others with MATS.

McNamara increased the confusion when he described the characteristics of the Airlift and Sealift package. He emphasized the industrial fund operations of MATS and MSTS, and the fact that

105. To cope with the increase in centralized decision-making, McNamara increased the number of Deputy Assistant Secretaries of Defense from eleven in 1961 to twenty-six in 1963.

106. The defense comptroller has been exceedingly important for several years, but never to the degree that has been so since 1961. See Ries, *op. cit.*, pp. 189–90.

107. *Polaris* might be considered a weapon more useful against urban centers than against other missiles, with *Minuteman* in the opposite category.

both agencies were available equally to all military customers.[108] This made some sense for MATS and MSTS, although one might argue that airlift and sealift are not wholly compatible; the amount of each depends more upon the assumptions the planner makes than upon comparisons between ships and planes. In any case, the Secretary's explanation did not include the reason for lumping TAC with MATS. TAC dealt only with the Army's forces, was not available to the Navy and Marines, and the functional tasks of TAC's transports more nearly resembled those of Army and Marine transports than MATS'. If, as McNamara asserted in 1961, STRICOM was the Army-Air Force equivalent of the Marine Corps, then STRICOM's airlift (TAC) would have been more at home in the same budgetary package as the remainder of STRICOM. The assignment of TAC's specialized wings to the generalized Airlift and Sealift package thus intensified the conflict between TAC and MATS, because the two commands became "alternatives" and "competitors." McNamara indicated as much when he said on one occasion that it was a "good question" whether TAC or MATS should control strategic airlift.[109] The corollary effect was to make it appear that the Army's transports were not competitive with TAC's.

Army airlift was in the General Purpose package, even though its mission (like that of Marine transports) was virtually homogeneous with TAC's. The Army-Air Force aviation dispute was real enough, however, even if it did appear that way in the package approach. The net effect, then, of the relationship between airlift functions and program packages was somewhat discouraging.

There is no evidence that comprehensive analysis preceded the assignment of airlift activities to the packages. Yet the new decision-making process (the selection between clearly defined alternatives) makes it as important, or more important, to group the activities properly as to make the final choices. The assumptions behind the packages are, in short, more important than any economic calculations leading to the budgetary decisions. If, as some argued, the packages acted to "bring clarity out of confusion," it was in some area other than airlift.[110]

108. U.S. Congress, House, Committee on Armed Services, *Military Posture Briefings, 1962*, Hearings, 87th Congress, 2nd Session (Washington: Government Printing Office, 1962), p. 3274.

109. *Military Procurement, 1963*, p. 150.

110. White, *op. cit.*, p. 46. The program packages were presented as a new approach, the first to overcome all of the many shortcomings of former

Like the three other decisions analyzed earlier, the adoption of program packages transferred old airlift controversies to new positions of prominence within the defense establishment. In trying to be decisive, the new administration came down firmly on both sides of difficult questions. Airlift was definable in many ways, each having its own champion. It was TAC versus MATS, the Army versus the Air Force, STRICOM and its paratroopers versus the remainder of the Army, and military leaders versus the civil service business managers. To make matters worse the new managerial framework pushed every decision to the top where it became a political commitment not reversible through any normal administrative process.

If one uses airlift as a point of departure for an analysis of defense policy formulation since 1961, the picture that emerges is somewhat different from the popular image of efficiency, economy, and a newly rationalized decision-making process. The trend toward autonomous defense agencies, or "bureaus," the formation of a new joint command which seeks its own strategic airlift, the rationale of the program packages, and the Army's own airlift operations, all seem contradictory to each other.

POTENTIAL CIVIL-MILITARY CONTROVERSIES

The post-1960 era of good feeling between the Defense Department and the airlines may well develop into a new controversy by the late 1960's. The dispute may occur if the Air Force acquires enough airlift of its own to reduce or eliminate any need for commercial assistance or, alternatively, if changes in military strategy and technology suddenly make obsolete the entire commercial fleet.

While the Defense Department did not release precise figures, the emphasis on the C-141 implied that many of them are in the military program.[111] MATS modernization has appeared to be fol-

procedures. During the 1961 hearings of the Jackson Subcommittee, the Eisenhower administration's Defense Comptroller (W. J. McNeil) disputed such implications and pointed out that budget categories can be developed in any number of ways. The Eisenhower administration, he said, had very nearly adopted a geographical approach in 1958; separate "packages" would have included the military expenditures in the Mediterranean, Europe, and the Pacific. The administration abandoned the idea because it seemed to offer no substantial improvement. *National Security Hearings*, pp. 1062, 1065.

111. Report on increases in the number purchased in *Aviation Daily*, August 21, 1961.

lowing the pattern that General Tunner originally outlined in 1960, thus indicating that several hundred of the aircraft are involved. Whatever the level of peacetime utilization at which they operate, the government will have a truly substantial airlift capability within the Air Force, and the military transports will be newer, larger, and more capable than the cargo jets the supplemental carriers (and some of the scheduled companies) have bought at government insistence. The companies may not be immediately willing to trade in these almost new jets for commercial versions of the C-141, even though the Air Force has announced that it will encourage the airlines to do just that.[112] Secretary McNamara, never involved in the detailed give-and-take with the airlines, complicated the matter further by announcing that it made little sense to rely on airlines whose only source of business is the government. Where airlines of this sort are involved, said McNamara, the wise course is for the government to buy its own aircraft.[113]

Foreign policy trends may well lead to gradual reductions in the U.S. forces maintained overseas and this, in turn, will reduce the airlift capacity needed on a day-to-day basis. The U.S. may withdraw troops either to protect its balance of international payments, or as the result of a decrease in tension, or both.[114] In any emergency situation, however, it seems sensible enough to expect that the government may need a great deal of airlift capacity. The fundamental problem may well remain the unwillingness of any agency to accept as necessary to military preparedness the peacetime operation of military and commercial transports with partial loads or with no loads at all. If both military and commercial transports must always fly in accordance with economic ground rules, some sort of dispute may be virtually inevitable.

There is another possible controversy between government and industry which is less likely than the one mentioned above, but

112. The Air Force told the carriers that their new jets were only "interim" models and that they should plan to replace them with C-141's. *Guarantee Loan Hearings*, pp. 154–55. Yet most companies were expending the depreciation schedules of their jets to indicate a life expectancy of fourteen years for those aircraft. Jets bought in 1963 would be only four years old when C-141's are available. *Airlift Magazine*, February, 1963, p. 35.

113. U.S. Congress, House, Committee on Armed Services, *Hearings before the Special Subcommittee on National Military Airlift*, 88th Congress, 1st Session (Washington, Government Printing Office, 1963), p. 5988.

114. Some troop withdrawals were announced in the spring of 1964, some tactical air units were withdrawn, and there were recurring reports of plans for more withdrawals in 1965. *New York Times*, April 11, 1964; *Journal of the Armed Forces*, September 26, 1964.

shifting strategies may bring it about. Both the Army and Air Force have had some interest in transports that can land and take-off without benefit of airfields. Transports of this type seem necessary to cope with the problems of isolated units dispersed within a tactical nuclear battlefield. The same types of aircraft have seemed attractive for use in areas such as Southeast Asia. Should the fascination with tactical nuclear war continue, the next step may be strategic transports with the ability to take-off and land vertically (VTOL). An Air Force study group already has described the need for such a plane; it would have a range of 10,000 to 12,000 miles, could carry both troops and equipment, could drop them by parachute or other means so as to avoid landing, and then return home without refueling.[115]

Whether the commercial fleet remains subsonic or takes another step ahead into supersonic jets, the likelihood is that the international and domestic airfield complex currently in use (with its normal expansion) will remain active for some time. If the U.S. adopts a strategy requiring strategic VTOL transports, the military and commercial fleets no longer will be compatible at all. This is because tactical nuclear wars may well call for larger armies and rear area facilities than needed in conventional wars.[116] If the government were to buy enough VTOL transports to cope with tactical nuclear war, there would be more than enough for conventional war. This would make commercial airlift superfluous to peacetime military needs, and useless in war because of the nature of the conflict.

McNamara has said that VTOL transports deserve a place in future U.S. military planning, but he did not make clear whether he was speaking of tactical or strategic airlift.[117] The need for long-haul VTOL transports, however, seems dubious on several grounds, despite its attractiveness to the Air Force study group. If tactical nuclear wars will call for larger numbers of troops and for sophisticated rear area logistic complexes to supply and repair nuclear weapons (and to care for battlefield casualties), there certainly

115. *New York Times*, February 16, 1964. This was described as the "most comprehensive look into the future" in many years, and there was no mention of a need for a supersonic military transport.

116. On the nature of tactical nuclear war, see B. H. Liddell Hart, *Deterrent or Defense* (New York: Frederick A. Praeger, 1960), pp. 74–82; James M. Gavin, *War and Peace in the Space Age* (New York: Harper & Brothers, 1958), pp. 150–51; Ridgway, *op. cit.*, p. 296; Maxwell D. Taylor, *The Uncertain Trumpet*, (New York: Harper & Brothers, 1959, 1960), p. 74.

117. *Military Posture, 1963*, p. 456.

should be enough room in those complexes for runways of some sort. If the aircraft are to operate at high utilization and deliver that many supplies, the ground handling system to support the transports alone will be quite substantial. The runways will be slightly more vulnerable to attack than the logistic complexes themselves, but strategic transports probably cannot operate into areas not relatively secure from attack.

Second, the notion of VTOL transports implies that huge aircraft will deliver supplies to units dispersed at many locations within the battle area. This would force the transports to make several landings in the combat zone, would make them increasingly vulnerable if only because of constant exposure, and would prevent high utilization. Third, theater commanders are likely to retain some sort of supply system of their own, whatever the sophistication of computerized logistic networks. Finally, there probably will continue to be such things as budgetary ceilings, and a huge VTOL fleet seems out of the question.

These conclusions do not rule out short-haul VTOL transports, another recommendation of the Air Force study group.[118] Development programs of this sort, moreover, seem compatible with commercial needs. The short-haul market between major cities may become quite prosperous for aircraft that do not need lengthy runways.[119] The military task for such aircraft appears to be distribution of supplies from overseas terminals to forward area battle units. Even this military need seems of modest size, however, for, if battle units are to depend completely upon air supply, they once again will have supply complexes large enough for runways.

The most hopeful development programs would seek to maintain some compatibility between the military and commercial fleets. Whatever the promise of advanced technology, there appears to be no overriding reason to dispose of that long-term relationship. Nor does it seem valid for the military to insist that its transports must get along without any bases at all between the U.S. and overseas destinations.

Singly or together, then, reduced U.S. troop commitments overseas, the continued buildup of military airlift capacity, and a shift

118. *New York Times,* February 16, 1964.

119. *Airlift Magazine,* February, 1963, pp. 19–20. A leading transportation expert adds that VTOL aircraft hold much promise for use in underdeveloped countries. Wilfred Owen, *Strategy for Mobility* (Washington: The Brookings Institution Transport Research Program, 1964), p. 104.

to a limited nuclear strategy could bring on a new domestic controversy involving the airlines, especially the remnants of the supplemental industry.

ORGANIZATIONAL SOLUTIONS FOR THE 1960'S

A broad view of military airlift experience suggests that its most distinguishing characteristic remains its newness. This is especially the case with strategic airlift, identified for so many years as only the equivalent of commercial airline operation. Strategic airlift is, in the 1960's, about where strategic bombardment was in the 1940's, for strategic troop deployment by air was unknown before 1960.

The combat organizations, traditionally and properly the leaders of the military services, have tried to force upon strategic airlift the combat rationale so familiar to them; this has been the case with both the Army and the Air Force, whose combat airlift traditions include only paratroopers. Both services have found it difficult to accept the notion that strategic airlift may not fit the traditional combat model and, for this reason, the first difficulty encountered in 1942 remains unresolved in the 1960's. Combat organizations, especially the airborne divisions, remain hesitant to agree that strategic airlift is best operated on a basis which prevents individual combat commanders from diverting the airplanes.

One encouraging development is the attempt to define with some precision the problem Admiral Radford suggested in 1957—the primary factor limiting the productivity of strategic airlift is the ability of route systems and facilities to accommodate aircraft. MATS calculated in 1963, for example, that the Pacific transport route can support only 112 aircraft each day. These must evenly cover the twenty-four hour period (as in airline-type operation), and the total presumably includes aircraft returning to the U.S. for another load.[120] Once this aerial pipeline is full, there simply is nothing more to be done in the way of supplying airlift. If the Air Force were to refine calculations of this sort (by indicating, for example, whether Wake Island calculations include the commercial transports who use that small airfield), the limitations of strategic airlift would be more easily discernible.

Strategic airlift and sealift are, for some time, likely to remain complementary to each other. This conclusion runs counter to McNamara's announcement that strategic airlift is almost automati-

120. *Military Air Transportation, 1963*, p. 170.

cally less expensive than sealift and should largely replace it.[121] Aside from the problem of route system capacity, at least two other questions are suggested by the Defense Department argument: (1) some items are so large that airlift will not be able to move them for some years to come; and (2) the basis on which the Defense Department decided that sealift is less expensive than airlift (no figures were released to prove the assertion).

If airlift and sealift are complementary, about all the Defense Department need do is determine on some general basis the amount of high-value, high-speed strategic airlift needed as a complement to sealift. Beyond that, nothing seems wrong with the management system developed in World War II. The JCS, or some lower level interservice committee, would have no difficulty in allocating the capacity among various claimants during wartime. The point to recall is that the reorganizations followed from the dubious assumptions of paratrooper operations. All of this suggests a logical sequence for some necessary reorganizational actions.

The Re-examination of Airborne Concepts / It does not appear necessary to await the creation of air assault divisions before evaluating the utility of paratrooper operations on a large scale. There is even some reason to argue, moreover, that the Army as a whole is less than an enthusiastic supporter of its two elite airborne divisions. The only form of battlefield airlift the Army has not volunteered to provide is that for the paratroopers, perhaps because of the high costs.

If massive airborne assaults are necessary to national strategy, alternative and less expensive means seem available. There should be no mistaking the central problem; if airborne divisions are needed and are to do their jobs, there is no choice but to let them have their own airlift. This being so, nothing should prevent the cost of the airlift from being added to the budget for those airborne divisions. This arrangement might be politically and militarily feasible if paratroopers were to use short-haul transports having only enough capability to perform the airdrop mission. Such aircraft would not need a very long range, would not be pressurized, and would not fly at jet speeds; they might then be inexpensive enough to be worth the cost. The idea of tying

121. Testimony in *Merchant Marine, 1962*, p. 88; Morton H. Halperin, *Limited War in the Nuclear Age* (New York and London: John Wiley & Sons, Inc., 1963), p. 123.

down transports that cost $6 million each while awaiting an airborne assault still remains dubious: ". . . The idea that you can fly troops for thousands of miles across the ocean, join up with other aircraft in neat formations over the assault area, in fair weather or foul, and then, after many hours of flying without landing, drop combat troops in an orderly fashion, ready to fight, makes little sense."[122] The existing structure easily could accommodate small-scale airborne assaults without requiring that massive numbers of aircraft remain constantly available for that purpose.

The Dissolution of STRICOM / A logical next step would be the dissolution of STRICOM on grounds that it is as superfluous as the previous experiments of that type. The command depends for its rationale upon a traditional airborne theory which the Defense Department has not implemented because of the cost of strategic transports. If STRICOM is maintained in being, however, the corollary requirements are to let it control its own strategic airlift, and then build another strategic fleet for the remainder of the armed forces. But even this would not make the command politically or geographically compatible with the U.S. position in world politics.

The Return to the Single System of Strategic Airlift / The third step would be to revert to a single system of strategic airlift, organized and managed in the manner first found necessary in World War II. The relatively finite limitations of available route systems and airfields make such a solution a practical requirement. It may be difficult to achieve because of the intense bitterness within the Air Force traceable to the establishment of STRICOM and the contradictory pressures that have resulted. Congressman Rivers, for example, has suggested on occasion that Congress remove MATS from the Air Force and assign it directly to the JCS.[123] More recently, his subcommittee indicated that it accepted Air Force control, subject to certain operational revisions.[124] Who

122. Tunner, *op. cit.*, p. 323. Alternatively, reserves could provide airlift.

123. Rivers apparently felt for some time that the Air Force simply did not appreciate strategic airlift and could not be trusted with it. He introduced legislation in 1962 to separate MATS from the Air Force; H. R. 12112, 87th Congress, 1st Session.

124. U.S. Congress, House, Committee on Armed Services, *Report of Special Subcommittee on National Military Airlift,* 88th Congress, 1st Session, Committee Print, (Washington: Government Printing Office, 1963), pp. 16–17, 22.

should control strategic airlift is part of a broader question involving defense organization in general and the specific subject of unification.[125]

The Merger of Tactical Airlift Systems / The fourth step would be merger of Army and Air Force tactical airlift programs. This refers only to the supply function of airlift; paratrooper airlift, if the U.S. needs it, should be left with those divisions. There probably is a need for much more tactical airlift than during World War II and the Korean War, even if the Army's more recent proposals are somewhat optimistic. The question of whether the Army or Air Force should do the job is not of immediate concern here; what is most important is that multiple loading and unloading of cargo is not likely to increase the efficiency of airlift within the combat zone.

Most of these actions would reverse post-1960 reorganizations which, on the whole, seem to have been made too quickly and with too little thought. One experienced observer remarked in February, 1961, that it was asking too much of Secretary McNamara and his associates to reappraise within four weeks the entire defense establishment and to recommend changes; snap judgments and ill-advised commitments were inevitable.[126] Most of the important decisions discussed here were taken within the first month or two after the 1961 inauguration (even if the Defense Department delayed announcing some of them).

The environment was not exactly the same, but the 1961 decisions caused some turns in airlift policy almost as abrupt as those of 1934. Apparently because of commitments to theoretical models, the administration's advisers failed to point out (or did not notice) that 1960 had been a year of substantial progress. The military problem was largely settled by 1961 and civil-military relationships within the Defense Department, as well as liaison between services, were better than they had been for some time. Joint strategic airlift exercises had begun, the Army was newly acquainted with long-haul airlift, and both planning and operations had proved

125. Something can be gained by leaving with each of the military services at least one important task associated with each of several strategies. This leaves each service with a voice in all strategies and does not stake service survival upon a single military function. Samuel B. Huntington, *The Common Defense* (New York and London: Columbia University Press, 1961), p. 423.

126. Hanson Baldwin in *New York Times,* February 1, 1961.

successful. The administration elected to quickly thrust aside those accomplishments and to use airlift as something of a political instrument. The administration thus managed to take the credit for the 1960 accomplishments of the congressional consensus.

[11]

International Airline Regulation in the 1960's: The Need for Re-examination

IF CONGRESSIONAL ACTION had by 1960 solved at least some of the military airlift problems (while leaving untouched the old paratrooper question), the same could not be said for international airline regulation. In the American context, the political and economic problems are the most important; any military solution is indeed fragile if strong nonmilitary policies do not accompany it. In this sense the long preoccupation with military questions only diverted attention from the crucial issues. In retrospect it is something of a mystery why so many individuals and groups looked to military traffic to solve a nonmilitary problem. This chapter returns, then, to the fundamental questions of whether and how to regulate the airline industry (especially the international companies).

The long-term U.S. difficulties are suggested by the contrast between the relatively steady decline of the U.S. international carriers with respect to those of other countries and the apparently outstanding success of U.S. promotion of export sales of transport aircraft. It is no exaggeration to conclude at the outset that both Congress and the president usually treat export promotion as a legitimate policy objective of the government, but look upon the problems of the international carriers as largely their own concern.

The appropriate agencies generally administer these two aspects of airlift policy in a compartmented fashion, apparently unaware that export decisions often cause a reaction within the airline industry. In an age of rapidly advancing technology, export promotion has come to include sales of long-haul jet transports to countries which must, in turn, compete in travel markets already occupied by U.S. carriers. Given the country's vested interest in the foreign carriers which buy the transports, the government finds its export and airline regulatory policies working at cross purposes.

As foreign companies have grown progressively stronger, the U.S. has found itself in an ever more uncomfortable position. Within the bilateral framework established in 1946, the U.S. gradually has reversed its original positions and has championed a more restrictionist policy toward foreign airlines. Yet the U.S. also remains an advocate of lower prices and some vague sort of unilateral regulation of international rates. What seems most needed is an approach not wedded to the economic theories of free competition that do not answer the hard questions of world politics. The most illustrative question is the one the Europeans had to face in the 1920's when the U.S. managed to avoid facing it: What must a country do if competition threatens to force its international airlines out of the market? In the absence of substantial variations to the old approaches, airlift policy may remain in something of a permanent impasse between government and industry.

THE DECLINE OF THE INTERNATIONAL CARRIERS AND THE STALEMATE ON MERGERS

The relative position of the U.S. international carriers has declined substantially for more than a decade. The industry's pessimistic forecasts, made during the controversy in the mid-1950's over foreign route concessions, have proved more accurate than the Eisenhower administration's sanguine estimates. In answering its critics, that administration had pointed to the American carriers' 65 per cent share of the commercial passenger traffic between the U.S. and other countries.[1] Even as late as 1957, the Celler Committee majority criticized what it saw as Pan American's dominance of international aviation.[2] Facing a bipartisan lack of sympathy, the airlines turned to the side issue of military business. It became even more clear by 1961, however, that the international problems had not vanished; despite substantial expansion in international air travel, the share of the U.S. companies had dropped even more rapidly (it had been at 70 per cent in 1951) to slightly

1. *Department of State Bulletin* 36 (June 24, 1957), 1012–13; also see Chapter 4.

2. U.S. Congress, House, Committee on the Judiciary, *Report of the Antitrust Subcommittee (Subcommittee No. 5) Pursuant to H. Res. 107 on Airlines*, 85th Congress, 1st Session, Committee Print (Washington: Government Printing Office, 1957), p. 263 (hereinafter *Celler Report*). The Republicans on the Committee dissented.

under 50 per cent.[3] In the rich transatlantic market, moreover, U.S. carriers handled only 37 per cent of the traffic in their competition against the eighteen foreign airlines flying those routes.[4] Those airlines extended their routes to interior cities (Chicago, Detroit, and Houston, among others), and their governments became more and more intractable in negotiation.[5] In a substantial reversal of its postwar positions, the U.S. turned to accusing foreign carriers of being too aggressive in competition.

Initially, of course, the U.S. was the champion of free competition; we insisted that each country determine for itself the number of trips its own airlines could operate.[6] As other nations recovered their economic strength, they permitted their airlines to increase their trip frequencies. The U.S. then attempted to invoke one of the secondary principles adopted at Bermuda; consultations would be in order if one party to a bilateral agreement believed the other's airlines were operating too many trips.[7] In short, the U.S. accused other countries of unfair competition (too much service to international passengers), an accusation about as far as imaginable from traditional notions of free competition. We argued with the Scandinavian countries and spent two years convincing India that it should reduce the trip frequencies of its national airline.[8] Indeed, the whole pattern of airlift negotiations now fits more than ever into broader patterns of international behavior.

Many of the world's airlines have created new organizations which can be described variously as consortiums, bloc airlines, pools, or commercial airlift power blocs. As individual countries take their airlines into such blocs, new and larger airlines are organized, facilities are consolidated, and member airlines and countries divide among themselves the total revenues. As in the case of the IATA, these new organizations are not traditional cartels, because the governments themselves own or control the airlines. These bloc airlines are more in the nature of politico-economic

3. The actual share was 49.6 per cent. It recovered slightly to 50 per cent in 1962 and 50.4 per cent in 1963. *ATA Facts and Figures*, 25th ed., 1964, p. 33.

4. U.S. Federal Aviation Agency, *Report of the Task Force on National Aviation Goals, Project Horizon* (Washington: Government Printing Office, 1961), p. 108 (hereinafter *Project Horizon*). The figure dropped somewhat in 1963. *Time*, May 3, 1963, p. 95.

5. International route map in *CAB Annual Report, 1961*.

6. See Chapter 4.

7. *Project Horizon*, pp. 111–17; *CAB Annual Report, 1961*, p. 37.

8. *Ibid.*, 1962, pp. 31–32.

alliances designed to extend the national power and prestige of member countries. In a sense, the alliances are technological necessities; these countries fly transports first designed for the long distances within the U.S., and they have no choice but to concentrate on international aviation. Several such alliances are in effect, each one seems to be increasing in internal cohesion, and the trend toward more such groupings seems clear enough.

One such alliance is the Scandinavian Airlines System (SAS) which, though treated as a single international airline, actually is a pool operated jointly by Denmark, Norway, and Sweden.[9] A second arrangement, if still a vague one, includes airlines in the British Commonwealth (the British, Indians, and Australians).[10] Several Latin American airlines have begun discussions looking toward some sort of agreement on pool operation.[11] Airlines of West Germany, France, Italy, and Belgium agreed in 1959 to form a co-operative, known as Air Union. Though limited at the outset to co-ordinating timetables and passenger reservation facilities, Air Union promises to evolve into a truly supranational airline, or so CAB Chairman Alan Boyd reported in 1964.[12] Eleven African countries have formed Air Afrique to meet their intercontinental airlift needs.[13] The most obvious of the consolidated airlines is the one belonging to the Soviet Union and its satellites.[14] The Soviet airline (Aeroflot) generally is recognized for what it is, but some U.S. traditions make it all too easy to look upon free world airlines as private enterprises disassociated from the foreign policies of their countries. Yet all of these politico-economic alliances are

9. This organization also has proved intractable in negotiations. *Ibid.*, 1961, p. 37.

10. *Wall Street Journal*, December 6, 1962.

11. *Ibid.*

12. *New York Times*, November 3, 1959; *Business Week*, December 1, 1962; Flight Forum, *The Future of the International Carrier* (Connecticut General Life Insurance Company, September, 1964), pp. 4–5 (hereinafter *International Carrier*). Indeed, one observer argued as early as 1956 that a European regional organization was absolutely necessary; multilateral planning was the only way to cope with the needs of continental air transport. Stephen Wheatcroft, *The Economics of European Air Transport* (Cambridge: Harvard University Press, 1956), pp. 311–31.

13. *International Carrier*, pp. 4–5.

14. Hans Heymann, "The Soviet Role in International Civil Aviation," Rand Research Memorandum RM 2213, 1958. Summaries of the Soviet experience are in Robert A. Kilmarx, *A History of Soviet Air Power* (New York: Frederick A. Praeger, Inc., 1962), pp. 270–73, and Asher Lee, *The Soviet Air Force* (New York: The John Day Company, 1962), pp. 172–86.

similar to each other, and they seem representative of global trends toward regionalism.

The consolidated airlift systems correspond to regional blocs of political and economic power and, in this sense, the European example may be the most significant. Air Union has paralleled the evolution of the European community. If Britain were to join the Common Market, the British airline might end its associations within the Commonwealth and join Air Union instead. Relationships between the U.S. and these bloc airlines also follow traditional behavior patterns of world politics, as in the case of the 1963 rate dispute.

When the CAB attempted to deny the IATA's new rate increases, the board confessed to a hope that the European countries and Britain would not take united action.[15] When the British threatened to deny landing rights to U.S. carriers who violated the approved rate schedule, the CAB hoped the French would not do the same. Taken together, U.S. disapproval and French disaffection might have persuaded the British to relent to American pressures and to defy the IATA by keeping international rates at their current levels. In the classic pattern of alliance behavior, the U.S. sought to use a Franco-American agreement to replace a loose Anglo-French alliance. The CAB Chairman has taken an optimistic view of the trend toward airline regionalism, but the CAB's actions in the 1963 rate dispute implied rather clearly that British membership in a unified airline (or, by extrapolation, the politico-economic unification of Europe) might not be in the best interests of the U.S.

Coincident with these international difficulties was a brief spurt of CAB support for airline merger proposals. There is no evidence that any sort of co-ordinated policy-making process preceded the CAB's mild overtures to the industry. Indeed, the board's activity probably was only an isolated attempt to find some solution to the problems of both domestic and international operation. CAB Chairman Alan Boyd let it be known in 1961 that he might give merger proposals a friendly reception. This reinforced the 1960 suggestions of Whitney Gilliland, then CAB Chairman, who made several speeches on the subject during that year.[16]

15. *New York Times,* May 16, 1963.

16. Boyd's views were interpreted at first as those of the Kennedy administration. Indications favored a speedup of a trend which included mergers of Colonial and Eastern in 1956, and United and Capital in 1961. Several observers predicted more mergers; W. A. Patterson of United predicted in 1959 that CAB policies would inevitably force companies to

The *Project Horizon* task force, appointed by the FAA in 1961 to study aviation policy, also encouraged mergers, but this group implied that only small airlines should take such action.[17] The new administration did not have to commit itself formally so that, on the evidence, one can only speculate as to what it wanted. The airline response, however, took the form of two proposals by the giants of the industry.

American Airlines and Eastern Air Lines advanced the first one in 1961; it would have consolidated two carriers whose routes were largely noncompetitive against each other. A CAB staff examiner recommended almost a year later that the board turn down the proposal because the new company would be too large and would dominate the industry.[18] After another six-month delay, the CAB tentatively agreed with its examiner, three Democrats on the board outvoting the two Republicans.[19] The second proposition had even broader implications, for it challenged the international policy framework constructed after World War II.

TWA and PAA asked permission in 1962 to form a new company, with PAA to provide the operating nucleus. Some observers interpreted the move as only the latest attempt by PAA to make itself the U.S. chosen instrument in world aviation.[20] The proposal would have merged PAA with its major domestic competitor in the rich North Atlantic market. If PAA were to become the only U.S. carrier in this largest of all markets, companies in other markets might follow this example and seek to merge with PAA. A TWA-PAA merger also would remove most remaining distinctions between domestic and international companies, because PAA would

---

propose mergers by 1961. Richard L. Worsnop, "Troubles of the Airlines," *Editorial Research Reports*, 1 (1962), 83; Samuel B. Richmond, *Regulation and Competition in Air Transportation* (New York and London: Columbia University Press, 1961), p. 74; *Aviation Daily*, September 21, 1959, August 31, 1960.

17. *Project Horizon*, pp. 181–82.

18. *Business Week*, January 21, December 1, 1962.

19. The CAB announcement described the rejection as "tentative." When American Airlines withdrew from the proposal, the CAB did not have to issue a formal explanation for the decision. *New York Times*, June 22, 1963.

20. *Ibid.*, December 22, 1962; *Wall Street Journal*, December 6, 1962. The merger reportedly followed an attempt by the CAB staff to adjust the TWA and PAA shares of the market so as to make them more nearly equal. John McDonald, "Pan Am: Peril and Opportunity," *Fortune*, 66 (August, 1962), 83, especially the summary of arguments for and against mergers on p. 86.

then be operating in both areas. The proposal, then, raised troublesome issues, and the CAB was almost relieved to discover that it did not have to produce a quick recommendation. The board announced it could not consider a merger until stockholders of both companies had approved it, but this could not be done until TWA's tangled financial problems were solved.[21] Given this impasse and TWA's suddenly improved position in 1963, the two companies terminated their agreement in October of that year.[22]

If the administration initially favored mergers, it probably had in mind consolidations of the types suggested by the *Project Horizon* group. Perhaps the hope was that the industry, divided into categories known as the "Big Four" and the "Little Seven," automatically would produce proposals for six or seven companies of the same size.[23] The administration appointed an interdepartmental committee to study the subject; it listed several questions it thought applicable to all proposed mergers, but it went further and voiced a general fear that mergers might inconvenience the traveling public through "reduced competition."[24]

The Kennedy administration appointed yet another study group to study the specific area of international air transport policy. While the $300,000 study was in preparation, the State Department advertised it as a new and definitive program.[25] Apparently in anticipation of a new approach the CAB, for the first time in memory, withdrew its request for authority to regulate international rates (this was before the 1963 rate dispute).[26] The final report, however, suggested only that the international industry be left about as it was: "Government flexibility in implementing international political and aviation policies would be reduced if the interests of any single carrier became too dominant a factor in U.S. aviation policy. . . . We should continue . . . a U.S. system in which one U.S. flag carrier has access to world markets . . . and

21. *Denver Post,* February 22, 1963.
22. *Aviation Daily,* October 17, 1963. There can be little doubt of TWA's amazing resurgence, especially in view of the company's tangled financial affairs. John McDonald, "The Epic Struggle for the Corporate Cockpit," *Fortune* (May, 1965), 106–11, 213–16.
23. The "Big Four" remain the four domestic carriers, American, Eastern, United, and TWA.
24. *Congressional Quarterly,* 21 (March 8, 1963), 294.
25. Statement of Assistant Secretary of State for Economic Affairs Edwin M. Martin, *Department of State Bulletin,* 45 (October 23, 1961), 687–88; *Ibid.,* 45 (November 13, 1961), 815.
26. *CAB Annual Report, 1962,* p. 8.

other U.S. carriers continue to be authorized to serve one or more areas of the world in overall competition with this carrier."[27] The president issued a letter of approval with release of the unclassified sections of the report, thus leading observers to conclude that the administration had taken this oblique route to indicate in advance its disapproval of the TWA-PAA merger.[28]

Some widely held economic and administrative theories still lead to the conclusion that the number of airlines should be kept about where it is. One economist, for example, ridicules the notion that fewer companies might provide both better competition and improvements in operation: "This is refuted by all evidence of conduct within the air transport industry, as well as by much evidence concerning unregulated industries in the American economy. And if this were not enough, the logic . . . implies that reducing the number of firms in an industry renders it more and more competitive until, with only two left, competition reaches its maximum effectiveness; no comment seems necessary."[29] Another observer advises the CAB to forget mergers and concentrate instead on "other issues where it can constructively and responsibly exercise its powers."[30] The international aviation problem, says Hans Heymann, is largely one of economic growth, probably solved best by organizing a new and more liberal international framework to replace the paralytic one-nation-one-vote principle of the IATA.[31] If economic forces had more maneuver room, says Heymann, rates would decrease while growth increased.[32] Heymann, like others, worries about unutilized capacity, known in 1963 as "excessive ca-

27. *White House Press Release*, April 24, 1963. The government released some excerpts and a presidential statement endorsing the report, but the document itself was kept secret.

28. *Statement on International Air Transport Policy*, White House Press Release (Mimeographed: April 24, 1963), pp. 11–12. The report itself has been kept classified.

29. Richard E. Caves, *Air Transport and its Regulators*, (Cambridge: Harvard University Press, 1962), p. 441.

30. Richard J. Barber, "Merging the Airlines: What Next?" *New Republic*, 141 (January 5, 1963), 11–13.

31. Hans Heymann, "Aviation and International Relations," Text of a talk given at the symposium on The Issues and Challenges of Air Transportation (Mimeographed: Hartford, Connecticut, November 2, 1961).

32. *Ibid.* Other observers agreed that economic growth was the main problem. *Time*, May 3, 1963. A variation of the argument put some emphasis upon U.S. recessions and implied that growth could overcome the slump. Oliver Lissitzyn, "The Participation of the United States in World Air Transport," *Journal of Air Law and Commerce*, 28 (Winter, 1961–62), 1–17.

pacity"; load factors of U.S. international companies stood at less than 53 per cent.[33]

These observations generally miss the point of the difficulties outlined earlier and overlook important evidence to boot. Demands for lower rates, for example, ignore the financial debacle that followed hard upon the CAB's optimistic price-cutting after World War II; in the international field, such demands still rest upon a premise that other nations will accept U.S. price fixing, and at a time when the country's relative strength is less than in the immediate postwar period.[34] As to competition, the twenty airlines that cross the Atlantic cannot be identified as evidence of a lack of competition. Even the airline industry's growth rate seems substantial enough; the international carriers doubled their operations between 1957 and 1962.[35] Constant references to the number of empty seats are as unsatisfactory as they were when the Hoover Commission asked the military to fill them. When one plane is filled, another is added to the schedule, the process repeats itself over and over again, and the load factor problem remains unsolved.

The principal failure, in short, may well be an intellectual one, for most economic and administrative theories apparently provide no conceptual framework suitable for regulating the airline industry or for relating aviation policy to foreign policy. Much of the remainder of this book is an attempt to outline such a framework and to show how the international industry might fit within it.

The administrators of government policy seem actively in search of a new theoretical framework. The CAB Chairman, avoiding use of the word "merger," speaks of excessive competition and the need to rationalize airlift capacity.[36] The Interstate Commerce Commission has approved several railroad mergers and bids fair to approve still more, thus indicating that the drive to reduce competition has taken firm hold in an industry importantly related to the airlines.[37]

33. *ATA Facts and Figures*, 25th ed., 1964, p. 20.

34. The issue has been confused by PAA's attempts to pose as the champion of lower rates. After the 1963 dispute, PAA announced it had asked for a 39 per cent reduction in the one-way fare to Europe. Yet the proposal specified the elimination of group and charter flights. PAA, moreover, had acquiesced in the IATA rate increases, had every reason to believe that drastic cuts were out of the question, and admitted privately that price-cutting on a substantial scale was not a practical approach. *Time*, July 5, 1963; *Denver Post* editorial, July 1, 1963; McDonald, *op. cit.*, p. 83.

35. International operations totaled 5,882 million revenue passenger miles in 1957, 10,138 million in 1962. *ATA Facts and Figures*, 24th ed., 1963, p. 20.

36. *International Carrier*, pp. 4–5.

37. *Time*, July 24, September 4, 1964; *Washington Star*, August 25, 1964.

THE CHAIN REACTION OF EXPORT PROMOTION

One way to begin is with an analysis of perhaps the most typical problem that any regulatory approach must deal with—that of export promotion. This single problem fans outward into most of the other areas that have troubled U.S. international airlift policy for so many years. Export sales began after World War II as a means of preserving a mobilization base, but the sales increasingly have contributed to the declining position of U.S. carriers. There are no clear limits to the application of the export promotion program, and it has reached a point where it contradicts both the *de facto* international regulatory framework and the U.S. policy of multiple competition on foreign routes. The contradictions seem traceable both to the export program's expansion and to advances in aeronautical technology. Ethiopia is a typical case in point.

At the end of World War II Ethiopia asked the U.S. to help it establish an airline that would meet the country's needs.[38] The State Department asked TWA to provide the management skills and technical assistance required. The program has been virtually continuous since that time, with TWA's participation financed through U.S. foreign aid programs and Ethiopia's new aircraft purchased with loans obtained through the U.S. export program. The airline has modernized itself, first with piston transports and more recently with long-haul jets.[39] The airline's new Boeing jets service long international routes which include several major cities, among them Frankfurt, Cairo, and Athens. The airline has advertised, presumably for the benefit of American tourists, that its transports still are flown by American pilots.[40]

Some underdeveloped countries have purchased older aircraft with the help of U.S. grants or loans. Air Laos, for example, was operating in 1961 with a fleet of thirteen Boeing four-engine Stratocruisers, one of the first transports produced after World War II.[41] Air Congo began flying between Leopoldville and New York in 1962 with a Boeing jet owned in part by the Belgian national airline.[42] All told between 1955 and 1960, fifteen countries

38. This account is from Theodore Geiger, *TWA's Services to Ethiopia* (Washington: National Planning Association, 1959), pp. 56 ff.

39. *New York Times*, November 7, 1962.

40. *Reporter*, 28 (February 14, 1963), 6.

41. U.S. Department of Commerce, Business and Defense Services Administration, *World Survey of Civil Aviation; Selected Asian Countries and Oceania* (Washington: Government Printing Office, 1961), p. 11.

42. *National Defense Transportation Journal*, 18 (July–August, 1962), 4.

obtained twenty-five Export-Import Bank loans to buy aircraft worth $173 million; in 1961 alone, the U.S. provided loans of $94 million to Brazil, Colombia, Ethiopia, France, India, and Japan.[43]

Most of these participating countries either are too underdeveloped to provide a domestic market for such huge transports (India), or they are too small in size to use the planes in anything but international travel (France and Japan). U.S. carriers already occupy the international markets in which these foreign carriers must operate if they are to pay off the loans. The U.S. thus acquires a vested interest in the competitors of its own airlines. Indeed, the loan itself seems to touch off a complicated series of cause-effect relationships seldom understood or examined in advance of Export-Import Bank decisions.

To begin with, the mobilization rationale makes it unnecessary to calculate the effect of any sale upon the international travel market. Nor can anyone predict in advance if other countries will ask for new routes or if the U.S. will grant them. The CAB-presidential decision-making process is virtually isolated from the export promotion structure. After the loan is made, however, the U.S. has little choice but to be generous in its bilateral negotiations with that country, as during the Eisenhower years. As the country becomes more troublesome to negotiate with, the CAB tries to restrict its operations. The CAB also increases mail payments to U.S. companies to compensate for the difficulties. The increase causes a rise in the postage rate, an increase in the Post Office budget and, then, complaints that the Post Office is guilty of mismanagement.

The U.S. carriers, of course, also turn to the Defense Department when the cause-effect pattern catches up with them. All of the companies in international competition, including the U.S. carriers, must engage in expensive advertising programs designed to lure passengers. This can be a difficult task in markets such as the Atlantic one, where twenty companies are in operation. The immediate result is what might be called extravagant expenditures for liquor and sumptuous meals. This gives each company a foothold in the market but also makes it impossible to lower international rates. If all of this ultimately requires that the government pay direct subsidies to the U.S. carriers, the companies and the CAB

43. Compiled from *Project Horizon*, p. 30; *Aviation Daily*, August 11, 1961; and U.S. Department of State, Agency for International Development, *The United States Program of Aviation Assistance to Less-Developed Countries*, Report to the Administrator, July, 1964, Appendix A-1 to A-14 (hereinafter *Aviation Assistance Report*).

stand accused of mismanagement. It is even possible that aircraft sales are a long-term detriment to the U.S. balance of payments, admitting that at the outset the sale provides an immediate help. Because the U.S. often provides the market, the final result is difficult to discern; the U.S. may well be repaying its own loans.[44]

Rather than being an unqualified success, export promotion has produced the equivalent of an international group of nonscheduled airlines. Export-Import Bank loans are quite similar to (1) the postwar loans that enabled the nonscheduled airlines to first organize themselves; (2) the merchant marine proposals that appear with some frequency; and (3) the tacit guarantees the Defense Department has used to encourage the purchase of new transports. In all of these instances, the government has established airlines in a manner that contradicts the international airline policy framework. The typical U.S. airline can purchase transports only when market estimates indicate that revenues will pay for the aircraft. These certificated airlines know from experience, moreover, that direct subsidy brings only direct criticism and, as a result, they cannot afford to be overoptimistic about the market.

When countries such as France and Japan buy American transports with U.S. loans, the major condition is that funds are not obtainable otherwise. Yet there surely is enough private capital in those countries to finance anything that is commercially promising. The situation with respect to underdeveloped countries is even more dubious.

The actual needs of the emerging countries appear to have been overlooked in the rush to help U.S. manufacturers by selling their transports. True enough, export sales have for some years been part of the Soviet-U.S. competition for the friendship of these nations.[45] Nor can it be doubted that the small nations have demanded the huge new jets that carry with them a form of national prestige. These jets do not, however, produce new capital within those countries; the internal "multiplier effect," whereby an injection of foreign aid produces far more additional capital than is represented by the original input of funds, simply is not at work. The new jets draw capital largely from outside the country; the ultimate

44. This disputes a generally held view that export sales always assist the balance of payments of the U.S. See, e.g., *Project Horizon*, pp. 129, 177; also Wilfred Owen, *Strategy for Mobility* (Washington: The Brookings Institution Transport Research Program, 1964), pp. 179–80.

45. The Cherington Report and the Elliott memorandum both took this approach. See Chapter 8.

effect is a multiplication of foreign aid but without any assurance that the additional aid will be spent as its donor might desire. It also is possible that the airlines of underdeveloped countries will for many years depend upon access to international markets that do not include the countries themselves. These airlines may become the real gypsies of international aviation; they seldom will land in their home countries.

It might be more reasonable to concentrate on selling aircraft that can directly benefit the internal development of such countries; short-haul transports, for example, could form the national transportation systems in countries without highways, railroads, or rivers.[46] Applied to such countries, then, the U.S. concentration on long-haul jets combines the characteristics of unregulated airline activity, concealed foreign aid not even understood by the donor country, and development programs that add little to the real wealth of the countries themselves.

If many economic theories have little to say about the U.S. international airlines other than to advocate more competition and more growth, these theories have virtually nothing to say about an export policy that relies upon defense needs. If the gaps in the intellectual framework are of this magnitude, can they be filled?

AIRLINE POLICY AND ECONOMIC THEORY: A REVISION TO TRADITIONAL APPROACHES

If a single important conclusion emerges from this book it is that the U.S. often has discarded reasonable solutions to airlift problems in favor of alternatives which overlook many of the important interests at stake. The first major case of this sort was the Roosevelt administration's dismissal of its predecessor's solution, even though that solution was the result of a decade of attempts to use more traditional approaches. The most important factor in

46. There are some active programs of this sort, and the FAA and the CAB are participants. For a description of activity in Latin America, see *Time*, July 31, 1964, p. 33. For a forthright criticism of these U.S. aid programs, see Hans Heymann, Jr., *Civil Aviation and U.S. Foreign Aid: Purposes, Pitfalls, and Problems for U.S. Policy* (Santa Monica: The Rand Corporation, R-424-RC, January, 1964), pp. 18–25. The State Department also has indicated that a major shift in emphasis may be in order. *Aviation Assistance Report*, p. 11. There has been no estimate, however, of the political impact within the U.S. of a decision to de-emphasize export promotion and its corollary benefits to U.S. manufacturers and workers.

the Roosevelt administration's decision was the absence of competitive bidding from government awards of air mail contracts. Insistence upon competitive bidding, then, led to the military accidents of 1934 and that same policy, until the CAB ruled it out in 1960, was at least partly responsible for later accidents. This same approach was the foundation for political support for the several attempts of small entrepreneurs to establish themselves as airline operators, and these attempts produced yet other distressing accidents.

If it is admittedly difficult to prove direct correlation between marginal operation and aircraft accidents, there is much economic evidence to demonstrate that optimistic rate slashes and multiplication of competition have not worked. The problems cannot be explained away by accusations that the airlines are guilty of mismanagement and incompetence, that the CAB is staffed only with political hacks in league with the airlines, that the president does not respect the independence of the CAB, or some combination of these contradictory allegations.

The international airlines initially were a joint effort by government and private enterprise, functioning as a team. It does not seem an exaggeration to say that much economic theory looks inward to the operation of a single society and does not recognize the possibility of such teamwork; the same economic theories seem never to have acknowledged the existence of such a partnership or its necessity as a means of furthering national objectives. Still without admitting that such a partnership was in being, the wartime administration used it to lay the foundation for the industry's postwar expansion, and all but forced the major domestic carriers into the world arena. With more companies in the international market, most of them primarily domestic companies whose international operations were a minor part of their total output, it once again was possible to perpetuate the fiction that the government has nothing to do with overseas operation.

The outcome of this behavior is that the government has used its authority to install the traditional model of economic activity—direct competition—but, because mythology prescribes that government actually has no such authority, it has not attempted to undo the mistakes. This attitude has produced almost the ultimate in confusion; the government insists on free competition even if it must force companies to remain in operation against their expressed desire not to do so.

Administrators of national policy approach their tasks with some general notion of what is legitimate and what is not. Definitions of both usually are provided by normative or empirical theories dealing with the phenomena under consideration. In the case of aviation policy, there is no lack of normative theories, those which prescribe how things *ought* to be; most theories tolerate only more competition and a constant lowering of prices.

There is, on the other hand, very little in the way of empirical theories, those which analyze things as they *are*. Any such theory, if it undertook to describe the CAB and aviation policy as at least a partially unique phenomenon not to be confused with a host of dissimilar regulatory processes, would be more than welcome. Despite the fact that airlines are totally subject to public regulation, they are forced (on normative grounds) into inefficient operational patterns and then criticized if they cannot achieve economies in operation.

New theoretical approaches are both timely and timeless because even though airlift problems of the 1960's demand attention, some of the proposals to be argued here were suggested as long ago as 1935. They seem to have been valid then, and this validity has not dissipated in subsequent years of technical advancement. Some of these approaches, moreover, are in accord with the new outlines of economic theory which promise ultimately to force significant modifications of traditional theory.

Airlines as a Combination of Oligopoly and Public Regulation / Even to those who favor free competition, it is now clear that for some years to come, a relatively small number of large airlines will dominate the market; new enterprises no longer can expect to enter the industry.[47] This is a significant admission, but it is based only upon the cost of new equipment; new jet transports cost approximately $5 million without flight crews, mechanics, or associated facilities. But the high costs are only part of the explanation; astute observers might have noticed in 1930 that the geographical future of transcontinental aviation inevitably would require large companies.

An even more significant reason for the small number of successful companies is the relative experience of the supplemental and established airlines. The supplementals, after all, have finan-

47. Caves, *op. cit.*, pp. 446–48.

cial backing in the form of the large corporations who engage in airlift as a sideline, but money is not enough. Simply stated, the established carriers have too much operational and management expertise, from mechanic to pilot to manager, for new companies to successfully overcome. It is difficult to escape the conclusion that when the airlines are criticized for mismanagement and incompetence, it is only because other explanations are not immediately available.

John Kenneth Galbraith pointed out in 1952 that most economists fail to recognize the dollar value of experience in any given industry. He noted, for example, that the government provided loans to postwar entrepreneurs seeking a place in the automobile industry. The experiment failed, despite substantial financial backing, largely because the newcomers could not quickly acquire enough experience to pose a major threat to the old companies.[48] Even allowing for CAB and congressional intervention, the same thing happened in the airline industry. This is not to say, however, that the government had a clearly defined objective in mind when it released surplus transports at reduced prices.

As a first step, then, we can argue that the airline industry is both an oligopoly and a regulated industry. There are relatively few companies, but the regulatory pattern forces them into direct competition against each other. It is necessary to this argument, then, to analyze the characteristics of these two forms of business activity. Oligopoly, for example, has proved historically troublesome for traditional economic theory.

Oligopoly as a Problem of Economic Theory / Many economists, even when they take into account the industries dominated by major producers, remain unwilling to stray very far from classic models of free competition. Because unregulated oligopolies display tendencies toward monopoly, the economists prescribe that oligopolies be as competitive as possible. What this means in specific situations is that economists argue against any reduction of the number of companies in an industry dominated by a few companies.

Gardiner Means cites the conclusion of some economists that it automatically makes more sense to have five instead of three com-

48. John Kenneth Galbraith, *American Capitalism: The Concept of Countervailing Power* (Boston: Houghton Mifflin Company, 1952), pp. 37–38.

panies producing spark plugs, despite the lack of proof that this is more efficient.[49] Means points out that economists have at least some obligation to examine the organizational imperatives of any given industry under consideration, as Means has done with the steel industry. He found, for example, that blast furnaces operate in groups which must include a minimum number of units if they are to be efficient. If anyone insists that the steel companies be reduced below that size, he forces a quantum increase in operating costs and prevents price reductions.[50] This rather precise operational factor easily can be overlooked in developing a theory which purports to cover all industries.

Economists, administrative lawyers, and political leaders have argued for many years that the best organization of the airline industry is the one which provides for the maximum number of companies. They consider mergers dangerous, criticize the CAB for not licensing new trunklines and, of course, they support a competitive system of unregulated carriers.[51] Yet the theory of monopolistic competition (which deals with oligopoly) does not contemplate a regulated industry necessary to public convenience, and the theory does not have to include any questions about subsidy. The traditional theories prescribe that the marginal company will disappear from competition, but this cannot be the case with the airlines. This has left the CAB with no guidelines other than to comply with as many political demands as possible, but this attempt has only caused the airlines to take on the worst characteristics of unregulated oligopolies.

The Phenomena of Competitive Waste and Excessive Advertising / What has been said thus far indicates that there is an important contradiction within most economic theories, and it prevents them from dealing successfully with oligopolies. The economist relies on price competition and presumably agrees that companies will drop out of the race if they cannot keep up. But the economist also insists that any reduction in the number of companies is dangerous, for it leads to monopoly. The contradiction thus requires that conditions remain static in an economic society that is supposed to be dynamic, and it makes it impossible for oli-

49. Gardiner C. Means, *Pricing Power and the Public Interest* (New York: Harper & Brothers, 1962), pp. 166–67.
50. *Ibid.*, pp. 182–83.
51. See, e.g., Caves, *op. cit.*, p. 441; Barber, *op. cit.*, pp. 11–13; *Celler Report*, pp. 268–70.

gopolies to engage in price competition in any real sense of the term. While these conclusions apply only to unregulated oligopolies, they are importantly related to the airline problem.

The managers of large enterprises tacitly agree that any use of price competition can lead only to disaster for those who play the game all the way through.[52] Oligopolies, almost by definition, are not marginal companies and, it might be added, the public interest often requires that such companies stay in business.[53] The managers thus adopt an informal convention which abolishes price competition as the principal weapon of economic warfare.[54] In its most sophisticated form, the convention has become known as administrative pricing, an arrangement whereby the major companies in a given industry set their prices without reference to day-to-day competitive pressures. In its most extreme form, the convention degenerates into old-fashioned price-fixing by direct consultation between the major companies.[55] In a pattern that is equally familiar in international politics, agreements between companies are designed to insure the survival of all of them. Having ruled out price wars, oligopolies turn to other forms of competition which preserve or extend the market of any individual company but not to the point of forcing another out of business.

The most familiar of these alternative forms of competition are the addition of nonrational capacity and the use of advertising on an extremely large scale, neither of which conforms to standard theory. Many industries adopt what Gardiner Means calls the

52. Galbraith, *op. cit.*, pp. 45–49.

53. This refers to the impact of the large concern upon the local community, and the community's feeling that the concern must be preserved.

54. Price warfare is a fearsome weapon in those industries where the convention is not observed. The merchant who starts the warfare is considered a traitor by his fellows. A typical example is the milk producer who suddenly lowers his prices. *Ibid.* I am indebted to Gardiner Means for pointing out that there is, indeed, no completely rational way of conducting price competition in such situations. Out-of-pocket costs often are only a small part of total economic costs and simply cannot serve as a pricing base. Price competition, in such circumstances, becomes an extremely upsetting form of psychological warfare.

55. There is much more to understand about this type of price-fixing. When the government seeks competitive bids from a group of large companies, for example, it is considered improper for the companies to act in collusion. Yet if one company consistently secures all of the contracts and succeeds in putting its competitors out of business, the resultant monopoly would get at least equal criticism. This is not to argue either for monopoly or for price-fixing, but only to point out that existing theory does not cope with the problem.

"principle of competitive waste"; the industries build many more distribution units than they can use with efficiency.[56] The typical example is the petroleum industry; the dominant companies construct new retail sales outlets on a ridiculously large scale. The street corner with four gasoline stations is a phenomenon known to all. The costs are quite high, especially when they include auxiliary services and prizes for customers. Truly massive advertising is quite familiar in industries such as tobacco, automobiles and, once again, gasoline.

If competition of this sort has no place in economic theory, the industries themselves look upon it as extremely wasteful and of little or no direct benefit to the consumer.[57] Inefficient retail capacity and advertising campaigns combine to prevent price reductions that might otherwise be attainable. The natural tendency of industries of this sort, adds Means, is to continually dissipate potential profits by building still more capacity; operating costs continue to rise and, while retail prices are reasonable with respect to costs, they are much higher than they ought to be. It is an odd twist that the airlines have proved that excess capacity and huge advertising campaigns are not limited to wholly private and unregulated enterprises. The industry has managed to achieve what might be called government-promoted competitive waste.

It may seem peculiar that airlines combine the distinguishing features of unregulated oligopolies with public regulation, but there is a logical enough explanation for the oddity. To Galbraith and Means, and to most others who deal with the entire national economy, the normal image of public regulation is the power company, the municipal bus company, or the telephone company, none of which operate within an over-all policy of direct competition.[58] In the industry under study here, the regulatory process prevents the attainment of efficiencies which the process is intended to provide.

The airline industry gradually has become similar to the petroleum industry and its gasoline stations. Each company provides its own crews, mechanics, equipment, ticket sellers, office space, baggage handlers, and all of the other appurtenances so indispensable to a complete airlift system. This inevitably has led to substantial increases in operating costs, a situation aggravated further by the

56. Means, *op. cit.*, pp. 213–18.
57. *Ibid.*, p. 205.
58. *Ibid.*, p. 186.

special nature of international operations. Some of the bases used around the world are not revenue-producers, and duplication of facilities is not offset by the hope of market expansion. Few, if any, paying passengers are available at Wake Island or the historically important Azores Islands in the Atlantic.

The obvious difficulties of facility duplication led the CAB in the 1950's to pressure the airlines to share equipment in the interest of "economies of operation," and it has since become commonplace for airline passengers to find that the airplanes on which they travel are flown by crews from more than one company.[59] The CAB-induced practice suggests this question: If facilities, crews, and airplanes are used jointly by several airlines, what is the nature of the choice that the consumer must make between companies?

Both domestic and international airlines also engage in service competition of the most expensive sort. Lavish provisions for meals, liquor, hotel accommodations around the world, first-run movies in flight, television at every seat, and countless other items lure passengers away from competitors or merely catch up to other advances.[60] The average citizen who looks for austere transportation at the lowest possible price simply has no chance to realize that hope.

Under the circumstances airlines have little choice but to increase their trip frequencies, because an important part of the service offered the passenger is that he not be forced to pass up any given airline because its airplanes have no empty seats. An airline hardly can afford to fill more of its seats than its competitors, and this factor has tended over the years to defeat every attempt to

59. *CAB Annual Report, 1962,* p. 27. Continental, one of the smaller trunklines, has found it necessary to share facilities with both Braniff and United. The former president of Eastern Air Lines has observed that sharing, even if theoretically efficient, goes against the theory of competition. On many occasions, he argues, the result is only dispute as to what critical equipment (e.g., loading steps) is available to which company at any one instant. Malcolm A. MacIntyre, *Competitive Private Enterprise under Government Regulation* (New York University: The Charles C. Moskowitz Lectures, School of Comerce, Accounts, and Finance, 1964), p. 35.

60. The costs of some of these services are quite substantial. American Airlines spent $2.5 million to install a new television and motion picture system in 47 of its jets, with the yearly operating costs of the system estimated at $1.7 million. *Newsweek,* October 19, 1964, p. 78. With U.S. carriers leading the way in this service, the Civil Aeronautics Board has rejected attempts by the International Air Transport Association to ban movies and television on international flights. The CAB decided that such services did not represent a "costly frill," and agreed with TWA that movies "ease the tension and tedium of long-haul air service." *Washington Post,* June 3, 1965.

overcome "unutilized capacity." In the international sphere, of course, this whole process of trip frequencies and empty seats has become a matter of intense negotiation. The CAB Chairman frankly admits that "excessive capacity" has become a household word in aviation, but the percentage of empty seats continues to increase.[61]

Because of the nature of the industry, the airlines gradually have become big spenders for direct advertising. Some unregulated oligopolies, e.g., steel, use a form of limited advertising directed at their relatively large customers. Airlines follow the trends set by cigarettes and automobiles; they aim at the entire population, especially at individuals who never have flown. The domestic companies spend three to six times as much for advertising as the average industry, and the international companies allocate 7 to 10 per cent of their total revenues for that purpose.[62]

Economists who accept the necessity for large airlines admit upon occasion that some routes produce so little traffic that it is plainly senseless to insist on multiple competition. They assert, on the other hand, that it is possible to devise a rational standard by which to distinguish between routes of that sort and other routes which can indeed support multiple competition. The CAB is advised, for example, to determine which routes are in each category.[63] The 1963 report on international air transport policy also concluded that the policy of multiple competition is "justified" on the heavy traffic routes.[64] The evidence demonstrates, however, that the CAB cannot find the rational standard, despite many years of efforts to do so.[65] And all of this is made worse by the fact that

61. *International Carriers,* p. 4.

62. Caves dismissed advertising as insignificant to the total problem, but added that its effect was unmeasurable. He admitted in any case that the industry is among the highest spenders on advertising. Caves, *op. cit.*, pp. 426–27. Also see his comment in *International Carriers,* p. 28. The airline industry thus manages to bring together some of the most challenging problems of economic theory, e.g., advertising and public regulation.

63. Caves, *op. cit.*, p. 447.

64. *Wall Street Journal,* April 24, 1963.

65. Caves, *op. cit.*, pp. 192–231. Caves did not say explicitly there is no rational standard, but his evidence indicates this to be the case. The *Project Horizon* study group did not make a specific recommendation, but it noted that one group had suggested that the standard for international operations be the level of traffic which would permit each of two carriers to operate a daily jet flight across any given route. Under these circumstances each company would have to install facilities to handle one flight at each base along the way, thus leaving those facilities idle for the remainder of each day. The suggestion, moreover, seems to have assumed

technological improvements in airplane size and speed tend to increase the capacity of airplanes themselves at a faster rate than the market can hope to expand. Faced with almost constant political pressure to increase competition while maintaining reasonable profit levels, the CAB produces decisions which cannot be sensibly explained; this results in the incoherent pattern mentioned in Chapter 5, with board members neither writing nor reviewing the official opinions.

Direct and multiple competition, excessive capacity, and large-scale expenditures on advertising make the airlines quite similar to unregulated oligopolies. The principal difference, on the other hand, is that a public agency sets airline prices and profits; they are not set by tacit convention or by direct consultation between companies. The difficulties of the regulatory process have persuaded many people over the years that the desirable objective is to turn the airlines into truly competitive enterprises by abolishing the CAB. As in the case of the supplemental airlines, the idea has been that open competition will bring lower prices. While it appears unlikely that this alternative can command serious attention, it is examined here because it has been popular on occasion. If the airline oligopoly were not subject to CAB regulation, how would it go about setting prices?

"Target Pricing": An Alternative to Regulation? / Those who support unregulated competition apparently conclude that airlines would reach their price levels in the classic manner. But it seems reasonably clear by now that oligopolies do not use the classic model of competition. Whether such companies tacitly agree, conspire together, or act in complete independence of each other, they arrive at prices through the use of a practice which Gardiner Means has chosen to call "target pricing."[66]

When giant companies use target pricing, Means argues, they arrange their operations so as to earn premium profits, defined

a static world. If new model aircraft prove twice as capable as those in use when routes are awarded, one airline becomes superfluous. But how is anyone to decide which airline is unneeded? *Project Horizon*, p. 120. For a brief review of the inconsistencies in decisions of recent years, see MacIntyre, *op. cit.*, pp. 6–10.

66. Of authors cited in this book, Caves and Mrs. Keyes take perhaps the strongest position in support of ending regulation. See especially Caves, *op. cit.*, pp. 448–49. "Target pricing" is developed in Means, *op. cit.*, pp. 232–48. This is Means's terminology, although he credits Donaldson Brown with an earlier term, "pricing for a target rate of return."

as profits that are higher than can be earned in the capital market. If the average return on capital investment, for example, is 6 per cent, an oligopoly might set a price which aims at a target of 12 to 15 per cent. The large companies use these premium profits, usually about double those of the capital market, to obtain expansion capital, thus avoiding any necessity to buy the capital in the money market. In a sense, the companies extract capital from their consumers, but the latter do not get shares of stock in return for their hidden investment. As Means sees it, target pricing is not a legitimate practice; it brings in profits above the market average, and his principal objective is to find ways of controlling the phenomenon of target pricing.[67]

The nature of these large companies makes it impossible for prices to change every day or so, in accordance with fluctuations in demand. The industries are geared to long-range development and production plans that force company operations into a continuous flow. The companies must ignore temporary changes in demand and, once made, price decisions must stand. Viewed in one context, target prices are a margin of safety. Viewed in another, they are part of a management art which consists of hitting the target and thus avoiding losses that might be traceable to competitive waste. Airlines, of course, operate on a continuous flow basis; the problems of rearranging schedules and printing rate lists make it impossible to respond to temporary changes in demand. If there were no regulatory agency, there is every reason to suppose that airlines would use target pricing.

For reasons mentioned earlier free competition would not bring more companies into the market, and it is even doubtful that direct competition could increase much beyond the point where it is now. The industry would incur the same expenditures as it does now. The CAB restricts domestic carriers to modest profits; target profits would be higher. If the complex and quasi-governmental IATA rate mechanism were to disappear from international aviation, the outcome would be difficult to predict. The point is that there is no reason to assume that the end of regulation would bring lower prices.

Airlift as an Example of Over-regulation / There is sufficient practical evidence to demonstrate that airlines, like other transportation units, would discard their most undesirable and unprofit-

67. Means, *op. cit.*, pp. 266–68.

able routes if they were left free to do so.[68] Commuter railroads and municipal bus companies, for example, often seek to terminate services on sparsely traveled routes. Public control remains necessary, then, to provide service to all who require it, to represent the U.S. in international negotiations, and to restrain profits. The policy of multiple competition, however, brings to the regulatory process some additional factors which tend to raise the price of tickets and thus offset the advantages of regulation.

Since it first combined multiple competition with the regulatory process, the CAB has involved itself in many activities which expand an already cumbersome procedural routine. Even when a regulatory agency manages a single public utility (natural monopoly), the process includes elaborate quasi-judicial procedures which take years to culminate in decisions. The costs of those procedures and the lawyers to carry them through are in all cases added to the public bill for the services in question.[69] But the logic of multiple competition forces the CAB to prolong its search for standards with which to choose between applicants. All of the companies are government-licensed, their pilots and mechanics are government-inspected, and their profit levels are controlled. About the only standard that might be used to choose between carriers is the level of auxiliary services provided passengers, and this would make little sense. The CAB, as part of its endless search, becomes involved in elaborate statistical analysis, tries to keep up with every detail of every company, and attempts to build a case that it can use to defend a decision in which at least one company must be a loser.

More and more lawyers examine and cross-examine witnesses, route cases take longer and longer to decide because any losing company must be afforded complete due process, and ticket prices absorb all of the costs. The CAB, immersed in statistics and procedural due process, loses sight of the broad national objectives it is supposed to advance.[70] The decisions themselves probably are the best available under the circumstances, even if they sometimes result from the ability of one company's lawyers to best another's.

68. Caves, *op. cit.*, pp. 326–29. It is worth recalling again that at the peak operation of the supplementals, they showed little interest in anything other than operation between major cities.

69. Means, *op. cit.*, pp. 191–92.

70. And there still are recommendations that the principal need is for more judicialization. Bernard Schwartz, *The Professor and the Commissions* (New York: Alfred A. Knopf, 1959), pp. 272–75.

One hopeful aspect of oligopolies in general is the ability of expert managers to make decisions without having to wait for a public agency to certify the decisions, even if industry managers must think about the government's reaction before raising prices. The regulatory process tends to prevent any sort of price experimentation. If one company asks permission to lower rates in an attempt to discover if this will attract new passengers, the CAB cannot accede to the experiment for fear it will damage other licensed operators.

In sum, the benefits of competition are lost because managers cannot make decisions, and the advantages of public regulation are thrown away in constant attempts to make the industry more competitive. The logical objective of a new empirical approach is to realize the benefits of both oligopoly and public regulation.

The Opportunity for Reorganization / The geographic characteristics of airlift systems still hold out the hope that something can be done to reverse the trend toward competitive waste in all of its forms, and to restore to the industry itself a greater share of the decision-making process. The acceptance of these objectives could open the way to a unique opportunity to put a giant enterprise to work in a manner that would more nearly realize the public interest. The new approach would accept the necessity for bigness, would restrain profits, and would reduce competitive waste.[71] There should be no reason to be overly fearful of bigness in an industry completely subject to public regulation.

The general rule for restructuring the industry would have to be that only one carrier operate across any particular route except when it is utterly impossible to avoid direct competition. The rule is advanced here primarily as an approach to international regulation, given the focus of this book. It would seem applicable to domestic regulation as well, although both geographic and competitive situations might dictate some modification. The objectives would be massive reductions in advertising budgets, logical consolidations of facilities, and, most important of all, a simplification of the regulatory process to the point where CAB policy-makers would be free of hopeless detail. Not bothered with the search for rational standards where none are available, the CAB

71. Means cites the example of David Lilienthal who served as head of the Tennessee Valley Authority, and then came to appreciate the usefulness of size. Means, *op. cit.*, pp. 251–52.

could avoid persistent temptation to corruption, could easily control profit rates, and could formulate policy by comparing the results achieved by carriers flying comparable routes. Such a solution might well dispose of some persistent myths.

It seems past the time to abandon the view that the airline industry is by its very nature the enemy of the public interest. It seems time to abandon unlimited faith in small business, to agree that the CAB is not the home of political conspirators or political incompetents but is instead trapped in an unworkable structure, and to admit that the regulatory process does not require that two contending companies be involved in each case. It is time, moreover, to appreciate that increased competition in a regulated industry can well be self-defeating. There is, in short, an impressive opportunity to comply with the plea of a Democratic president, made when opponents challenged his economic policies: ". . . We must move on from the reassuring repetition of stale phrases to a new, difficult, but essential confrontation of reality. . . . The great enemy of the truth is very often not the lie . . . but the myth—persistent, persuasive, and unrealistic."[72]

The theoretical framework outlined here would require the reinstitution of some sort of government-industry partnership, if only to the extent of a new and mutual respect. Indeed, sensible route structures probably are unattainable in any other fashion. The argument advanced here puts major emphasis on the notion that the way to organize a regulated industry is to begin with the most efficient operating structure imaginable and then depend upon the regulatory process to overcome the evils of unregulated monopoly. The importance of partnership lies in the fact that all of the expertise is with airline management. If route systems are to be realistic, those who must operate them are necessary to the policy process.

It would be admittedly difficult to repeat, in the domestic arena, Postmaster General Brown's 1930 conferences. Aside from the emotional residue which still remains, the 1938 legislation authorized existing trunklines to continue in business. The international routes, on the other hand, are directly traceable to presidential authority over foreign and military affairs, even if this fact often is forgotten. The argument here is that presidential authority is more than sufficient to reorganize the international airlines. As

72. Speech by President Kennedy, reproduced in *New York Times*, June 12, 1962.

described thus far, however, this theoretical framework does not have an international dimension; it must now be added.

AIRLINE POLICY AND INTERNATIONAL POLITICS: AN ADDITION TO THE FRAMEWORK

The theoretical framework being argued here rests upon a premise that there is a relationship between the airline industry, foreign policy, and military strategy. Both government and industry have recognized this relationship on many occasions in the past and, in fact, the relationship persists on a day-to-day basis in such crisis areas as Berlin and Southeast Asia. National decisions concerning the international airline structure and export promotion have been relatively few in number, but all of them have been made on grounds of national defense.

The U.S. first participated in international aviation largely in response to what was perceived as a German threat in Latin America. Similar considerations accompanied the later expansion into both the Pacific and the Atlantic, and the government maintained the partnership until multiple competition changed the atmosphere. This historical association between airlift and the national interest suggests that the traditional approaches have overlooked important considerations.

The first of these is that instruments of national policy and power are measurable for the most part only in relative terms; absolute measurements are much less significant. To use a simple analogy, the size and composition of any military force depends largely upon the size and composition of the forces maintained by likely military opponents and the perceived degree of the threat. It is much less important, then, to look at the absolute growth of U.S. international carriers (especially when the growth is artificially stimulated by military traffic) than to look at the standing of those carriers with respect to their foreign competitors.

A second consideration is that international politics imposes definite limits upon any theoretical framework. The transatlantic market, for example, simply cannot be organized for maximum efficiency, or for the lowest price, because a place must be found for at least some of the twenty airlines now operating in that area. The international airline system is likely to retain at least some of the characteristics of competitive waste, but it still seems possible to make useful adjustments. But it will not do to argue that the U.S. should drive down international prices, by fiat if necessary.

Yet another consideration is the apparent obsolescence of the mobilization policy which still underlies the large scale promotion of export sales. The policy emerged when the U.S. visualized a new World War II. Such a war, so it was thought at the time, would be a long one, would require the U.S. to reinvade Western Europe, and would call for at least the sustained level of production attained in World War II.[73] Industrial capacity seemed as valuable as military forces in being, and export promotion seemed a useful way to keep alive an important aircraft industry. Organized labor liked the policy, if for different reasons, and the policy also fit into the long-term U.S. dedication to production. The Morrow Board of 1925, it will be recalled, viewed aircraft production and export sales as important aspects of aviation policy, and at a time when national defense seemed less important.

What follows are the specific steps that seem necessary to structure a new framework for international airlift regulation and to establish some sort of policy equilibrium. That none exists now seems clear enough; national defense needs and the economic difficulties of aircraft manufacturers still are generally thought legitimate reasons for government spending, but subsidy to airlines is something else again.

The Airline Industry and the Export Program / Assume for the moment that all of the commercial passengers and cargo airlifted to and from the U.S. traveled only on U.S. international airlines and that all of the bases and facilities used along the way were under U.S. control. This would represent the maximum extension of the commercial airlift component of U.S. national strength, and there would be direct relationships to military strength as well. Under the conditions postulated, there probably would be no direct military threat to the routes and bases used by U.S. carriers and, moreover, the same companies easily could provide assistance to the military in times of crisis.

If the reverse condition is assumed, foreign airlines would airlift all of the passengers and cargo and, for this reason, those countries

73. For the thinking behind U.S. strategy up to 1953, see James E. King, Jr., "NATO: Genesis, Progress, Problems," in Gordon Turner and Richard D. Challener (eds.), *National Security in the Nuclear Age* (New York: Frederick A. Praeger, Inc., 1960), pp. 149–51.

probably would control the routes and bases as well. The economic and military presence of those countries would extend to U.S. boundaries, and this would restrict the U.S. to domestic commercial operations. As a first step, then, governmental decision-making should include a look at any trend toward either of the two extremes. This would require the government to admit that it would not permit U.S. international carriers to be forced out of the market, and this alone would be the most explicit acceptance of responsibility since 1930. While this might account for the airlines, it remains to allow for the manufacturers and their employees.

If the world's international airlines operated nothing but U.S. transports, this would represent the maximum projection of U.S. industrial strength in this field. In the opposite case, all of the world's airlines would fly transports built outside the U.S. The relative positions of the U.S. carriers and the relative success of the export program can be combined in various theoretical ways. In one situation, for example, the U.S. would maufacture all of the transports, but only foreign airlines would fly them.

The principal necessity is to include both of these dimensions in the decision-making process, and to balance one against the other. The maintenance of a healthy manufacturing industry might upon occasion have to give way to the necessity of retaining U.S. access to routes and bases, because the extension of U.S. commercial and military strength would depend upon both. As stated thus far, this framework bears some resemblance to earlier theories of international affairs.

The Relevance of Admiral Mahan / The nature of military airlift and international airlines as components of national power seems similar in many respects to Admiral Alfred T. Mahan's theories of sea power.[74] National power, he argued, depends upon military and economic dimensions of naval strength—the Navy and the Merchant Marine. To Mahan a weakness in either dimension signified a deterioration in U.S. strength.

It was Mahan's great contribution to bring together the separate considerations of geography, economics, and military strategy. He

74. This comparison with Mahan is based upon William E. Livesey, *Mahan on Sea Power* (Norman: University of Oklahoma Press, 1947), pp. 48–49, 241–44, 277; Margaret Tuttle Sprout, "Mahan: Evangelist of Sea Power," in Edward Mead Earle (ed.), *Makers of Modern Strategy* (Princeton: Princeton University Press, 1943), pp. 415–45, especially 415–24.

added, moreover, that air might one day surpass water as the great medium of transportation between nations, and students of Mahan conclude that he probably would have become concerned about bases, routes, and commerce for ships of the air. His theories seem quite applicable to U.S. international airline systems, especially when one includes the bases, facilities, and other appurtenances so important to airline routes. To argue for the relevance of Mahan's sea power theories, however, is not to argue that they be used now in the manner his followers used them at the time.

The political dimension of Mahan's approach was a form of maritime colonialism; the U.S. acquired and retained overseas colonial bases. This dimension of Mahanism has been obsolete since World War I, but recognition of this deficiency does not invalidate everything he said. In geopolitical terms, U.S. economic strength still rests to some extent upon the military security of routes and bases above ocean surfaces and upon them.

Application to the Airlines / What policy-makers need is a set of relatively simple rules of thumb that account for the balance of U.S. political, economic, and military interests in this field. Even the highlighting of issues might serve to avoid some disastrous results. If the U.S. becomes the least important customer of the countries who control important bases, they may restrict U.S. access to those bases. Typical examples are the Azores and the bases in eastern Canada which are closest to Europe.

While exact lines are difficult to draw, the current U.S. position in the transatlantic market seems well below any level that can be thought sufficient. The U.S. provides more than half the commercial passengers, but U.S. carriers airlift only about a third of the total. One logical rule might be that the U.S. regard as dangerous any situation in which its carriers' share of the market is less than the amount of U.S. traffic in that market. It would be a matter of policy to restore the balance any time it fell below that level, and the government would conduct negotiations with that objective clearly in view. After restoring the balance, the government would consider airline questions in conjunction with other foreign economic policies.

The airlines might object to this aspect of the international relationship, but route awards and landing rights are logical trade-offs for specific aspects of trade and tariff policies, export promotions, and similar economic policies. If such an arrangement ultimately might limit the expansion of U.S. carriers, it would at least guarantee their minimum level of participation. Even as outlined here, however, implementation of the policy would cause sharp conflicts with other countries.

The current U.S. position in the transatlantic market probably requires abrupt corrective actions, perhaps even the withdrawal of concessions already granted to airlines of other countries. There seems little choice but to restrict the operations of foreign airlines whose governments cannot retaliate to the same degree. While this is not a specific suggestion, Australia is a typical example. If the U.S. denied the Atlantic market to Australia, that country would deny landing rights to U.S. carriers. The resultant dispute would be undesirable, but the alternative of a permanent deterioration in the U.S. position is much more undesirable. If a rule of this sort could help in CAB-White House decisions on route awards, however, it would not solve the export questions.

Application to the Export Program / A revised export policy would have to be integrated with the other dimensions of foreign economic policy and be related to a more comprehensive mobilization policy than the one still in use. Both requirements seem attainable, and without any reduction in the number of airplanes produced and sold, even if fewer would be sold abroad.

Two rules of thumb seem necessary. In the case of underdeveloped countries, the emphasis should shift from long-haul jets to aircraft of shorter range usable within the countries themselves. In the case of developed countries, a revised policy would prohibit export sales when market analyses indicated that the airplanes would reduce the participation of U.S. carriers below the levels outlined earlier. These analyses could not be purely economic in character. The U.S. would have to assess the probability that the purchaser country would ask for new routes and the probable U.S. response. Had a policy of this sort been in effect during the 1950's and early 1960's, the U.S. probably would not have sold as many aircraft abroad.

Mobilization theory still rests upon maximum aircraft production, but it prescribes no relationship between that production and possible U.S. needs for emergency airlift capacity. The air merchant marine programs tried to deal with the relationship, but they did not prescribe that any of the airlift capacity be kept in a truly reserve status, i.e., that it not be used during peacetime. Even if mobilization theory must call for keeping factories in production at some reasonable level, there could be a purely domestic solution to the problem.

Relevant historical experience suggests that in any substantial emergency, the national transportation capacity will be insufficient if it already is operating at full tilt when the emergency begins. Domestic transportation needs probably will increase, as will the demands for strategic airlift. If both international and domestic carriers operated at something less than maximum utilization in peacetime, this would provide for emergency expansion. Instead of flying at rates of ten to twelve hours per day, carriers might operate at five to eight hours. One outcome would be a Civil Reserve Air Fleet of much greater utility than the imaginary fleet maintained since 1952.

It appears at first glance that an arrangement of this sort would increase airline operating costs. The actual increase would be very small, perhaps nonexistent, if funds now used for competitive waste were diverted to the more utilitarian solution of providing transports for emergency use. A mobilization approach of this variety would put into the hands of U.S. carriers the transports normally sold abroad. The over-all regulatory process would include a prohibition against excessive operation by U.S. carriers. Perhaps the result might be called a regulated air merchant marine of known capacity, as opposed to the unregulated auxiliaries proposed over the years.

If such arrangements seem the only way to provide for a sensible reserve airlift capacity, they also would be quite difficult to administer. There would have to be agreements with crews and mechanics, providing for employment at less than peak capacity during normal times. Any program which involves full pay for partial work, and in an occupational group already well paid, could not help but produce political dispute. This suggests that a substantial part of U.S. strategic airlift capacity would have to be military, if only on grounds of political feasibility; it should be less difficult to fly military transports at less than peak capacity during peacetime.

This new framework accounts thus far for an economic model of the entire airline industry, a way of approaching the international issues, and the organization of a legitimate reserve capacity for emergency use. It is time now to carry the approach one step further, and suggest the precise nature of an international airlift structure that might offer a solution.

THE THEORETICAL FRAMEWORK APPLIED: REGIONALISM AND THREE INTERNATIONAL AIRLINES

The reorganization of U.S. carriers that is suggested here recognizes that government participation was the most important ingredient present at the birth of those airlines and has not since diminished in importance. Only traditional economic theories prevent the government from taking the decisions to correct the most obvious deficiencies.

The structure of three regional airlines outlined in the remainder of this chapter conforms to the pattern of U.S. international airline expansion in the 1930's, formed the basis of the 1935 recommendations of the Federal Aviation Commission, was vaguely mentioned during World War II by President Roosevelt, parallels the structure found necessary for efficient operation of strategic military airlift, was specifically suggested at the end of World War II by individuals who understood the airline business but had no direct interest in how the airlines would be organized and, finally, is designed to avoid the pitfalls of the single chosen instrument. The three regional airlines are best described in terms of geography, the nature of airlift systems in general, and their prospective size with respect to prominent domestic companies.

The geographic position of the U.S. in the world airlift market and the initial expansion of PAA in the 1930's indicated from the start that the international carriers would extend outward from the country in three general directions. While their size is not uniform, there are in fact three major regional markets for U.S. carriers—Europe, Latin America, and the Pacific. In practically all international operations aircraft have left the U.S., traveled into one of these regional markets, and then have returned along the same route. When one company (PAA) has operated in all three markets, it has organized semi-autonomous regional subsystems within the company.

A reorganization of the existing international companies would produce three new companies, each comparable in size to the largest of the domestic trunklines (see Table VII). The reorganization thus would take into account the requirements of rationality and efficiency, an objective wholly compatible with public regulation. This would hold to a minimum the problem of competitive waste, and the obvious relationship between these reorganized airlines and the national interest might even overcome the problem of excessive advertising.

Table VII • The Proposed Regional Airline Systems Overseas Revenue Traffic in Billions of Ton-Miles (1964)

Present Company	New Atlantic	New Pacific	New Latin America
Alaska		.01	
American			.02
Braniff			.02
Caribbean			.01
Delta			.01
Eastern	.05		.05
Northwest		.10	
Panagra			.04
PAA-Alaska		.01	
PAA-Pacific		.40	
PAA-Atlantic	.50		
PAA-Latin America			.30
Trans-Caribbean			.04
TWA	.30		
United		.10	
Western			.01
Seaboard World	.07		
Pacific Northern		.02	
Totals for new systems	.92	.64	.50
Added military traffic	.10	.15	
	(1.02)	(.79)	(.50)

Domestic Trunk Lines as of 1964

American	1.00	Continental	.20
Eastern	.60	Delta	.40
TWA	.80	National	.20
United	1.00	Northeast	.10
Braniff	.20	Northwest	.20

Pan American Total

1.2

Source: Compiled from charts in U.S. Civil Aeronautics Board, *Monthly Report of Air Carrier Traffic Statistics* (Mimeographed: Washington, November, 1964). Military traffic estimated from charts in U.S. Civil Aeronautics Board, *Supplement to the Monthly Report of Air Carrier Traffic Statistics* (Mimeographed: Washington, December, 1961), and U.S. Department of Defense, Military Air Transport Service, *Airlift Service Management Report July 1961–June 1962.*

Three new regional systems quite clearly would be U.S. representatives in those areas of the world and would be identifiable as something other than purely private enterprises. The current status of the international carriers suggests that it may not be too long before the U.S. will resort to patriotic appeals to its travelers, encouraging them to give their business to U.S. companies.[75] Perhaps the greatest advantage of the reorganization, however, would be the return of a government-industry partnership (no matter how vague) and an end to the policy paralysis caused by multiple competition.

With only one U.S. company operating into each major area, the divergent interests of competing companies no longer would hinder the U.S. negotiating position. As outlined in Chapter 4, multiple competition has made it virtually impossible for other countries to grant concessions which seem to them to be double those granted by the U.S. The same policy has made it too convenient for U.S. negotiators to explain away concessions on grounds that U.S. carriers cannot agree among themselves.

Because the regional companies would not operate in the domestic market, it would be possible to restore and re-emphasize the essential differences between domestic and international airlines. Because these differences have been obscured for so long a time, it will be hard for many to accept the possibility that it may not be necessary for each U.S. traveler anywhere in the world to have a choice between two U.S. companies. Those who object to departing from tradition, however, should be prepared to accept the regional reorganization, if only because the alternatives to it seem much worse.

The Alternatives to Regionalism / If the regional approach outlined here is unacceptable, only two alternatives seem available. A brief examination of these alternatives leads to the conclusion that neither offers much promise.

The maintenance of multiple competition into the 1970's seems to depend upon U.S. ability to keep international companies in operation against their wishes. There are important political reasons for hoping that the U.S. will not attempt this approach. Inter-

75. Indications are that an economic cold war already is underway. TWA began a "fly American" program in 1963, designed to insure "defense of the U.S. dollar." The airline claimed also that U.S. carriers had more experienced crews, precision maintenance, and superior equipment. The foreign reaction was immediate. *Aviation Week*, November 25, 1963. More recently, President Johnson urged U.S. businessmen to patronize U.S. airlines as one means of preserving the dollar balance. *Time*, February 26, 1965. Foreign carriers then implied they would reduce their orders for U.S. jet airliners. *Newsweek*, April 12, 1965.

national enterprises finding themselves in this position might well be tempted to make undesirable arrangements with foreign competitors. And even the possibility that international companies will play some part in national emergencies makes it undesirable to continue indefinitely the mutual hostility between them and the government.

The second alternative promises, if anything, even more damaging consequences. Because it seems unlikely that the U.S. can force companies to remain in operation, PAA ultimately may become the single U.S. company in the world airlift market. TWA, the first company to ask for a significant merger, is the company that led the 1945 fight to establish the domestic operators in the international markets; TWA, moreover, remains the PAA's only U.S. competitor for passenger traffic in the Atlantic market. Any merger of this sort would be a major step toward a U.S. chosen instrument, but indirect competition between permanent regional carriers would tend to prevent the complacency often thought to accompany a chosen instrument.

It thus seems necessary to undertake the reorganization effort if the U.S. is to head off these undesirable alternatives. Even the CAB has begun to consider a rearrangement whereby TWA and PAA would operate in different zones within the larger Atlantic market, e.g., TWA might have Spain while PAA might operate in Italy.[76] This tentative CAB approach is a hopeful beginning but, if the board agrees to the necessity for an area approach, there is no reason to stop short. The CAB approach would separate revenue-producing zones that are quite close together (Spain, Italy) while forcing two passenger carriers (not to mention cargo airlines) to fly side-by-side across thousands of miles of routes that do not produce revenue.

The Method for Installing the System / National defense needs and other aspects of foreign policy make every international decision subject to presidential approval, and the president has the authority to organize the international airlines in any way he deems

76. *New York Times*, July 20, 1963; McDonald in *Fortune, op. cit.*, p. 84. It is conceded here that a detailed study might reveal that a two-company solution (Atlantic and Pacific) might be more efficient. It is even possible that the relationships between supersonic speeds, time, and passenger availability may force supersonic transports to travel around the world in one direction rather than flying back and forth between continents. *International Carrier*, p. 15.

necessary. This statement, while quite forthright, is not intended as an exercise in legalistic hairsplitting. Indeed, if the president does not have the authority, why should CAB recommendations be sent to him for approval? Nor is it intended to minimize the political difficulties attendant to such a reorganization. An appropriate political climate would be the first order of business.

The supplementals apparently have lost enough support in Congress to eliminate them as a potent political force. Whatever support remains probably would evaporate during the dialogue, assuming that their actual status would become better known. The most troublesome political problem might come from within the Defense Department. As any government organization head opposes the loss of functions assigned him, the Defense Department's traffic managers might be reluctant to agree to the removal of the supplementals from their position of subservience to a prospective new transportation agency. This opposition could be quite strong, because the status of a new agency and the size of its overhead managerial complex would depend upon the number of airlines employed by the Defense Department, not necessarily upon the aggregate size of all of those airlines.

The three new companies could be set up through a consolidation of the assets currently in use in the international field. Companies now engaged in international operations would have to sell their equipment to the three new companies, probably with some government assistance. The government, moreover, would have to insure that the dispositions were equitable ones, because public policy established all of the present companies. The government could ease the transition by having aircraft currently engaged in military contract operations carry those arrangements with them into the new companies. This also would make it clear to the military leaders that they had incurred no loss in airlift capacity. It is not necessary to propose any more detail, because government and industry would have to use a pragmatic approach which allowed for adjustments but not for subversion of the three-company approach.

After establishment of the three companies, a suitable regulatory process could include a joint committee made up of the CAB, the Departments of State, Defense, and Commerce, and the three new companies. This would be an unusual combination of industry, the executive branch, and a regulatory agency, but it would emphasize the dual position of the CAB, would discard the notion that regulation of international airlines is not a matter for the executive branch, and would recognize the necessity of partnership in the international field.[77] The CAB would be a minority member of the committee, thus making it clear that the principal issues at stake are a part of foreign and military policy. Indeed, it might even be preferable not to include the CAB, if only to make it completely clear that the objectives of domestic and international aviation are not the same.

The System in Operation: Civil-Military Partnership / How the new three-company system would operate in practice can be shown by an analysis of some of the specific problems that lie ahead. The reorganization would, for one thing, correspond to the military system adopted in World War II on grounds of necessity and efficiency. The strategic military airlift organization has contained in recent years only two regional sub-organizations, for the Atlantic and Pacific areas, principally because of reduced military requirements in the Latin American area. Military airlift history suggests, therefore, that regional companies could reinstate a government-industry partnership in the most meaningful sense of the term.

The regional companies would fit into the national defense picture in a much better fashion than ever has been possible with the vast numbers of large and small airlines awarded government contracts. The supplementals always have been available to take government business, but this responsiveness has not been translatable into operational flexibility at the working level. It often is necessary, for example, to subordinate military requirements to the operational inflexibility of the company that has only one or two aircraft and can fly them only on one fixed schedule. These technical considerations, substantial as they are, need not be set out in detail, but the Berlin Airlift was a significant example.

Had only one U.S. commercial company been operating in the Atlantic area, that company might well have been usable in the Berlin Airlift. The military could have dealt effectively with an or-

77. Landis, for example, proposed that the CAB conduct international negotiations, with assistance from the State Department. This would make the international and military considerations secondary to those emphasized by an independent regulatory agency oriented principally toward domestic economic conditions. *Landis Report*, p. 44. There seems little choice but to reject any position of primacy for the CAB in the international field.

ganization that could treat as an integrated whole the pilots, mechanics, aircraft, and facilities in the area. Arrangements of this sort could be highly beneficial in national interest operations requiring an approach not ostentatiously military in nature, e.g., the Hungarian evacuation of 1956 and the Cuban affair of 1962. Nor should it be forgotten that Pan American is an important ingredient of current U.S. operations into Berlin. Because of their one-airplane, one-crew aspects, the small companies simply cannot be used with much effectiveness in situations of this sort.

The System in Operation: The Supersonic Transport / The relationship between the production of transport aircraft and their sale to both U.S. and foreign airlines shortly will take on a shape that is unusual in the country's airlift history. Government normally has provided the necessary funds for the research and development of new models of transport aircraft, principally through expenditures for military needs. Aircraft manufacturers usually have funneled all of this information into the production of civil transports, a typical example being the Boeing 707 as the by-product of military purchases of B-47 and B-52 bombers.[78]

This arrangement proved eminently workable even if largely a matter of happenstance, because the military has awarded its contracts solely on the basis of military needs. The over-all result, however, has been to keep the government out of the decision-making process with respect to the construction and sale of commercial transports.

The next technological advancement in the commercial industry probably will be the long-haul supersonic transport capable of traveling up to 2,000 miles per hour. Both the Eisenhower and Kennedy administrations, however, virtually halted military development projects in the field of manned aircraft of comparable size.[79] All that remained was a skeletonized research program involving two experimental B-70 bombers; this may produce by the late 1960's some information important to a supersonic transport (SST) development but not enough to permit the government to avoid a

78. See remarks in U.S. Congress, House, Committee on Armed Services, *Hearings before Special Subcommittee on National Military Airlift*, 86th Congress, 2nd Session (Washington: Government Printing Office, 1960), pp. 4425–26.

79. *Time*, June 14, 1963.

direct and full-fledged commitment if the U.S. is to produce such an aircraft.[80]

President Kennedy announced in mid-1963 that the administration would join with industry in a "partnership" to produce the SST.[81] Making clear that the partnership would involve principally the aircraft manufacturers, he specified that the government would provide them with credit assistance but would not pay out direct subsidies to any airline operating the SST.[82] The announcement turned out to be only the introduction to a new series of difficulties and disagreements.

The president initially turned the project over to the FAA but with important elements of the research program consigned to the National Aeronautics and Space Administration (NASA). After several months of design competition, the FAA awarded study contracts to four manufacturers to refine airframe and engine designs. The four companies, Boeing, Lockheed, General Electric, and United Aircraft, had been selected by a 210-man government team which also had taken into account recommendations from the airlines.[83] The government let these contracts in the hope of finding a way to overcome the important technical problems that threaten the SST. In the absence of some new solution, for example, the sonic boom may be unbearable to those on the ground.[84]

It is somewhat difficult to trace the government's involvement thus far in the SST program because so much of it remains secret. The president chose a commencement address at the Air Force Academy to make a hasty announcement of the SST, apparently prompted by the knowledge that an Anglo-French consortium to build such a transport was well under way and that Pan American already had ordered some of its SST's.[85] The U.S. program initially required that the manufacturers provide 25 per cent of the development costs, expected to amount to a total of $1 billion.[86] The president also appointed a special Cabinet Committee,

80. *New York Times*, May 12, 1964; *Aviation Daily*, May 25, 1964.

81. *New York Times*, June 6, 1963.

82. *Ibid.*, June 15, 1963.

83. *Baltimore Sun*, June 3, 1964; *Time*, June 12, 1964.

84. *Ibid.*, July 24, 1964.

85. Jerry Hannifin and Charles J. V. Murphy, "We're Losing the Supersonic Transport Race," *Fortune*, 69 (February, 1964).

86. *Ibid.* The *Fortune* researchers ascribe the formula to Defense Secretary McNamara, who already had ruled against any military need for the SST.

headed by Vice-President Johnson, and a special study group, chaired by Eugene R. Black, former head of the World Bank, to study the difficult questions of financing the development of the SST.

The Black Report was not made public, but its principal recommendations appeared in several locations. The report suggested: (1) transfer of the SST program from the FAA to an independent agency reporting directly to the president; (2) slowing the pace so as to provide for more cautious development, especially extensive research financed, at least in large part, by military appropriations; and (3) government financing of 90 per cent of the development costs.[87] This seemed on the surface a discouraging report, but it actually was quite close to what both the manufacturers and the airlines thought practicable. The proposal for 90 per cent financing seemed realistic, as did the recommendation for cautious prototype development rather than early commitment to a production model. The Black approach would result in a close approximation of the method used for civil transport development over the years. The historic U.S. approach has been somewhat unplanned, in that manufacturers and airlines have been left to their own devices except for government support of military research. The Black Report nevertheless led to a slowdown, if not a complete halt, in the U.S. SST program.

Rather than transferring the program to another agency, President Johnson appointed yet another Advisory Committee, headed this time by Defense Secretary McNamara.[88] At the group's first meeting, McNamara secured assent to his proposal that all proceedings be kept secret, and very little concrete information has been available since.[89] The increasing difficulties of the Anglo-French program have tended to dissipate some of the urgency felt in 1963 when the president announced the U.S. program.[90] The European program has suffered steadily rising costs and unforeseen design problems, and the Anglo-French agreement may not weather these problems or associated political conflicts.[91] The Sovi-

87. *New York Herald Tribune*, February 13, 1964.
88. Mimeographed White House Order, April 1, 1964.
89. *New York Times*, May 3, 1964.
90. *Space/Aeronautics Magazine*, August, 1964.
91. *Ibid.* For the detailed arrangements between France and Britain, see Hannifin and Murphy in *Fortune*, *op. cit.* As with the future of Air Union, the SST consortium may be a casualty of other problems in the Atlantic alliance.

ets have chimed in also and have announced that they expect to produce the first SST.[92]

The net effect of all of this recalls some pertinent U.S. aviation history. Whatever the background and experience of any Secretary of Defense he is, of necessity, interested primarily in defense matters and his staff assistants, whether in uniform or in mufti, are similarly oriented to military considerations. The current U.S. approach, then, puts the SST in the hands of the Defense Department, and this is at least a partial reversal of a policy dating back to the 1920's, to the effect that civil and military aviation should be kept apart. The Defense Department reportedly has ruled against the military need for either the SST or the research associated with it, even though the Black Report recommended military support of the research.[93] The president's Advisory Committee has kept secret a program having rather substantial commercial and foreign policy interests, and there is little evidence that the distinctively different questions of foreign and domestic operations have been treated in that fashion.

It is far too early to tell if the supersonic transport program is both desirable and necessary, or if the long-term U.S. lead in civil aviation is in serious danger. Under the most optimistic circumstances the Anglo-French airplane will be in production in the late 1960's, with a U.S. model following in the early 1970's. Some specific difficulties seem highly possible, however, and the U.S. would do well to attempt to guard against them.

The first of these is the continued reluctance to consider making both the airlines and the manufacturers partners of the government in solving the problems that lie ahead. The government may opt for some specific model of the SST, even though no federal agency has any recognizable expertise in the actual problems that face the airlines. The FAA's only previous experience with attempting to develop an airplane was the C-141, and the industry withdrew from that program.[94] The CAB, the only agency with anything approaching detailed knowledge, has been ruled out of consultation

92. *New York Times*, July 8, 1964.

93. Hannifin and Murphy, *Fortune*, *op. cit.* A modest study program has been kept going under the supervision of ex-Air Force generals newly appointed to the FAA and apparently content to accept Defense Department decision-making. The program cannot be called a bold one, since it may not lead even to a prototype. *Aviation Week*, August 31, 1965.

94. See Chapter 10.

on the traditional grounds that regulation is an independent function, even in the international field.

A second difficulty could follow directly from the first. If the long-haul airlines, especially the international ones, are not fully in accord with the government's choice of transports, they may buy a foreign-built SST while the government promotes the export sale of the U.S. model.[95] If the government uses guaranteed loans to sell its SST's abroad, a vicious cycle will be in operation. The U.S. will have a vested interest in the foreign airlines, any administration will award concessions to protect its investment, it will criticize U.S. carriers for buying their aircraft abroad, and bitterness between government and industry will reach an all-time high. If this happens, it will be traceable to the perennial failure to resolve the old question of direct and indirect subsidy, first raised in the 1920's.

Even if the government continues to include the airlines in at least some of the consultations leading to ultimate production of the SST, direct competition between carriers can prejudice the results and produce yet a third difficulty. Companies in direct competition against each other naturally will tend to argue for design characteristics that favor their individual route structures. It seems at least possible that three regional companies not in direct competition against each other could agree upon a single design that would serve the needs of all and would restore some meaning to the term "partnership."

It seems necessary, in summary, that some attempt be made to head off these difficulties before they become far too real. The minimum needs are acceptance that the international carriers must stay in operation as a matter of national policy, that sensible mergers or combinations reduce competitive waste while guarding against a single chosen instrument, and that direct subsidy of the international carriers is as legitimate as the indirect subsidy of export sales. A regional airline system would be helpful in meeting all of these needs.

CONCLUSION

To arrive at the theoretical framework and the policy suggestions in this chapter, and in the preceding one, it has been neces-

95. PAA, TWA, Continental, and American have ordered seventeen of the Anglo-French jets. Hannifin and Murphy, *Fortune, op. cit.*

sary to retrace national airlift policy to its very beginnings so as to put into proper perspective the contemporary policies seemingly traceable only to World War II. The broad view suggests that U.S. airlift history contains only a small number of military and commercial experiments with a relatively new phenomenon. The 1960's seem an appropriate time to select from among these experiments a revised policy that can advance the national interest at a time when the structure of world politics is less than ever that of two giant power blocs in direct confrontation. If it is important to maintain the strength of the free world, it seems equally important that the U.S. maintain its own economic position within the free world.

Bibliography

Bibliography

A. BOOKS

Adams, Walter, and Horace M. Gray. *Monopoly in America.* New York: The Macmillan Company, 1955. 221 pp.

Almond, Gabriel A. *The American People and Foreign Policy.* New York: Frederick A. Praeger, Inc., 1960. 269 pp.

Arnold, H. H. *Global Mission.* New York: Harper & Brothers, 1949. 626 pp.

Arnold, Major General H. H. and Lt. Col. Ira C. Eaker. *This Flying Game.* New York and London: Funk & Wagnalls Company, 1936. 307 pp.

Bennett, Lowell. *Berlin Bastion.* Frankfurt/Main, Germany: Fred Rudl, 1952. 185 pp.

Brereton, Lewis H. *The Brereton Diaries.* New York: William Morrow and Company, 1946. 450 pp.

Brodie, Bernard. *Strategy in the Missile Age.* Princeton: Princeton University Press, 1959. 423 pp.

Burden, William A. M. *The Struggle for Airways in Latin America.* New York: Council on Foreign Relations, 1943. 245 pp.

Burns, James MacGregor. *The Deadlock of Democracy: Four-Party Politics in America.* Englewood Cliffs, N. J.: Prentice-Hall, Inc., 1963. 388 pp.

Caves, Richard E. *Air Transport and Its Regulators.* Cambridge: Harvard University Press, 1962. 479 pp.

Cleveland, Reginald M. *Air Transport at War.* New York and London: Harper & Brothers, 1946. 324 pp.

Cooper, John C. *The Right to Fly.* New York: Henry Holt and Company, 1947. 380 pp.

Craven, Wesley Frank, and James Lea Cate (eds.). *Plans and Early Operations, January 1939 to August 1942.* (*The Army Air Forces in World War II*, Vol. I) Chicago: The University of Chicago Press, 1948. 788 pp.

———. *Europe: Argument to V-E Day. January 1944 to May 1945.* (*The Army Air Forces in World War II*, Vol. III.) Chicago: The University of Chicago Press, 1951. 948 pp.

——. *Services Around the World.* (*The Army Air Forces in World War II*, Vol VII.) Chicago: The University of Chicago Press, 1958. 667 pp.

Cushman, Robert E. *The Independent Regulatory Commissions.* New York: Oxford University Press, 1941. 780 pp.

Davison, W. Phillips. *The Berlin Blockade: A Study in Cold War Politics.* Princeton: Princeton University Press, 1958. 423 pp.

Dearing, Charles L., and Wilfred Owen, *National Transportation Policy.* Washington: The Brookings Institution, 1949. 459 pp.

Deitchman, Seymour J. *Limited War and American Defense Policy.* Cambridge: The MIT Press, 1964. 273 pp.

Eisenhower, Dwight D. *Crusade in Europe.* Garden City: Permabooks, 1952. 573 pp.

Frederick, John H. *Commercial Air Transportation.* Homewood, Illinois: Richard D. Irwin, Inc., 1955. 547 pp.

Friendly, Henry J. *The Federal Administrative Agencies.* Cambridge: Harvard University Press, 1962. 180 pp.

Fuller, Major General J. F. C. *The Second World War.* New York: Duell, Sloan and Pearce, 1949. 431 pp.

Futrell, Robert Frank. *The United States Air Force in Korea, 1950–1953.* New York: Duell, Sloan and Pearce, 1961, 774 pp.

Galbraith, John Kenneth. *American Capitalism: The Concept of Countervailing Power.* Boston: Houghton Mifflin Company, 1952. 217 pp.

Gann, Ernest K. *Fate is the Hunter.* Greenwich, Conn.: Fawcett Publications, Inc., Crest Reprint, 1962. 352 pp.

Gauvreau, Emile, and Lester Cohen. *Billy Mitchell.* New York: E. P. Dutton, 1942. 303 pp.

Gavin, James M. *Airborne Warfare.* Washington: Infantry Journal Press, 1947. 186 pp.

——. *War and Peace in the Space Age.* New York: Harper & Brothers, 1958. 304 pp.

Geiger, Theodore. *TWA's Services to Ethiopia.* Washington: National Planning Association, 1959. 80 pp.

Goldberg, Alfred (ed.). *A History of the United States Air Force 1907–1957.* Princeton: D. Van Nostrand Company, Inc., 1957. 277 pp.

Halperin, Morton H. *Limited War in the Nuclear Age.* New York and London: John Wiley & Sons, Inc., 1963. 191 pp.

Hammond, Paul Y. *Organizing for Defense.* Princeton: Princeton University Press, 1961. 403 pp.

Hart, B. H. Liddell. *Deterrent or Defense.* New York: Frederick A. Praeger, Inc., 1960. 257 pp.

——. *Strategy.* Rev. ed. New York: Frederick A. Praeger, Inc., 1960. 420 pp.

Harvey, Frank. *Strike Command.* New York: Duell, Sloan and Pearce, 1962. 212 pp.

Higham, Robin. *Britain's Imperial Air Routes 1918 to 1939.* London: G. T. Foulis & Co., Ltd., 1960. 407 pp.

Hitch, Charles J., and Roland N. McKean. *The Economics of Defense in the Nuclear Age.* Cambridge: Harvard University Press, 1960. 422 pp.

Huntington, Samuel P. *The Common Defense.* New York: Columbia University Press, 1961. 500 pp.

International Security: The Military Aspect. Report of Panel II of the Special Studies Project, Rockefeller Fund. Garden City: Doubleday & Company, Inc., 1958. 64 pp.

Janowitz, Morris. *The Professional Soldier.* Glencoe, Illinois: The Free Press, 1960. 464 pp.

Kaufmann, William W. *The McNamara Strategy.* New York, Evanston, and London: Harper & Row, 1964. 339 pp.

Keyes, Lucile Sheppard. *Federal Control of Entry Into Air Transportation.* Cambridge: Harvard University Press, 1951. 405 pp.

Kilmarx, Robert A. *A History of Soviet Air Power.* New York: Frederick A. Praeger, Inc., 1962. 359 pp.

Kintner, William R., in association with Joseph I. Coffey and Raymond J. Albright. *Forging A New Sword.* New York: Harper & Brothers, 1958. 238 pp.

Kissinger, Henry A. *Nuclear Weapons and Foreign Policy.* New York: Harper & Brothers, Published for the Council on Foreign Relations, 1957. 455 pp.

———. *The Necessity for Choice.* Garden City, New York: Doubleday & Company, Inc., Anchor Edition, 1962. 387 pp.

Krislov, Samuel, and Lloyd D. Musolf (eds.). *The Politics of Regulation* (Boston: Houghton Mifflin Company, 1964), 261 pp.

LaFarge, Oliver. *The Eagle in the Egg.* Boston: Houghton Mifflin Company, 1949. 320 pp.

Lee, Asher. *The Soviet Air Force.* New York: The John Day Company, 1962. 288 pp.

Levine, Isaac Don. *Mitchell: Pioneer of Air Power.* Cleveland and New York: World Publishing Company, 1943. 420 pp.

Lincoln, George A., William S. Stone, and Thomas H. Harvey (eds.). *Economics of National Security.* New York: Prentice-Hall, Inc., 1950. 601 pp.

Livesey, William E. *Mahan on Sea Power.* Norman: University of Oklahoma Press, 1947. 334 pp.

Lissitzyn, Oliver James. *International Air Transport and National Policy.* New York: Council on Foreign Relations, 1942. 478 pp.

McGeary, M. Nelson. *The Development of Congressional Investigative Power*. New York: Columbia University Press, 1940. 172 pp.
MacIntyre, Malcolm A. *Competitive Private Enterprise under Government Regulation*. New York University: The Charles C. Moskowitz Lectures, School of Commerce, Accounts, and Finance, 1964. 61 pp.
Malkin, Richard. *Boxcars in the Sky*. New York: Import Publications, Inc., 1951. 282 pp.
Marx, Daniel, Jr. *International Shipping Cartels*. Princeton: Princeton University Press, 1953. 323 pp.
Means, Gardiner C. *Pricing Power and the Public Interest*. New York: Harper & Brothers, 1962. 359 pp.
Millis, Walter. *Arms and Men*. New York: Mentor Books, 1958. 342 pp.
——. with Harvey C. Mansfield and Harold Stein. *Arms and the State*. New York: Twentieth Century Fund, 1958. 436 pp.
Mitchell, William. *Winged Defense*. New York and London: G. P. Putnam's Sons, 1925. 261 pp.
Moore, Samuel Taylor, *U.S. Airpower*. New York: Greenberg, 1958. 196 pp.
Morgenthau, Hans J. *Politics Among Nations*. 2nd ed. New York: Alfred A. Knopf, 1956. 600 pp.
New Frontiers of the Kennedy Administration: The Texts of the Task Force Reports Prepared for the President. Washington: Public Affairs Press, 1961. 185 pp.
Owen, Wilfred. *Strategy for Mobility*. Washington: The Brookings Institution Transport Research Program, 1964. 249 pp.
Richmond, Samuel B. *Regulation and Competition in Air Transportation*. New York and London: Columbia University Press, 1961. 309 pp.
Ridgway, Matthew B., and H. H. Martin. *Soldier: The Memoirs of Matthew B. Ridgway*. New York: Harper & Brothers, 1956. 371 pp.
Ries, John C. *The Management of Defense* (Baltimore: The Johns Hopkins Press, 1964) 221 pp.
Rowat, Donald C. (ed.). *Basic Issues in Public Administration*. New York: The Macmillan Company, 1961. 500 pp.
Salomon, Leon I. (ed.) *The Independent Federal Regulatory Agencies* (New York: H. W. Wilson Co., 1959), 195 pp.
Schlesinger, Arthur M. *The Coming of the New Deal*. Boston: Houghton Mifflin Co., 1958. 669 pp.
Schwartz, Bernard. *The Professor and the Commissions*. New York: Alfred A. Knopf, 1959. 275 pp.

Shepley, James, and Clay Blair, Jr. *The Hydrogen Bomb.* New York: David McKay Company, Inc., 1954. 244 pp.
Smith, Henry Ladd. *Airways.* New York: Alfred A. Knopf, 1942. 430 pp.
——. *Airways Abroad.* Madison: University of Wisconsin Press, 1950. 355 pp.
Spanier, John W. *American Foreign Policy Since World War II.* New York: Frederick A. Praeger, Inc., 1960. 234 pp.
Speas, R. Dixon. *Technical Aspects of Air Transport Management.* New York, Toronto, London: McGraw-Hill Book Company, Inc., 1955. 316 pp.
Spencer, Francis A. *Air Mail Payment and the Government.* Washington: The Brookings Institution, 1941. 402 pp.
Taylor, Maxwell D. *The Uncertain Trumpet.* New York: Harper & Brothers, 1959. 1960. 203 pp.
The Berlin Airlift. Air Ministry and the Central Office of Information. London: His Majesty's Stationery Office, 1949. 63 pp.
The Memoirs of Field-Marshal The Viscount Montgomery of Alamein, K. G. Cleveland and New York: The World Publishing Company, 1958. 508 pp.
Thompson, Captain Annis G. *The Greatest Airlift.* Tokyo: Dai-Nippon Printing Company, 1954. 404 pp.
Tunner, William H. *Over the Hump.* New York: Duell, Sloan and Pearce, 1964. 340 pp.
Urquhart, Major General R. E., with Wilfred Greatorex. *Arnhem.* New York: W. W. Norton & Company, Inc., 1958. 239 pp.
Wheatcroft, Stephen. *The Economics of European Air Transport.* Cambridge: Harvard University Press, 1956. 358 pp.
Wilmot, Chester. *The Struggle for Europe.* New York: Harper & Brothers, 1952. 766 pp.
Zacharoff, Lucien. *The World's Wings.* New York: Duell, Sloan and Pearce, 1946. 310 pp.

B. PUBLICATIONS OF THE GOVERNMENT

U.S. Air Coordinating Committee. *Civil Air Policy.* Washington: Government Printing Office, 1954.
U.S. Civil Aeronautics Board. *Annual Reports, 1942–1962.* Washington: Government Printing Office, 1943–1963.
U.S. Commission on Organization of the Executive Branch of the Government. *Business Enterprises.* Washington: Government Printing Office, 1955.
——. *Business Organization of the Department of Defense.* Washington: Government Printing Office, 1955.

———. *Final Report to the Congress.* Washington: Government Printing Office. 1955.

———. *Subcommittee Report on Business Enterprises of the Department of Defense.* Washington: Government Printing Office, 1955.

———. *Subcommittee Report on Transportation.* Washington: Government Printing Office, 1955.

———. *Task Force Report on Military Procurement.* Washington: Government Printing Office, 1955.

———. *Transportation.* Washington: Government Printing Office, 1955.

U.S. Congress, House. *Airline Equipment Reinvestment Program.* Communication from the President, 85th Congress, 2nd Session, House Document 430. Washington: Government Printing Office, 1958.

———. *Report of the Air Coordinating Committee, Calendar Year 1948.* 81st Congress, 1st Session, House Document 59. Washington: Government Printing Office. 1949.

———. *Report of the Air Coordinating Committee, Calendar Year 1949.* 81st Congress, 2nd Session, House Document 476. Washington: Government Printing Office, 1950.

———. *Report of the Air Coordinating Committee, Calendar Year 1950.* 82nd Congress, 1st Session, House Document 55. Washington: Government Printing Office, 1951.

———. *The Transportation System of Our Nation.* Message from the President of the United States. 87th Congress, 2nd Session, House Document 384. Washington: Government Printing Office, 1962.

———. *United States Defense Policies Since World War II.* Prepared by the Legislative Reference Service. Library of Congress, 85th Congress, 1st Session, House Document 100. Washington: Government Printing Office, 1957.

———. *United States Defense Policies in 1958.* Prepared by the Legislative Reference Service. Library of Congress, 86th Congress, 1st Session, House Document 227. Washington: Government Printing Office, 1959.

———. *United States Defense Policies in 1959.* Prepared by the Legislative Reference Service. Library of Congress, 86th Congress, 2nd Session, House Document 432. Washington: Government Printing Office, 1960.

———. *United States Defense Policies in 1960.* Prepared by the Legislative Reference Service. Library of Congress, 87th Congress, 1st Session, House Document 207. Washington: Government Printing Office, 1961.

———. *United States Defense Policies in 1961.* Prepared by the Legislative Reference Service. Library of Congress, 87th Congress, 2nd Session, House Document 502. Washington: Government Printing Office, 1962.

———. *A Compilation of Material Relating to United States Defense Policies in 1962.* Prepared by the Legislative Reference Service, Library of Congress, 88th Congress, 1st Session, House Document 155. Washington: Government Printing Office, 1963.

———. *United States Defense Policies in 1963.* Prepared by the Legislative Reference Service, Library of Congress, 88th Congress, 2nd Session, House Document 335. Washington: Government Printing Office, 1964.

U.S. Congress, House, Committee on Appropriations. *Department of the Air Force Appropriations for 1956.* Hearings before Subcommittee, 84th Congress, 1st Session. Washington: Government Printing Office, 1955.

———. *Department of the Army Appropriations for 1956.* Hearings before Subcommittee, 84th Congress, 1st Session. Washington: Government Printing Office, 1955.

———. *Department of Defense Appropriations for 1956.* Hearings before Subcommittee, 84th Congress, 1st Session. Washington, Government Printing Office, 1955.

———. *Department of the Air Force Appropriations for 1957.* Hearings before Subcommittee, 84th Congress, 2nd Session, Washington, Government Printing Office, 1956.

———. *Department of the Army Appropriations for 1957.* Hearings before Subcommittee, 84th Congress, 2nd Session. Washington: Government Printing Office, 1956.

———. *Department of Defense Appropriations for 1957.* Hearings before Subcommittee, 84th Congress, 2nd Session. Washington: Government Printing Office, 1956.

———. *Department of Defense Appropriations Bill for 1957.* 84th Congress, 2nd Session, House Report 2104. Washington: Government Printing Office, 1956.

———. *Department of the Air Force Appropriations for 1958.* Hearings before Subcommittee, 85th Congress, 1st Session. Washington: Government Printing Office, 1957.

———. *Department of Defense Appropriations for 1958.* Hearings before Subcommittee, 85th Congress, 1st Session. Washington: Government Printing Office, 1957. Part 2.

———. *Department of Defense Appropriations Bill for 1958.* 85th Congress, 1st Session, House Report 841. Washington: Government Printing Office, 1957.

——. *Department of the Air Force Appropriations for 1959.* Hearings before Subcommittee, 85th Congress, 2nd Session. Washington: Government Printing Office, 1958.
——. *Department of Defense Appropriations for 1959.* Hearings before Subcommittee, 85th Congress, 2nd Session. On Advanced Research Projects Agency, *et al.* Washington: Government Printing Office, 1958.
——. *Department of Defense Appropriations Bill for 1959.* 85th Congress, 2nd Session, House Report 1830. Washington: Government Printing Office, 1958.
——. *Department of Defense Appropriations for 1960.* Hearings before Subcommittee, 86th Congress, 1st Session. Washington: Government Printing Office, 1959. Parts 1–6.
——. *Department of Defense Appropriations Bill for 1960.* 86th Congress, 1st Session, House Report 408. Washington: Government Printing Office, 1959.
——. *Department of Defense Appropriations Bill for 1960.* 86th Congress, 1st Session, House Report 743. Washington: Government Printing Office, 1959.
——. *Supplemental Appropriations Bill for 1960.* 86th Congress, 1st Session, House Report 943. Washington: Government Printing Office, 1959.
——. *Department of Defense Appropriations for 1961.* Hearings before Subcommittee, 86th Congress, 2nd Session. Washington: Government Printing Office, 1960. Parts 1–7.
——. *Department of Defense Appropriations Bill for 1961.* 86th Congress, 2nd Session, House Report 1561. Washington: Government Printing Office, 1960.
——. *Department of Defense Appropriations Bill for 1961.* 86th Congress, 2nd Session, House Report 2040. Washington: Government Printing Office, 1960.
——. *Department of Defense Appropriations for 1962.* Hearings before Subcommittee, 87th Congress, 1st Session. Washington: Government Printing Office, 1961. Parts 2, 3, 6.
——. *Department of Defense Appropriations for 1963.* Hearings before Subcommittee, 87th Congress, 2nd Session. Washington: Government Printing Office, 1962. Parts 2–3.
——. *Department of Defense Appropriations for 1964.* Hearings before Subcommittee, 88th Congress, 1st Session. Washington: Government Printing Office, 1963. Parts 1–3.
U.S. Congress, House, Committee on Armed Services. *Investigation of National Defense, Phase II.* Hearings before Subcommittee, 85th Congress, 2nd Session. Washington: Government Printing Office, 1958.

——. *Investigation of National Defense, Phase II.* 85th Congress, 2nd Session, Committee Print. Washington: Government Printing Office, 1958.

——. *Adequacy of Transportation Systems in Support of the National Defense Effort in Event of Mobilization.* Hearings before Subcommittee, 86th Congress, 1st Session. Washington: Government Printing Office, 1959.

——. *Hearings before Special Subcommittee on National Military Airlift.* 86th Congress, 2nd Session. Washington: Government Printing Office, 1960.

——. *Report of Special Subcommittee on National Military Airlift.* 86th Congress, 2nd Session, Committee Print. Washington: Government Printing Office, 1960.

——. *Authorizing Appropriations for Aircraft, Missiles, and Navy Vessels for the Armed Forces.* Hearings, 87th Congress, 1st Session. Washington: Government Printing Office, 1961.

U.S. Congress, House, Committee on Armed Service. *Military Posture Briefings.* Hearings, 87th Congress, 1st Session. Washington: Government Printing Office, 1961.

——. *Military Construction Authorization, Fiscal Year 1963.* Hearings, 87th Congress, 2nd Session. Washington: Government Printing Office, 1962.

——. *Hearings before Special Subcommittee on Defense Agencies.* 87th Congress, 2nd Session. Washington: Government Printing Office, 1962.

——. *Report of Special Subcommittee on Defense Agencies.* 87th Congress, 2nd Session, Committee Print. Washington: Government Printing Office, 1962.

——. *Richmond Air Crash, November 8, 1961.* Hearings before Subcommittee, 87th Congress, 2nd Session. Washington: Government Printing Office, 1962.

——. *Report of Special Subcommittee to Investigate Crash of Imperial Airlines, Inc., Airplane, Richmond, Va., November 8, 1961.* 87th Congress, 2nd Session, Committee Print. Washington: Government Printing Office, 1962.

——. *Military Posture Briefings.* Hearings, 87th Congress, 2nd Session. Washington: Government Printing Office, 1962.

——. *Hearings before the Special Subcommittee on National Military Airlift.* 88th Congress, 1st Session. Washington: Government Printing Office, 1963.

——. *Report of the Special Subcommittee on National Military Airlift*, 88th Congress, 1st Session, Committee Print. Washington: Government Printing Office, 1963.

——. *Military Posture Briefings.* Hearings, 88th Congress, 1st Session. Washington: Government Printing Office, 1963.

U.S. Congress, House, Committee on Commerce. *Federal Aviation Act of 1958*. 85th Congress, 2nd Session, House Report 2360. Washington: Government Printing Office, 1958.

———. *Supplemental Air Carriers*. 87th Congress, 1st Session, House Report 1177. Washington: Government Printing Office, 1961.

U.S. Congress, House, Committee on Government Operations. *Military Air Transportation*. Hearings before Subcommittee, 85th Congress, 2nd Session. Washington: Government Printing Office, 1958.

———. *Military Air Transportation*. Twenty-eighth Report. 85th Congress, 2nd Session, House Report 2011. Washington: Government Printing Office, 1958.

———. *Military Air Transportation*. Hearings before Subcommittee, 86th Congress, 1st Session. Washington: Government Printing Office, 1959.

———. *Military Air Transportation*. Tenth Report of the Committee, 86th Congress, 1st Session, House Report 1112. Washington: Government Printing Office, 1959.

———. *Military Air Transportation*. Hearings before Subcommittee, 86th Congress, 2nd Session. Washington: Government Printing Office, 1960.

———. *Military Air Transportation*. Hearings before Subcommittee, 87th Congress, 1st Session. Washington: Government Printing Office, 1961.

———. *Military Air Transportation, 1963*. Hearings before Subcommittee, 88th Congress, 1st Session. Washington: Government Printing Office, 1963.

U.S. Congress, House, Committee on the Judiciary. *Monopoly Problems in Regulated Industries*. 4 vols. Hearings before Subcommittee, 84th Congress, 2nd Session. Washington: Government Printing Office, 1956.

———. *Report of the Antitrust Subcommittee (Subcommittee No. 5)*. Pursuant to H. Res. 107 on Airlines, 85th Congress, 1st Session. Washington: Government Printing Office, 1957.

———. *The Airlines Industry*. 85th Congress, 2nd Session, House Report 1328. Washington: Government Printing Office, 1958.

U.S. Congress, House, Committee on Merchant Marine and Fisheries. *Review of Merchant Marine Policy, 1962*. Hearings before Subcommittee, 87th Congress, 2nd Session. Washington: Government Printing Office, 1962.

U.S. Congress, House, Select Committee of Inquiry into Operations of the United States Air Services. *Testimony of Major Edward V. Rickenbacker, et al.* Hearings, 68th Congress, 2nd Session. Washington: Government Printing Office, 1925.

——. *Inquiry into Operations of the United States Air Services.* 68th Congress, 2nd Session, House Report 1653. Washington: Government Printing Office, 1925.

U.S. Congress, Senate, Committee on Armed Services. *Study of Airpower.* 2 vols. Hearings before Subcommittee, 84th Congress, 2nd Session. Washington: Government Printing Office, 1956.

——. *Military Procurement Authorization, Fiscal Year 1962.* Hearings, 87th Congress, 1st Session. Washington: Government Printing Office, 1961.

——. *Military Procurement Authorization, Fiscal Year 1963.* Hearings, 87th Congress, 2nd Session. Washington: Government Printing Office, 1962.

U.S. Congress, Senate, Committee on Appropriations. *Department of Defense Appropriations for 1955.* Hearings before Subcommittee, 83rd Congress, 2nd Session. Washington: Government Printing Office, 1954.

——. *Department of Defense Appropriations for 1958.* Hearings before Subcommittee, 8th Congress, 1st Session. Washington: Government Printing Office, 1957.

——. *Department of Defense Appropriations Bill for 1958.* 85th Congress, 1st Session, Senate Report 543. Washington: Government Printing Office, 1957.

——. *Department of Defense Appropriations for 1959.* Hearings before Subcommittee, 85th Congress, 2nd Session. Washington: Government Printing Office, 1958.

——. *Department of Defense Appropriations Bill for 1959.* 85th Congress, 2nd Session, Senate Report 1937. Washington: Government Printing Office, 1958.

——. *Supplemental Appropriations for 1960.* Hearings before Subcommittee, 86th Congress, 1st Session. Washington: Government Printing Office, 1959.

——. *Department of Defense Appropriations Bill for 1960.* 86th Congress, 1st Session, Senate Report 476. Washington: Government Printing Office, 1959.

——. *Department of Defense Appropriations for 1960.* Hearings before Subcommittee, 86th Congress, 1st Session. Washington: Government Printing Office, 1959.

——. *Department of Defense Appropriations for 1961.* Hearings before Subcommittee, 86th Congress, 2nd Session. Washington: Government Printing Office, 1960.

——. *Department of Defense Appropriations Bill for 1961.* 86th Congress, 2nd Session, Senate Report 1550. Washington: Government Printing Office, 1960.

———. *Department of Defense Appropriations for 1962.* Hearings before Subcommittee, 87th Congress, 1st Session. Washington: Government Printing Office, 1961.

———. *Department of Defense Appropriations for 1963.* Hearings before Subcommittee, 87th Congress, 2nd Session. Washington: Government Printing Office, 1962.

U.S. Congress, Senate, Committee on Commerce. *To Create the All-American Flag Line, Inc.* Hearings, 79th Congress, 1st Session. Washington: Government Printing Office, 1945.

———. *Creating the All-American Flag Line, Inc. and to Assure the United States World Leadership in the Field of Air Transportation.* 79th Congress, 1st Session, Senate Report 805. Washington: Government Printing Office, 1945.

———. *Consolidation of International Air Carriers.* Hearings before Subcommittee, 80th Congress, 1st Session, Washington: Government Printing Office, 1947.

———. *Development of Civilian Aircraft Adaptable to Auxiliary Military Service.* Hearings before Subcommittee, 80th Congress, 2nd Session. Washington: Government Printing Office, 1948.

———. *Air-Line Industry Investigation.* 5 vols. Hearings, 81st Congress, 1st and 2nd Session. Washington: Government Printing Office, 1949, 1950.

———. *Improvement of Procedures for the Development of Foreign Air Commerce.* 85th Congress, 1st Session, Senate Report 119. Washington: Government Printing Office, 1957.

———. *Study of Military Air Transport Service.* Hearings before Subcommittee, 85th Congress, 2nd Session. Washington: Government Printing Office, 1958.

———. *Government Guarantee of Loans to Certain Air Carriers.* Hearings before Subcommittee, 86th Congress, 2nd Session. Washington: Government Printing Office, 1960.

———. *Freedom of Communications.* Speeches, Remarks, Press Conferences, and Statements of Senator John F. Kennedy, August 1 through November 7, 1960. 87th Congress, 1st Session, Senate Report 994. Washington: Government Printing Office, 1961.

———. *International Air Transportation Problems.* Hearings before Subcommittee, 87th Congress, 2nd Session. Washington: Government Printing Office, 1961.

———. *Aircraft Guarantee Loan Program.* Hearings before Subcommittee, 87th Congress, 2nd Session. Washington: Government Printing Office, 1962.

———. *Regulation of Supplemental Air Carriers.* Hearings before Subcommittee, 87th Congress, 2nd Session. Washington: Government Printing Office, 1962.

U.S. Congress, Senate, Committee on Government Operations. *Organizing for National Security*. Hearings before Subcommittee, 87th Congress, 1st Session. Washington: Government Printing Office, 1961.

U.S. Congress, Senate, Committee on the Judiciary. *Report on Regulatory Agencies to the President-elect*. 86th Congress, 2nd Session, Committee Print. Washington: Government Printing Office, 1960.

U.S. Congress, Senate, Congressional Aviation Policy Board. *National Aviation Policy*, 80th Congress, 2nd Session, Senate Report 949. Washington: Government Printing Office, 1948.

U.S. Congress, Senate, Select Committee on Small Business. *Role of Irregular Airlines in United States Air Transportation Industry*. Hearings before Subcommittee, 82nd Congress, 1st Session. Washington: Government Printing Office, 1951.

——. *Report on Role of Irregular Airlines in United States Air Transportation Industry*. 82nd Congress, 1st Session, Senate Report 540. Washington: Government Printing Office, 1951.

——. *The Role of Competition in Commercial Air Transportation*. Submitted by the Civil Aeronautics Board. 82nd Congress, 2nd Session, Committee Print Number 9. Washington: Government Printing Office, 1952.

——. *Future of Irregular Airlines in United States Air Transportation Industry*. Hearings before Subcommittee, 83rd Congress, 1st Session. Washington: Government Printing Office, 1953.

——. *Future of Irregular Airlines*. 83rd Congress, 1st Session, Senate Report 822. Washington: Government Printing Office, 1953.

——. *Materials Relative to Competition in the Regulated Civil Aviation Industry*. 84th Congress, 2nd Session. Washington: Government Printing Office, 1956.

U.S. Department of the Army, Office of the Chief of Military History. *United States Army in the Korean War. South to the Naktong, North to the Yalu (June–November 1950)*, by Roy E. Appleman. Washington: Government Printing Office, 1961.

——. *United States Army in World War II. Logistical Support of the Armies. May 1941–September 1944*. Vol. I. Washington: Government Printing Office, 1953.

——. *September 1944–May 1945*. Vol. II. Washington: Government Printing Office, 1959.

U.S. Department of Commerce, Business and Defense Services Administration. *World Survey of Civil Aviation, Africa*. Washington: Government Printing Office, 1960.

——. *World Survey of Civil Aviation, Japan.* Washington: Government Printing Office, 1960.
——. *World Survey of Civil Aviation. Selected Asian Countries and Oceania.* Washington: Government Printing Office, 1961.
U.S. Department of Commerce, Civil Aeronautics Administration. *Statistical Handbook of Civil Aviation, 1948.* Washington: Government Printing Office, 1949.
U.S. Department of Commerce. *Federal Transportation Policy and Program.* Washington: Government Printing Office, 1960.
U.S. Executive Office of the President. *Aviation Facilities Planning.* Final Report by the President's Special Assistant. Washington: Government Printing Office, 1957.
U.S. Federal Aviation Agency. *Report of the Task Force on National Aviation Goals: Project Horizon.* Washington: Government Printing Office, 1961.
U.S. Interstate Commerce Commission. *Annual Reports, 1942–1946.* Washington: Government Printing Office, 1943–1947.
U.S. President's Air Policy Commission. *Survival in the Air Age.* Washington: Government Printing Office, 1948.
U.S. President's Committee on Administrative Management. *Report with Special Studies.* Washington: Government Printing Office, 1937.
U.S. *Report of the President's Aircraft Board.* Washington: Government Printing Office, 1925.

C. OTHER GOVERNMENT PUBLICATIONS AND PUBLICATIONS OF MISCELLANEOUS ORGANIZATIONS

Air Transport Association of America. *Facts and Figures About Air Transportation.* 23rd ed. Washington: 1962. 35 pp.
——. *Facts and Figures About Air Transportation.* 24th ed. Washington: 1963. 35 pp.
——. *Facts and Figures About Air Transportation.* 25th ed. Washington: 1964. 35 pp.
The Future of the International Carrier. Flight Forum, Connecticut General Life Insurance Company, 1964. 31 pp.
Heymann, Hans, Jr. *Civil Aviation and U.S. Foreign Aid: Purposes, Pitfalls and Problems for U.S. Policy.* R-424-RC. Santa Monica: The RAND Corporation. January, 1964.
——. *The Soviet Role in International Civil Aviation.* Research Memorandum RM-2213. Santa Monica: The RAND Corporation, 1958.
U.S. Civil Aeronautics Board. *Report of Task Group 2 on Aviation Mobilization.* Mimeographed: Washington, October 25, 1959.

——. *Monthly Report of Air Carrier Traffic Statistics, December 1961*. Washington: 1962.

——. *Monthly Report of Air Carrier Traffic Statistics, November 1964*. Washington: 1965.

——. *Subsidy for United States Certificated Air Carriers*. Washington: January, 1962.

——. *Supplement to the Monthly Report of Air Carrier Traffic Statistics*. Quarter ended December 31, 1961. Washington: 1962.

U.S. Department of Commerce, Defense Air Transportation Administration. *Report of the Clemson Committee*. Mimeographed: Washington, 1952.

——. *History of the Defense Air Transportation Administration to June 30, 1953*. Mimeographed: Washington, 1953.

U.S. Department of Commerce and Department of Defense. *Report to the Air Line Presidents on Civil Aviation Mobilization*. Washington: 1952.

U.S. Department of Defense. *The Department of Defense Plan for the Civil Reserve Air Fleet*. Washington: 1952.

——. *The Role of Military Air Transport Service in Peace and War*. Washington: 1960.

U.S. Department of Defense, Military Air Transport Service. *Civilian Pilots in Combat Areas*. Paper prepared by Historical Division. Mimeographed: February 4, 1959.

——. *Airlift Service Management Report*, July, 1958-June, 1959.

——. *Airlift Service Management Report*, July, 1959-June, 1960.

——. *Airlift Service Management Report*, July, 1960-June, 1961.

——. *Airlift Service Management Report*, July, 1961-June, 1962.

U.S. Department of the Air Force. *Reed Committee Report on MATS*. Mimeographed: Washington, 1960.

——. *USAF Airborne Operations, World War II and Korean War*. Washington: USAF Historical Division Liaison Office, 1962.

——. *Air Supply in the Burma Campaigns*, by Dr. Joe G. Taylor, USAF Historical Division, Research Studies Institute, Air University, 1957.

U.S. Department of State. Agency for International Development. *The United States Program of Aviation Assistance to Less-Developed Countries*. Report to the Administrator. Washington: 1964.

U.S. Executive Office of the President, National Security Resources Board. *Report on Utilization of Airlines for Wartime Airlift and Proposals to Aid Expansion of the Civil Air Fleet*. Mimeographed: Washington, 1950.

D. PERIODICALS

Baldwin, Hanson W. "The McNamara Monarchy," *Saturday Evening Post*, 236 (March 9, 1963), 8–12.

Barber, Richard J. "Merging the Airlines: What Next?" *New Republic*, 141 (January 5, 1963), 11–13.

Bendiner, Robert. "The Rise and Fall of the Nonskeds," *Reporter*, 16 (May 30, 1957), 29–34.

Bernstein, David. "Our Airsick Airlines," *Harper's*, 198 (May, 1949), 64–73.

Caputo, Vincent F. "Remembrance of the Year Past," *National Defense Transportation Journal*, 18 (March–April, 1962), 22–27.

Damon, Ralph S. "Airline Outlook," *Air Affairs*, 3 (December, 1950), 459–68.

Frederick, John H. "American Air Cargo Development," *Air Affairs*, 2 (Autumn, 1947), 79–100.

"Freight by Air," *Fortune*, 35 (May, 1946), 134–39 ff.

Godehn, Paul M., and Frank E. Quindry. "Air Mail Contract Cancellations of 1934 and Resulting Litigation," *Journal of Air Law and Commerce*, 21 (Summer, 1954), 253–86.

Gross, Robert E. "The Transport Plane in Peace and War," *Air Affairs*, 3 (December, 1950), 469–80.

Hannifin, Jerry, and Charles V. Murphy. "We're Losing the Supersonic Transport Race," *Fortune* 69 (February, 1964) 18–22, 168–72.

Hessler, William H. "America's Merchant Marine of the Air," *U.S. Naval Institute Proceedings*, 85 (May, 1959), 48–63.

Huntington, Samuel P. "Interservice Competition and the Political Roles of the Armed Services," *American Political Science Review*, LV (March, 1961), 40–52.

Jones, Lieutenant Douglas N. "Economics of Logistic Support," *Air University Quarterly Review*, XIII (Fall, 1961), 10–21.

Kraar, Louis. "The Two Lives of Robert McNamara," *Life*, 22 (November 30, 1962), 94–102.

Lindsey, John M. "The Legislative Development of Civil Aviation: 1938–58," *Journal of Air Law and Commerce*, 28 (Winter, 1961–62), 18–45.

Lippman, Walter. "The Senate Inquisition," *Forum*, 84 (1930), 129–32.

Lissitzyn, Oliver. "The Participation of the United States in World Air Transport: Statistics and the National Interest," *Journal of Air Law and Commerce*, 28 (Winter, 1961–62), 1–17.

Little, Virginia. "Control of International Air Transport," *International Organization*, 3 (February, 1949), 29–40.

McDonald, John. "The Epic Struggle for the Corporate Cockpit," *Fortune*, 71 (May, 1965), 106–11, 213–16.

——. "Pan Am: Peril and Opportunity," *Fortune*, 66 (August, 1962), 83–88.

"Merger and Monopoly in Domestic Aviation," *Columbia Law Review*, 62 (May, 1962), 851–83.

Murphy, Charles J. V. "The Berlin Airlift," *Fortune*, 38 (November, 1948), 89–93.

Nigro, Edward H. "Early Troop Carrier Operations in Korea," *Air University Quarterly Review*, VII (Spring, 1954), 86–98.

"Notes," *Harvard Law Review*, 75 (June, 1962), 575–89.

Putnam, Carleton. "Airline Outlook," *Air Affairs*, 2 (Winter, 1949), 491–504.

"The Airline Squeeze," *Fortune*, 35 (May, 1947), 117–20 ff.

Tunner, Lieutenant General William H. (ret.). "Strategic Airlift," *Air University Quarterly Review*, 12 (Winter and Spring, 1960–61), 104–19.

Tunner, Major General William H. "Technology or Manpower," *Air University Quarterly Review*, 5 (Fall, 1952), 3–19.

"What's Wrong With the Airlines?" *Fortune*, 32 (August, 1945), 159–64.

White, General Thomas D. (ret.). "Strategy and the Defense Intellectuals" *Saturday Evening Post*, 236 (May 4, 1963), 10–14.

White, Theodore H. "Revolution in the Pentagon," *Look*, 27 (April 23, 1963), 31–49.

Worsnop, Richard L. "Troubles of the Airlines," *Editorial Research Reports*, 1 (1962), 83–100.

E. ESSAYS AND ARTICLES IN COLLECTIONS

Arnold, General Henry H. "Air Power and the Future," from the Third Report to the Secretary of War by the Commanding General of the Army Air Forces, November 12, 1945. Reproduced in Emme, Eugene M., *The Impact of Air Power*. Princeton: D. Van Nostrand Company, Inc., 1959. Pp. 303–15.

Earle, Edward Mead. "Adam Smith, Alexander Hamilton, Friedrich List: The Economic Foundations of Military Power," in Earle, Edward Mead (ed.), *Makers of Modern Strategy*. Princeton: Princeton University Press, 1943. Pp. 117–54.

Holborn, Hajo. "Moltke and Schieffen: The Prussian-German School," in Earle, Edward Mead (ed.), *Makers of Modern Strategy*. Princeton: Princeton University Press, 1943. Pp. 172–205.

King, James E., Jr. "NATO: Genesis, Progress, Problems," in Turner, Gordon B., and Richard D. Challener (eds.), *National*

Security in the Nuclear Age. New York: Frederick A. Praeger, Inc., 1960. Pp. 143–74.

Kuter, General Laurence S. "The Berlin Airlift," Excerpts from an address before the Institute of Aeronautical Sciences, New York, January 24, 1949. Reproduced in Emme, Eugene M., *The Impact of Air Power*. Princeton: D. Van Nostrand Company, Inc., 1959. Pp. 377–83.

Snyder, Glenn H. "The 'New Look' of 1953," in Schilling, Warner R., Paul Y. Hammond, and Glenn H. Snyder, *Strategy, Politics, and Defense Budgets*. New York and London: Columbia University Press, 1962. Pp. 379–524.

Sprout, Margaret Tuttle. "Mahan: Evangelist of Sea Power," in Earle, Edward Mead (ed.), *Makers of Modern Strategy*. Princeton: Princeton University Press,1943. Pp. 415–45.

Turner, Gordon B. "Air and Sea Power in Relation to National Power," in Turner, Gordon B., and Richard D. Challener (eds.), *National Security in the Nuclear Age*. New York: Frederick A. Praeger, Inc., 1960. Pp. 227–52.

Warner, Edward. "Douhet, Mitchell, Seversky: Theories of Air Warfare," in Earle, Edward Mead (ed.), *Makers of Modern Strategy*. Princeton: Princeton University Press, 1943. Pp. 485–503.

F. UNPUBLISHED MATERIALS

Caroom, Hiram C. "Management Responsibilities and National Defense Interests: An Analysis of the Practicability of the U.S.-Flag International Passenger Airlines' Transporting More Military Traffic in Peace and War." Unpublished Ph.D. dissertation, Harvard University, 1959.

"Civil Airlift and the National Defense. How the Scheduled Airlines Contribute to the Nation's Security." Pamphlet prepared by the Air Transport Association of America, 1962.

Douglas Aircraft Company, Press Release, June 11, 1962. Mimeographed.

Einzig, Robert S. "CAB Control Over Domestic Routes and Competition, 1938–52." Unpublished Ph.D. dissertation, University of Michigan, 1953.

Excerpts from speech by Senator A. S. Monroney, Wings Club Banquet, New York, October 19, 1959.

Fowle, Herbert L., Jr. "A Critical Analysis of the Airlift Service Industrial Fund." Unpublished Master's thesis, University of Texas, Austin, 1962. 73 pp.

Heymann, Hans, Jr. "Aviation and International Relations." Text of a talk given at the symposium, "The Issues and Challenges

of Air Transportation." Mimeographed: Hartford, Connecticut, November 2, 1961.
Letter from American Airlines to Secretary of Commerce, June 2, 1959.
Letter from Civil Aeronautics Board Chairman Whitney Gilliland to Defense Secretary Thomas S. Gates, June 24, 1960.
Memorandum from William Y. Elliott to Vice-President Richard M. Nixon, February 11, 1958.
"Outline of U.S. Airline Responsiveness in Emergencies." Memorandum prepared by Air Transport Association of America, 1960. Mimeographed.
Legere, Lawrence J. "Unification of the Armed Forces." Unpublished Ph.D. dissertation, Harvard University, 1950.
U.S. Civil Aeronautics Board Press Release 59–21, June 11, 1959.
U.S. Department of Defense Press Releases, 1286–56, December 13, 1956; 1006–61, September 19, 1961.
U.S. Department of the Air Force, Unclassified briefing sheet prepared for House Committee on Armed Services, July 6, 1962.
U.S. Executive Office of the President. Bureau of the Budget. Letter to the President, signed by Defense Secretary Thomas S. Gates, and Budget Director Maurice H. Stans. Mimeographed: Washington, December 13, 1960.
——. Statement on International Air Transport Policy. Press Release, April 24, 1963.
U.S. Industrial College of the Armed Forces, "Problems Involved in Mobilization of Civil Aviation," Speech of Brigadier General Milton W. Arnold (ret.), February 19, 1953. Mimeographed.
U.S. National Archives. *Special File #2. Correspondence Pertaining to Merger of Air Transport Command and Naval Air Transport Service, 1947–48.*
U.S. President's Advisory Committee on Government Reorganization, "The President and the Regulatory Commissions" (Mimeographed: Washington, 1961).

G. NEWSPAPERS AND MAGAZINES

Aeroplane and Commercial Aviation News, September 17, 1964.
Air Force Magazine, October, 1959.
Air Force Times, June 30, 1962.
Airlift Magazine, June, September, October, 1959; March, 1960; February, 1963.
Armed Forces Management, April, 1963.
Army Information Digest, May, 1962.
Army Magazine, November, 1960; April, 1962.
Army-Navy-Air Force Journal, July, 1959–September, 1964.

Army-Navy-Air Force Register and Defense Times, December 12, 1959.
Aviation Daily, May, 1959–September, 1965.
Aviation Week, January, 1949–August, 1965.
Baltimore Sun, March, 1960–June, 1964.
Business Week, March, 1961–November, 1962.
Chicago Sun-Times, April 26, 1962.
Chicago Tribune, January 22, 1961.
Christian Science Monitor, November, 1959–February, 1961.
Congressional Quarterly, June 24, 1962.
Congressional Quarterly Almanac, 1961.
Congressional Quarterly Weekly Report, March 8, 1963.
Congressional Record, January, 1941–December, 1961.
Daily Oklahoman (Oklahoma City), February 28, 1960.
Daily Press (Newport News, Virginia), May 21, 1960.
DATA Magazine, November, 1962.
Denver Post, November, 1959–July, 1963.
Department of Defense Directives, 5160.12, January 31, 1956; 5160.2, December 7, 1956.
Department of State Bulletin, 1943–61.
Flying, May, 1962.
Forbes Magazine, June 15, 1959.
Interavia, January, 1962.
Journal of Commerce, February 10, 1960.
Los Angeles Times, July 25, 1961.
National Defense Transportation Journal, July–August, 1962.
Navy Times, October 21, 1959.
New Republic, May 25, 1963.
Newsweek, April 12, 1965.
New York Herald Tribune, June, 1959–February, 1964.
New York Times, November, 1945–July, 1964.
Reporter, February 14, 1963.
St. Louis Globe-Democrat, July 31, 1959.
St. Louis Post-Dispatch, July, 1959–October, 1963.
Time, March, 1960–February, 1965.
Wall Street Journal, July, 1959–April, 1963.
Washington Daily News, November, 1959–October, 1964.
Washington Post, November, 1959–June, 1965.
Washington Star, October, 1963–August, 1964.

Index

Index

D

T

www.ingramcontent.com/pod-product-compliance
Ingram Content Group UK Ltd.
Pitfield, Milton Keynes, MK11 3LW, UK
UKHW040841030726
472823UK00001B/103